News, Numbers and Public Opinion
in a Data-Driven World

News, Numbers and Public Opinion in a Data-Driven World

Edited by
An Nguyen

BLOOMSBURY ACADEMIC
LONDON • NEW YORK • OXFORD • NEW DELHI • SYDNEY

BLOOMSBURY ACADEMIC
Bloomsbury Publishing Plc
50 Bedford Square, London, WC1B 3DP, UK
1385 Broadway, New York, NY 10018, USA

BLOOMSBURY, BLOOMSBURY ACADEMIC and the Diana logo
are trademarks of Bloomsbury Publishing Plc

First published 2018
Paperback edition first published 2019

Cover design by Jason Anscomb

A catalogue record for this book is available from the British Library.

A catalog record for this book is available in the Library of Congress.

ISBN: HB: 978-1-5013-3035-3
PB: 978-1-5013-5400-7
ePDF: 978-1-5013-3037-7
eBook: 978-1-5013-3036-0

Typeset by Deanta Global Publishing Services, Chennai, India

To find out more about our authors and books visit
www.bloomsbury.com and sign up for our newsletters.

Contents

List of Contributors

Muhammad Idrees Ahmad, PhD, is lecturer in digital journalism at the University of Stirling, UK, and the author of *The Road to Iraq: The Making of a Neoconservative War* (2014). He writes for many international media outlets and has appeared as an on-air analyst on Al Jazeera, the BBC, RAI TV, Radio Open Source and several Pacifica Radio channels.

Stuart Allan is professor and head of the School of Journalism, Media and Cultural Studies, Cardiff University, UK. His current science-related research examines the evolving interconnections between citizen science and citizen journalism.

Charles R. Berger is professor emeritus in the Department of Communication, University of California, Davis, USA. He is a former editor of *Human Communication Research* and co-editor (with Sandra Ball Rokeach) of *Communication Research* as well as a fellow and past president of the International Communication Association and a National Communication Association Distinguished Scholar. He has devised theories concerned with the explanation of communication under uncertainty (Uncertainty Reduction Theory), message production processes (Planning Theory) and, most recently, narrative impact (Story Appraisal Theory).

Renata Faria Brandão is a postdoctoral research associate at King's College London. She recently completed a PhD at the University of Sheffield, UK, under the Brazilian government's Science Without Borders scholarship scheme. She earned her BA cum Laude with Honours in communication with minor in international relations in 2011 and an MA in global journalism. Her research explores how statistical data are used to articulate narratives and shape news discourses of science.

Coy Callison (PhD, Alabama, 2000) is professor of public relations and associate dean of graduate studies in the Texas Tech College of Media and Communication, USA. His research focuses on using experimental methods to investigate message and source factors influencing the effects of persuasive messages and recall of information. He has published more than thirty-five peer-review articles, presented more than sixty papers, and has received research funding from various state and national agencies.

Stephen Cushion is a reader at the Cardiff School of Journalism, Media and Cultural Studies, Cardiff University, UK. He has published many journal articles on issues in journalism, news and politics, three sole-authored books, *News and Politics: The Rise of Live and Interpretive Television News* (2015, Routledge), *The Democratic Value of News: Why Public Service Media Matter* (2012, Palgrave) and *Television Journalism* (2012,

Sage) and co-edited (with Justin Lewis) *The Rise of 24-Hour News: Global Perspectives* (2010, Peter Lang).

Sharon Dunwoody (PhD, Indiana University) is emeritus professor of journalism and mass communication at the University of Wisconsin–Madison, USA. She studies the construction of science and environmental messages, as well as the ways in which individuals use such messages to inform their judgements about science issues. She is a fellow of the American Association for the Advancement of Science, the Society for Risk Analysis, and the Midwest Association for Public Opinion Research.

Rhonda Gibson (PhD, Alabama, 1993) is associate professor of journalism in the School of Media and Journalism at the University of North Carolina at Chapel Hill, USA. Her research focuses on the effects of exemplification in journalism on issue perception and the effects of images of sexual minorities in the media. Her research has been published in *Journalism & Mass Communication Quarterly, Communication Research, Newspaper Research Journal,* and *Journalism & Mass Communication Educator*, among others. She is working on a book examining the changing communication strategies and public opinion regarding same-sex marriage.

Robert J. Griffin (PhD, University of Wisconsin–Madison) is emeritus professor in the Diederich College of Communication at Marquette University in Milwaukee, Wisconsin, USA. He has focused much of his teaching and research on communication about environment, energy, health, science and risk. He is a recipient of Marquette's Faculty Award for Teaching Excellence, awarded to the university's top educators. He is a fellow of the American Association for the Advancement of Science and the Society for Risk Analysis.

Yael de Haan (PhD, Amsterdam 2012) is senior researcher at the University of Applied Sciences in Utrecht, the Netherlands. She analyses shifts in journalism practice in a changing media environment. She has published in several books and peer-reviewed journals such as *Digital Journalism, European Journal of Communication* and *Journal of Media Innovation*.

Sanne Kruikemeier is assistant professor of political communication at the Amsterdam School of Communication Research (ASCoR) at the University of Amsterdam, the Netherlands. Her research mainly focuses on the content and effects of online communication in a political context. She is currently co-chair of the political communication division of the Netherlands-Flanders Communication Association.

Willem Koetsenruijter is an assistant professor at Leiden University in the Netherlands, where he is a faculty member since 2000. He started his career as an editor and writer for magazines. He has published work about numbers in the news and about visual language, including *Visual Language: Perspectives for both Makers and Users* (co-authored), *Methods of Journalism Studies* (co-authored with Tom Van Hout), and *Numbers in the News* (sole-authored in Dutch).

Brendan Lawson is a PhD researcher at the University of Leeds, UK, where he is exploring the role of quantifiable information in humanitarian communication. He has a BA in History from the same university and a MA Media in Development from the SOAS, University of London.

Sophie Lecheler is professor of communication science with a focus on political communication at the Department of Communication at the University of Vienna, Austria. Her work has been published in a wide range of international journals, such as *Communication Research, Journal of Communication, New Media & Society, Journalism, Communication Theory, Communication Monographs, Journalism Studies,* and *International Journal of Press/Politics.*

Justin Lewis is professor of communication and dean of research for the College of Arts, Humanities and Social Sciences, Cardiff University, UK. His books, since 2000, include *Constructing Public Opinion* (2001, New York: Columbia University Press), *Citizens or Consumers: What the Media Tell Us About Political Participation* (2005, Open University Press), *Shoot First and Ask Questions Later: Media Coverage of the War in Iraq* (2006, Peter Lang), *Climate Change and the Media* (2009, Peter Lang) and *Consumer Capitalism: Media and the Limits to Imagination* (2013, Polity).

Seth C. Lewis, PhD, is the founding holder of the Shirley Papé Chair in Emerging Media in the School of Journalism and Communication at the University of Oregon. His award-winning research explores the social implications of media technologies for the dynamics of media work and innovation, particularly in the case of journalism and its digital transformation. He presently focuses on three areas: the interplay of humans and machines in news; the role of reciprocity in the changing dynamics among journalists, audiences, and communities; and the social dimensions of journalism and its boundaries. Drawing on a variety of disciplines, theories and methods, Lewis has published some 50 peer-reviewed articles and book chapters, covering a range of sociotechnical phenomena – from big data, coding, and open-source software, to social media, APIs, and digital audience analytics.

Wiebke Loosen, PhD, is a senior researcher at the Hans Bredow Institute for Media Research and a lecturer at the University in Hamburg, Germany. She has made various contributions to theoretical and empirical research in the field of journalism's transformation in the internet age, including topics such as online/digital journalism, cross media journalism, and the 'social mediatization' of journalism. In her current research, she studies the changing nature of the journalism-audience relationship, data journalism, and the emerging startup culture in journalism.

Jairo Lugo-Ocando is associate professor in the School of Media and Communication at the University of Leeds, UK. He is author of three monographs, one edited book, eighteen peer-reviewed journal articles and eleven book chapters together with several public and private sector reports. He has been the principal investigator for a series of grants and fellowships funded by Nuffield Foundation (UK), Carnegie Trust for the Universities of Scotland (UK), the TODA Institute (Japan), AHRC-CDA (UK), John F. Kennedy Library (USA), among others.

Scott Maier, PhD, is professor at the University of Oregon School of Journalism and Communication, USA. He is also US project director of the European Journalism Observatory. For his research on media accuracy, Maier received (with Philip Meyer) the Sigma Delta Chi Award for Research about Journalism. A twenty-year newspaper and wire-service veteran, Maier was co-founder of CAR Northwest, an industry–academic partnership providing training in computer-assisted reporting to newsrooms and journalism classrooms. His research interests include online news coverage, newsroom numeracy, media accuracy, and technological change management.

Kevin McConway is emeritus professor of applied statistics at the Open University, UK. His research and teaching has been in statistical theory and in the application of statistics in health and in the life sciences. More relevantly to this paper, he has a long-standing interest in and experience of the use and presentation of statistics in the media. He was the Open University's main academic liaison with the BBC Radio 4 programme *More or Less* for eleven years and he works with the UK's Science Media Centre to provide briefing for journalists on stories with statistical aspects.

An Nguyen is associate professor of journalism at Bournemouth University, UK. He has authored, in the English language, two books and about forty research papers and reports in several areas: digital journalism, news consumption and citizenship, citizen journalism, science journalism, and news and global development. His latest work includes *Data and Statistics in Journalism and Journalism Education*, a special issue of *Journalism: Theory, Practice and Criticism* (co-edited, 2016) and *Developing News: Global Journalism and the Coverage of 'Third World' Development* (co-authored, 2017) – both with Lugo-Ocando.

Julius Reimer, MA, is a junior researcher at the Hans Bredow Institute for Media Research in Hamburg. He studied communication studies, economic policy and sociology at the University of Münster (Germany) and at the Università della Svizzera italiana in Lugano (Switzerland) and was previously a junior researcher at TU Dortmund University's Institute of Journalism. His research focuses on participation and transparency in journalism as well as (personal) branding for journalists and new ways of organizing news work (e.g. freelancer networks, startups).

Fabienne Crettaz Von Roten, PhD in mathematics, is assistant professor and director of the Observatoire Science, Politique et Société at the University of Lausanne, Switzerland. Her research offers a symmetrical view of the relations between science and society: public perceptions of science and technology and scientists' engagement with society. In particular, she analyses the problems associated with statistics in society and in science. She has authored several books, book chapters, articles in journals in these fields.

Gerard Smit is a visual consultant who was a lecturer and researcher at the University of Applied Sciences in Utrecht, the Netherlands. A philosopher, he worked as an editor for professional magazines, co-authored books at the intersection of psychology and philosophy, and taught investigative journalism, philosophy, and information visualization. He conducted research on the applicability of visual storytelling for journalism.

Renee van der Nat is PhD candidate and teacher at the University of Applied Sciences Utrecht, the Netherlands. Her PhD research focuses on new storytelling practices in journalism, particularly where journalism, storytelling and technology intersect. She has worked on a research project on the production and reception of news visualizations.

Hong Tien Vu is assistant professor in the Williams Allen White School of Journalism at the University of Kansas, USA. His research focuses on international communication, development communication and changes in newsroom practices amid the rise of technological innovations. His research has been published in *Journalism & Mass Communication Quarterly; Journalism Studies; Journalism; Asian Journal of Communication;* and *Journal of Information Technology & Politics.* He has won a number of prestigious awards for his academic excellence and research potential at the University of Kansas and the University of Texas at Austin.

Oscar Westlund, PhD, is Associate Professor at University of Gothenburg, Sweden, and Adjunct Professor at Volda University College, Norway. He has long employed various methods in his award winning interdisciplinary research into the production, distribution and consumption of news via different platforms. He has published more than 100 articles, book chapters and reports in several languages. With Seth Lewis, he won the ICA's 2016 Wolfgang Donsbach Outstanding Journal Article of the Year Award. He is the incoming Editor-in-Chief of Digital Journalism (January 2018), serves eight editorial boards, and currently co-edits special issues for journals such as *New Media & Society* and *Media and Communication.* Throughout 2017–20 he leads a research project called the epistemologies of digital news production, funded by the Swedish Foundation for Humanities and Social Sciences.

Holger Wormer is an ordinary professor of science journalism and head of the Science Journalism Department TU Dortmund University, Germany. He studied chemistry and philosophy in Heidelberg, Ulm and Lyon, France, and was science editor at the German nationwide newspaper Sueddeutsche Zeitung between 1996 and 2004. He holds several journalism awards, including the Journalist of the Year award in the science journalism category. His research focuses mainly on quality and ethics of science, medical and environmental reporting. His latest book, *Endlich Mitwisser!* was 'Science Book of the Year' in 2012 in Austria. In 2014, his department started Germany's first systematic university course in data journalism.

Foreword

Stuart Allan, *Cardiff University, UK*

'The modern fortune teller is the statistician,' Walter Lippmann observed in his highly influential newspaper column in April, 1935. He continued:

> This suits the temper of our age in which the greatest triumphs of man [*sic*] have been achieved through the use of exact quantitative measurement, and surely it is an excellent thing that men [and women] are learning to count the facts instead of asserting them.

Still, he cautioned, certain pitfalls were apparent.

> Just because statistics seem so cold and so impersonal, they tend to tyrannize over reason and common sense and men [and women] will often believe on the authority of a statistical curve something which cannot be true according to all the rest of their knowledge (Lippmann 1935, 14).

Lippmann, one of the leading journalists and political commentators of his day in the United States, proceeded to argue that in order to overcome the 'disposition to believe the incredible' based on 'statistical interpretation', it is crucially important to avoid making hasty generalizations from 'unanalysed figures'. Further dangers lurked in the use of statistics, he warned, including making unwarranted speculations regarding extrapolated trajectories over time, or gratuitous comparisons reliant on sets of facts drawn from dissimilar contexts. 'The best statisticians are very skeptical,' he surmised. 'They respect their tools but they never forget that they are tools and not magic wands and divining rods.'

Newspaper readers – then as now, we may add – would do well to remember Lippmann's advice to 'learn from the statisticians how to be thoroughly skeptical', particularly when the statistics in question invite counter-intuitive conclusions. This is a theme raised in his column previously, when he expressed his concern that so 'much of our current discussion has to do with statistics and graphs and broad general tendencies and movements and forces and what not, that it gets to be tiresome and unreal'. In the absence of 'something personal and definite to deal with', readers were left with 'abstractions and impersonal things' inhibiting their understanding (1933, 18). In striving to chart a more positive, productive way forward, Lippmann had argued in his earlier book *Liberty and the News* for journalists to become 'patient and fearless men [and women] of science', prepared to delve deeply 'to see what the world really is'. Good reporting, he contended, 'requires the exercise of the highest of scientific virtues', which are 'the habits of ascribing no more credibility to a statement than it warrants, a nice sense of the probabilities, and a keen understanding of the quantitative importance of particular facts' (1920, 82–3). Easier said than done, of course. More than a question of journalism training being compelled to improve the teaching and learning of numeracy skills, as welcome as this would be, he believed it would necessarily entail a critical reorientation of journalism's routine, everyday engagements in the mediation of statistical data.

Considered from the vantage point of today's 'post-truth' politics, when the authority of 'experts' advancing claims made on quantitative evidence seems increasingly open to criticism, if not outright cynicism, Lippmann's analysis of how journalism can become mired in accusations of manipulating numbers in the service of opinion will be sure to resonate. Indeed, apparent from the outset of An Nguyen's edited collection *News, Numbers and Public Opinion in a Data-Driven World* is the perception that journalism is consistently falling short in fulfilling its public responsibilities where the reporting of such evidence is concerned. This volume's contributors, from an array of complementary perspectives, assess what has gone wrong, why this is the case, and how best to improve matters in the public interest. Not only does one recognize how the challenges unfolding for journalism in the digital era of big data have been reverberating since Lippmann's day, this awareness of the fluid vicissitudes of contingency alerts us to real prospects for future change. Statistical claims can only 'tyrannize over reason and common sense', to repeat his apt turn of phrase, to the extent any insistence on their intrinsic truth-value is allowed to pass unopposed, their presumed authority aligned with accustomed rituals of assured expertise. This is not to deny that when confronted with the uncertainties of calculating future trends the desire for recourse to predictable patterns can be understandable, but it makes for neither good science nor rigorous news reporting.

In learning to live with complexity, journalists and their citizen counterparts must continue to innovate through experimentation, and in so doing strive to secure new strategies for bringing to light causal forces and imperatives otherwise obscured in a superficial swirl of data. Bearing these priorities in mind, one appreciates how *News, Numbers and Public Opinion in a Data-Driven World* succeeds in pinpointing exigent issues recurrently eluding sustained attention and critique. Moreover, I would suggest, its chapters taken together represent an important intervention, one certain to inspire us to recraft anew a pragmatic agenda for reform.

References

Lippmann, W. (1920), *Liberty and the News*. New York: Harcourt, Brace and Howe.
Lippmann, W. (1933), 'Personal devils'. *The Boston Globe*, 14 December, p. 18.
Lippmann, W. (1935), 'Elusive curves'. *The Boston Globe*, 13 April, p. 14.

Introduction

Exciting Times in the Shadow of the 'Post-Truth' Era: News, Numbers and Public Opinion in a Data-Driven World

An Nguyen, *Bournemouth University, UK*

Exciting times

Not so long ago, it would have sounded bizarre to predict that journalism – the profession that often 'gets a bad press' for its permanent hostility to and poor use of numbers – would soon engage with numbers and statistics as an essential daily routine. A browse through online and print news pages today, however, can give an immediate sense of something of an ongoing transformation: from the common most-read/most-viewed lists to the increasingly rich and diverse menu of fancy infographics, interactive data maps and so on, data are now behind many things journalists are producing and delivering to news consumers. As one enters a newsroom, that transformation becomes even more readily observable. Computer screens are jam-packed with web metrics (audience analytics), with continuous streams of such data being beamed and projected in real time onto large and prominently positioned 'performance panels' (Nguyen 2013). Editors keep 'a hawk eye' on those click and view figures behind their most-read/most-viewed/most-shared lists to make news judgement 'on the fly' around the clock (MacGregor 2007), with many having established the routine of beginning news meetings with a rundown of audience data. Here and there, an innovation mood can be felt, with newspeople across soft and hard content areas being busy experimenting with new data-driven storytelling tools. The physical space of some newsrooms has been reconfigured accordingly to accommodate new operations and teams, such as audience strategies (or audience development) and data journalism, and new workflows are created for non-news professionals – such as statisticians, data scientists, programmers and app developers – to sit and work alongside traditional newspeople.

Such are only a few sketches of the change that digital technologies are bringing to journalism's relationship with numbers and statistics. On the one hand, as seen above, the accelerating datafication of society – that is, the use of digital technologies to collect, record, store, monitor and quantify human behaviours and activities for various social, economic and political purposes – has resulted in the intensive use of audience metrics in the journalistic gatekeeping process. On the other, with the world's move to transparency and open-access data in the digital environment and the recent introduction of simple, user-friendly data analysis and visualization software, the relatively (not entirely) new genre of 'data journalism' has enjoyed quite a surge, especially since the beginning of this decade. Both have been hailed as essentials for journalism to go into the future. Improving the use of metrics in the newsroom, for example, was seen as 'very important' by 76 per cent – and 'somewhat important' by another 21 per cent – of senior news executives and digital strategists around the world who responded to a study by Newman (2016). Another recent survey put data journalism as the third most sought additional training area by American journalists – below video shooting/editing and social media engagement but above twenty other categories (Willnat and Weaver 2014). Tim Berners-Lee, the inventor of the World Wide Web, declared in 2010 that 'data-driven journalism is the future':

> Journalists need to be data-savvy. It used to be that you would get stories by chatting to people in bars, and it still might be that you'll do it that way some times. But now it's also going to be about poring over data and equipping yourself with the tools to analyze it and picking out what's interesting. And keeping it in perspective, helping people out by really seeing where it all fits together, and what's going on in the country.[1]

Similarly, Clay Shirky (2014) points out that, unlike 'the old "story accompanied by a chart" (that) was merely data next to journalism, increasingly, the data *is* the journalism'. Hence, the first thing that Shirky (2014) advised newspaper journalists to do to save their job and their profession is to 'get good with numbers'. If journalists want to stand out in this brave new world, in his views, they will have to learn to code, take online classes in statistics, and familiarize themselves 'with finding, understanding, and presenting data'.

It is, in short, exciting times ahead.

With this potential future in mind, however, one ought to ask some hard questions. For this 'quantitative turn in journalism', as Petre (2013) calls it, to make sense and to benefit us all in the long term, it needs to be understood and explored in the historical context in which it takes place. However powerful this quantitative turn is, its socio-historical context will play a critical role in determining the extent to which its power can be unleashed and materialized for the sake of a better informed, more self-governed citizenry. Here, I would propose an examination from the perspectives of three key stakeholders in news production and communication: the public, statistical sources, and journalists as the middleperson between them.

Numbers as a staple of modern life

Let us start with the public. The first thing is to go beyond the hype to recognize that statistics have been a critically important part of our daily life for a very long time. As

---

[1]   *The Data Journalism Handbook*, http://datajournalismhandbook.org/1.0/en/introduction_2.html.

early as 1903, when statistics had not even established itself as an academic discipline, Herbert G. Wells had already predicted in *Mankind in the Making* that statistical skills would soon become as essential as reading and writing for efficient citizenship:

> The great body of physical science, a great deal of the essential fact of financial science, and endless social and political problems are only accessible and only thinkable to those who have had a sound training in mathematical analysis, and the time may not be very remote when it will be understood that for complete initiation as an efficient citizen of one of the new great complex world-wide states that are now developing, *it is as necessary to be able to compute, to think in averages and maxima and minima, as it is now to be able to read and write.* (Wells 1903, 189, my emphasis)

Such prophecy would have materialized over the course of the twentieth century, with statistics having now reached a point where they are 'part of the fabric of the contemporary world' (Nguyen and Lugo-Ocando 2016, 4). Numbers occupy a pervasive position in modern life because almost every key aspect of it – from the quality of the air we breathe to the national leader we choose – is numerically measured and represented in one way or another. In this 'hyper-numeric world', as Andreas and Greenhill (2010, 1) point out from a political perspective, 'if something is not measured, it does not exist' and 'will not be recognized, defined, prioritized, put on the agenda, and debated'. The age of 'big data' means numbers will exercise more power over the way we live and work – but we should not lose sight of the fundamental fact that they, as Lorenzo Fioramonti (2014) puts it, have long 'ruled the world'. In development policy-making, for example, statistics such as GDP or poverty threshold have become an indispensable tool to summarize, justify and/or rubberstamp policies across the globe:

> One extra dollar a day per capita, and the funding for a series of hospitals in India might stop on its track. A dozen more deaths of a certain disease, and the World Health Organisation declares an epidemic. For many in the world, these numbers mean the difference between life and death, poverty and prosperity, or happiness and misery. For others, who have seen Gross Domestic Product grow for decades and their lives remain as miserable as ever, these statistics mean absolutely nothing. (Lugo-Ocando and Nguyen 2017, 43)

As such, statistics have become a constant staple of our daily news diet (Blastland and Dilnot 2008; Maier 2002). A recent content analysis of British broadcasters' news output, for example, found at least one reference to statistics in more than a fifth of the sample, especially among health, politics, business and economics coverage (Cushion et al. 2016). And although there are legitimate concerns that ordinary citizens do not have the knowledge and skills to handle so much numerical information (Utts 2002, 2010), the broad assumption remains that they rely on data reported in the news to exercise some control over the social world, to make rational choices over public affairs and ultimately to maintain their influence over public institutions.

The age of dubious numbers and 'statistical bullshit'

This leads us to the second stakeholder: the people and institutions that produce and/or circulate statistics in the public sphere. The establishment in every society understands the power of numbers very well and tries its utmost to use that power to their own advantage. Today it is rare to find a resourceful political, economic or social institution that does not resort to statistical information as a crucial tool to shape public opinion. Indeed, this is a 'chicken and egg' relationship: as statistics are too central to daily life and daily news, they force such institutions to continuously measure and collect data for just everything within their operational scope to defend, legitimize and/or promote what they do or want to do (Best 2001).

One outcome of this is a deluge of bad statistics being circulated in the public sphere, thanks to both traditional media platforms and, more recently, the very helpful hand of social media and their echo chambers. Many of these numbers are deliberate attempts to win the public debate with dubious and/or misinterpreted numbers – such as the now-famous misleading £350 million per week that the Boris Johnson, Michael Gove and their Brexit followers claimed the UK had to pay for EU membership a way to persuade UK voters to vote to leave EU during the EU Referendum campaign in 2016. Some others are deliberately manipulated or even fabricated to serve some political or commercial interests. All these belong to what Tim Harford (2016) calls 'the rise statistical bullshit – the casual slinging around of numbers not because they are true or false, but to sell a message'. In the world of social media, as Harford notes, such bullshit 'spreads easily these days (because) all it takes is a click', with the bullshitter caring nothing about its truthfulness but its ability to catch attention and/or stir emotion. The bullshitter could even be more dangerous than the liar, because at least the latter cares about and defers to the truth.

That is what some have called the post-fact, post-truth era. Indeed, journalism embarks on its new quantitative journey at a very troubling time for numerical facts, a time overshadowed by the rise of fake news, the emotionally charged populist movements behind Brexit and Donald Trump's controversial election, and the increasing power of the few private companies that control big data through algorithms. Instead of helping to serve as stable reference points and to settle arguments, as sociologist William Davies (2017) puts it in the *Guardian*, statistics have lost their authority:

> Rather than diffusing controversy and polarisation, it seems as if statistics are actually stoking them. Antipathy to statistics has become one of the hallmarks of the populist right, with statisticians and economists chief among the various 'experts' that were ostensibly rejected by voters in 2016. Not only are statistics viewed by many as untrustworthy, there appears to be something almost insulting or arrogant about them. Reducing social and economic issues to numerical aggregates and averages seems to violate some people's sense of political decency.

The situation has reached such a critical point that the highly respected University of Washington has recently started a 10-week course named 'Calling bullshit in the age of Big Data', which they offered both face-to-face to 160 students and online to the general public (through recorded lectures/seminars and readings).

The crucial but largely forfeited role of journalism

In this increasingly chaotic world, journalists must be adept at using quantitative data to filter the true from the false and ultimately to lead the public in the right direction. Indeed, even without entering the post-truth era of fake news and filter bubbles, one would have expected that role long ago: if statistics can be worse than lies and damn lies, to use that famous old saying, then the logical outcome would be that journalism, the profession that purports to tell the truth and expose the untruth, would have long found in them a fertile land for deep-digging and world-changing investigation. The somewhat odd concurrent rise of big data and 'post-truth' can only highlight that critical importance of journalism in exposing misleading, dubious numbers that are central to public life.

On the other hand, from a more positive side, journalists need to be statistically competent because there is also a great deal of good data and statistics which, if used well as an authoritative form of evidence, could assist them greatly in searching the truth and helping the public to make sense of the complex world around them. This is by no means a new argument: as Stuart Allan says in the Foreword, it was indeed a central point that the great American thinker, Walter Lippmann, made about journalism in the early part of the twentieth century. Not long after H. G. Wells said the above, Lippmann argued ardently in *Public Opinion* for the integration of scientific rigour into journalism, calling newspeople to bring to their truth-seeking and truth-telling function the objective method of science, which is 'based on exact record, measurement, analysis and comparison' (Lippmann 1920, 138). 'It does not matter that the news is not susceptible of mathematical statement,' he said. 'In fact, just because news is complex and slippery, good reporting requires the exercise of the highest of the scientific virtues' (1922, 49). This means, as he pointed out in an earlier book, *Liberty and the News*, journalists would have to possess the ability to '(scribe) no more credibility to a statement than it warrants, a nice sense of the probabilities, and a keen understanding of the quantitative importance of particular facts' (Lippmann 1920, 82).

Flash forward a century and, as hinted at the outset, the ability to work with numbers is still not only an intellectual luxury but sometimes a cursed subject in the news industry. Many journalists simply cannot settle themselves with the idea of scrutinizing or putting statistics to good use in news production: why do they have to deal with numbers, some might ask, while they come into journalism to work with words and to avoid all those eye-numbing figures? 'Journalism is one of the few professions that not only tolerates general innumeracy, but celebrates it,' said Aron Pilhofer, executive director for the digital at the *Guardian*. 'I still hear journalists who are proud of it, even celebrating that they can't do math, even though programming is about logic. It's hard to get a journalist to open up a spreadsheet, much less open up a command line. It is just not something that they, in general, think is held to be an important skill' (as quoted in Howard 2014, 46). Meanwhile, statistical training has been largely ignored by both universities and industry accreditation bodies – and the recent rise of big data has not changed that mindset in any substantial way (Howard 2014, Nguyen and Lugo-Ocando 2016).

In that context, it is perhaps not surprising that nearly two-thirds of 4,300 sampled references to statistics in the above study of UK broadcasters' news output were either vaguely mentioned or were clearly presented but with little context, while only 4.2 per cent of them were challenged in some way (Cushion et al. 2016). By the same token, early research into data journalism has found it to be done on a superficial level, in limited scopes and formats, with 'as much decorative as informative' impact (Knight 2015, 55). Neither is it to see that the key drivers of most data journalism projects are not traditional newspeople but those with non-journalism skills and experiences, such as programmers, developers, designers and producers (Tabary, Provost and Trottier 2016). Some influential names that have triggered the interest in data journalism, such as Nate Silver or Ezra Klein, had never worked as journalists before entering the news scene (Nguyen and Lugo-Ocando 2015).

All this begs the question: could journalists ever 'get good with numbers' – to not only make the most from the potential of 'data journalism' and audience metrics but also, and more importantly, to effectively handle, use and communicate data and statistics in general?

My personal answer is a resolute yes – and it must be a resolute yes. The traditional ignorance of statistics in journalism training and education is based on too many incorrect assumptions and prejudices about journalism's relationship with numbers. As Sisi Wei, a news app developer at *ProPublica*, puts it:

> I wish that no j-school ever reinforces or finds acceptable, actively or passively, the stereotype that journalists are bad at math. All it takes is one professor who shrugs off a math error to add to this stereotype, to have the idea pass onto one of his or her students. Let's be clear: journalists do not come with a math disability. (quoted in Howard 2014)

Which is quite true: research has shown clearly that beginning journalism students have no less ability to deal with numbers than those in other disciplines – including those in the natural sciences (Dunwoody and Griffin 2013; Maier 2002; Nguyen and Lugo-Ocando 2016). As Pilhofer (ibid.) commented: 'Journalism programs need to step up and understand that we live in a data-rich society, and math skills and basic data analysis skills are highly relevant to journalism.' All that is needed is the right statistical training approaches that (a) place a total confidence in the ability of journalism students to handle numerical and statistical information and (b) equip them with the knowledge and skills not only to source, verify and scrutinize data but also to effectively present and deliver them to the lay public. The ultimate aim of any such training is for journalists to be able to help citizens to relate themselves to the world beyond them and to vanquish the true from the false, the gold from the bullshit. It would require not just statistical skills but also in-depth knowledge and understanding about the socio-psychological interaction between news, numbers and public opinion and its socio-political impact in today's data-driven world. It is time for media and journalism scholars, statisticians and the news profession to get together to develop a strong knowledge base for this task.

This book's intervention

It is in that spirit and towards that aim that this book comes together. While there is no shortage of practical guidelines and textbooks on statistical skills for journalists and the general public, scholarly research into the area remains relatively scarce and, at best, scattered. How do news, numbers and public opinion interact with each other? Is it really the case, as common wisdom tells us, that most citizens and journalists do not have the necessary cognitive resources to critically process and assess data and statistics? What strategies do journalists use to gather and communicate statistical data in their stories? What skills do they need to deal effectively with the daily influx of numbers in the resource-poor and deadline-driven newsroom? What sort of formats and frames do they often use to represent different types of data? What are the potential socio-psychological and political effects of such formats and frames on public information, knowledge acquisition and attitude formation? What good practices are there to promote and what bad ones to avoid – and how?

Gathering some of the most established scholars as well as emerging names in both statistics and media and journalism studies, this book offers a comprehensive, authoritative, first-of-its-kind collection of past and present research into the issues above. As represented in the word cloud below (generated from the book's manuscript

Figure 0.1 A simple word cloud featuring the key themes and topics of this book (created from its book manuscript).

with the free but powerful tools on wordclouds.com), our focus will be on how journalism makes use of data and statistics, how such numerical information is told in news stories, and how all this affects different types of audiences and publics. Towards that purpose, the book is divided into three sections, examining how numbers are received, treated and handled by at both ends of the news communication process – the journalist in the newsroom and the news consumer in the public at large – with a particular attention to their effects on public reasoning and deliberation.

Section 1: Data and statistics in news production

The first section consists of seven chapters that offer theoretical and empirical insights into how and why data and statistics are gathered, used and represented in certain ways in the news. Chapter 1 is a personally informed account of an engaged statistician with (often frustrating) experience with journalists in Switzerland. Fabienne Crettaz von Roten starts with an inventory of common statistical errors in the news before outlining three key roles that, in her view, journalists should assume but have largely failed to fulfil in the statistics – society relationship – namely verifying and scrutinizing statistics, communicating statistics in an accurate manner, and educating the public about statistics. She identifies the lack of statistical training in journalism education as part of the problem and concludes with some thoughts on potential solutions for that shortage, with particular attention to whether to teach statistical reasoning as a separate course or as embedded elements across different courses in the journalism curricula.

Chapter 2 offers some empirical insights of the overall quality of statistical news reporting, based on results from the aforementioned BBC Trust-funded review of statistics in the UK broadcast media. Stephen Cushion and Justin Lewis look deeply into the current state of everyday reporting of statistics across radio, television and online news platforms, both with a quantitative overview and two qualitative case studies. Although they found that 'statistics were routinely invoked in vague and imprecise ways, with limited context and explanation about their meaning or method behind them', some rare good-practice examples that they identified suggested that journalists would do a far better service to the public by rethinking and reinterpreting their (sometimes problematic) impartiality ethos in a way that enables and prepares them to scrutinize and challenge statistics whenever necessary.

Chapter 3 provides another insightful case study, examining the uncritical news reporting of body-count statistics during the US-led drone war in Afghanistan in the past fifteen years. Muhammad Idrees Ahmad, like Cushion and Lewis above, demonstrates how journalists, overshadowed by their 'strategic ritual of objectivity' and constrained by the difficulty to find accurate numbers about drone casualties, disregard questions about the validity of official figures from the military. In reporting drone-induced body counts to a remote passive public without knowing, for example, they included more terrorists or more civilians, journalists have greatly contributed to a long-running political and military attempt to use dubious and imprecise casualty statistics to legitimize the drone war. The consequence has been a false public image of drones as a panacea for security problems, leading to a dramatic shift in public attitudes towards extrajudicial killing, with dangerous consequences.

Chapter 4 examines the political communication of statistics in the context of news coverage of poverty. At the heart of Jairo Lugo-Ocando and Brendan Lawson's critique is the diminishing ability of journalists to critically analyse and counter data-based hegemonic discourses, leaving too much space for political sources – such as government officials, multilateral governance bodies, NGOs and others – to spin over poverty concepts and numbers. The consistent failure of the media around the world to report and expose PR stunts around poverty statistics, according to the authors, can be attributed to not only the lack of an agreed definition of poverty between its key stakeholders but also, again, to the lack of statistical preparation among newspeople and decreasing resources in the newsroom. The result is an over-reliance on official sources, who are allowed to get away easily with their own, often flawed and ideologically driven, explanatory frameworks. With that, Lugo-Ocando and Lawson conclude the purported function of journalism as a mediator in development policy formulation and a watchdog of democratic decision-making has been ineffective.

Chapter 5 moves the discussion to another area where statistics are an inevitable part of the day: science journalism. Renata Brandao and An Nguyen start with a review of journalism's traditional deference to statistics from cultural and epistemological perspectives before citing a number of good reasons to hope that science journalism might be an exception to that relationship. However, their ensuing review of the past and present of science journalism – particularly its traditional cheer-leading approach to science, with its focus on promoting science literacy and the consequent failure to equip lay people with the ability to critically engage with and question science – suggests that that hope might be more like a dream. Employing a secondary content analysis of 1,089 science articles in four leading UK and Brazil broadsheets, they found evidence that although journalists did not deprive readers of the opportunities to understand science statistics in the social context, these statistics were presented mostly as unquestionable and undisputed facts, without the essential methodological information that readers need to critically understand, judge and, if necessary, raise their voices over science developments.

Chapter 6 explores the latest trend in the news – numbers relationship, namely the evolving shape of the new genre of data journalism. Using a content analysis of projects nominated for the Global Editors Network's annual Data Journalism Awards (DJA) between 2013 and 2015, Julius Reimer and Wiebke Loosen explore if, and how, data journalism at its finest has changed over the years in terms of, among others, data sources and types, visualization strategies, interactive features, topics, and types of nominated media outlets. Their results suggest that although data journalism is increasingly personnel-intensive and progressively spreading around the globe, the set of structural elements and presentation formats that the DJA-nominated pieces are built upon remain rather stable and restricted to relatively simple forms. In addition, although half of the sampled projects assume a watchdog role (i.e. with elements of criticism or calls for public intervention), this role is limited as journalists rely mainly on publicly available, thus institutionally friendly, datasets. There is, however, evidence that they are increasingly looking to non-institutional data sources for their stories.

Chapter 7 deals with a very different but fast-growing use of statistics in the newsroom: the use of digital audience-tracking data in news production and distribution. Against the backdrop of the traditionally powerless position of the audience in the journalistic gatekeeping process, An Nguyen and Hong Tien Vu review the recent unprecedented penetration of web metrics into the newsroom, examining its key social and technological drivers as well as its potential impacts on news decision-making processes and journalists' work ethos and autonomy. As news moves from being exclusively 'what newspapermen make it' to also something that the crowd wants it to be, the authors demonstrate that newsroom processes and relationships are being transformed in ways that invite more misgivings and reservations than hopes and innovations. In observing the latest developments, however, the authors conclude with a more positive and optimistic note, introducing a range of promising audience data policies and measures that might help journalists to go beyond head counting and get more deeply into news engagement metrics to harness their power for a better and more sustainable future for journalism.

Section 2: Data and statistics in news consumption

The second section examines data and statistics from the news audience's perspective, with five richly insightful chapters on how statistical information in the news interacts with individuals' reasoning, knowledge acquisition and attitude/belief formation in their public and private life.

Scott Maier starts with a thought-provoking chapter that challenges our common beliefs about the power of numbers in the news. Statistics are to trigger rational action but their impact on human emotion could be big and undesired. Maier proves it through the case of news about humanitarian crises, where experimental psychology research shows that statistical information can actually diminish empathy and discourage humanitarian response. This 'psychophysical numbing' effect (lives are valued less as their numbers increase) was tested and confirmed in his research, including a year-long examination of reader responses to the use of personified numbers by *New York Times* columnist Nicholas Kristof and an experimental study on how numbers-based stories differ from other storytelling forms in their influences on reader emotions. The cautionary lesson he offers: the news media might select some statistics to raise awareness but should not rely solely on them to drive responses to calamity and mass injustice. Embedding personal stories in numerical news reporting might be the way, he concludes.

That, however, comes with potential issues that Chapters 9 and 10 deal with. In Chapter 9, Charles Berger offers an authoritative review of research into the effects of a regular use of anecdotal cases and quantitative data to depict threatening trends (e.g. traffic facilities, crime and health problems) in the news. He begins with a discussion of how news consumers, in the absence of relevant statistical data, use mental shortcuts and heuristics – such as extraordinary cases and non-representative anecdotes that are regularly provided in the news to illustrate hazards and threats – to make wrong inferences about the general trends/risks that such anecdotes depict. Even when journalists present quantitative data in lieu of anecdotes, however, there persist many problems, especially with respect to their widespread, advertent or inadvertent, use of mere frequencies rather than base-rate data in depicting threatening trends. Through an analysis of some recent examples in the US media and a thorough review of experimental research, Berger shows how such statistical depiction may encourage news consumers to form distorted estimates of their risk of exposure to hazards and threats. This, as he reviews, leads to many individual and social consequences in a public with generally low numeracy, such as excessive anxiety (constituting a public health problem), attenuated achievement expectations and increased risk aversion.

In Chapter 10, Rhonda Gibson and Coy Callison delve even more deeply into the rational and emotional responses of news consumers to numeric information and exemplar cases, with attention to how these responses vary across different quantitative literacy levels and different numerical presentation formats. Following a brief introduction to exemplification theory, which predicts how base-rate data and exemplars in news reports affect audience perception of the reported issues, the authors review research that measures the effectiveness of various data formats – such as percentages, fractions, probabilities, and primary versus secondary ratios. The second part of the chapter addresses more recent research into (a) the effects of an individual's quantitative literacy level on his/her willingness and ability to process numeric information in news reports, (b) the rational impressions and affective reactions that news users with different numeric ability form after exposure to statistical information and/or exemplifying cases about risk-related topics, (c) the subsequent effects on those individuals' levels of empathy with victims, their personal risk assessment as well as their assessment of risk to others. The chapter ends with useful recommendations for journalists and others who create statistical messages designed to inform the public.

If Chapters 9 and 10 raise issues about the potential undesired effects of statistical information in the news, Chapter 11 seeks an explanation for the so-called number paradox: the fact that the media use a lot of numbers, even though it is known that news consumers do not usually recall, understand or appreciate such numbers in news articles. Willem Koetsenruijter first reviews the relevant research literature to demonstrate that the impact of numbers on the way news consumers recall and perceive news facts might not be as strong as expected. Assuming that reporters are reasonable actors with purposes behind everything they do in news making processes, the author hypothesizes that journalists use numbers as a rhetorical device to appear credible, that is, they use the authoritative status of statistics as objective and scientific facts to boost their ethos in the mind of news consumers. Koetsenruijter presents a series of experiments designed to test this hypothesis in which he found enough evidence to conclude that the use of numerical information did increase the perceived credibility of the news both in newspapers and on television.

Chapter 12 reports an innovative, comprehensive mixed-method study on how news consumers use and evaluate infographics, the major distinctive form of the new data journalism genre. Yael de Haan, Sanne Kruikemeier, Sophie Lecheler, Gerard Smit and Renee van der Nat first conduct an eye-tracking study to explore the extent to which news consumers pay attention to and use data visualizations in three different news modalities (a print newspaper, an e-newspaper on a tablet and a news site in the Netherlands). Based on the results of this part, they then use the same material to conduct focus groups and an online survey to investigate whether news consumers appreciate and value data visualizations. The results show that news consumers do read news visualizations, regardless of the platform on which the visual is published. In addition, visualization elements are appreciated, but only if they are coherently integrated into a news story and fulfil a function that can be easily understood. Given the scarcity of research into this area, the chapter will provide the first comprehensive picture of the usefulness of infographics in the news and contribute to a growing literature on alternative ways of storytelling in journalism today. Its innovative mixed-method approach also suggests some fresh ideas for future research into the use and effect of new storytelling formats in journalism and beyond.

Section 3: Agenda for the future

The third section offers some forward-looking perspectives on what can be done to improve the status quo of the journalism, statistics and society relationship and how we might theoretically and empirically approach to improve this relationship in the age of big data.

In Chapter 13, Kevin McConway, an academic adviser to BBC Radio 4's More or Less show and the UK Science Media Centre, offers a useful mix of history, theory and personal experience from working with journalists to discuss how the interaction between statisticians and journalists can be more fruitful. Although they do similar things at the macro level, he argues, fundamental differences in the way each profession works exist and need to be mutually understood and acknowledged if they are to work effectively with each other. Reflecting on the strange case of news about potential links between mobile phone use and the risk of brain tumours, McConway goes on to outline the key differences between the two worlds, calling on statisticians to understand journalism in at least three respects: its timescales (deadline pressures), its agenda (news values and editorial policy), and its pressure to personalize stories from numbers. His 'take-home message': statisticians need journalists as they do not have the strengths of journalists – namely telling stories to the right audience in a short space. Therefore, they 'should not simply blame journalists for getting things wrong' but 'must help them to get things right' by being 'proactive in making known to journalists what we do, and why and how we do it'.

Chapter 14 looks to the future with a perspective from inside the newsroom, examining the potential role of science journalism as a possible bridge between data and journalism. Holger Wormer charts the recent rise of data journalism – from the tendency to tell stories in 'charticles' in the late 2000s to the growing use of colourful and interactive graphics in online media – as an extension of the media's traditional 'love affairs' with numbers. At the same time, he argues, there is not any real evidence that journalists in general have significantly improved their knowledge on data and statistics in recent years. Neither is it clear whether data journalism projects are more about a truthful appearance than substance, that is, whether they can offer something really meaningful from a statistician's point of view and be better than the simplistic 'he-said-she-said' journalism formula. In exploring these issues, Wormer argues that one possible bridge to connect both sides could be the traditional cadre of systematically educated science journalists who are familiar with journalistic needs and skills (between news selection, classical investigation and storytelling) but also with basic principles of statistics and scientific methods. From this perspective, science journalism education strategies may be not only a key to improve the strange relationship between journalists and statistics but also predestined for a 'second generation data journalism'.

Chapter 15 focuses on the current place of statistical reasoning in journalism education to propose some strategic directions for the teaching of these skills in the future. Robert Griffin and Sharon Dunwoody showed from findings of twin surveys with US journalism chairs in 1997 and 2008 that a clear majority valued statistical reasoning skills as a competitive advantage for their students in the job market. The level of statistical reasoning instruction, however, remained relatively low and saw only a small improvement between the two waves of the study. Although these surveys date from, respectively, one and two decades ago, the authors yield some useful and generalizable points that bear fresh implications for today and the future. For journalism education to be at the helm in the 'next big thing' of data journalism and to equip journalists with evaluative capacities to align truth claims with evidence to suppress fake news, universities will need to address the many factors that the surveys found to be behind the past and present low-key profile of statistical training in journalism curricula – namely the perceived unwillingness and inability of students to acquire statistical reasoning, the shortage of faculty with relevant expertise, the tightness of existing journalism curricula and other structural factors, and the lack of reward for faculty entrepreneurship (i.e. attempts to bring statistical reasoning into their classes). 'But getting there', they warn from their decadal surveys, 'will be challenging' because of the very slow pace of changes in academic institutions and requires much entrepreneurship on part of journalism chairs themselves.

The last chapter shifts the focus to journalism research, with Oscar Westlund and Seth Lewis offering a set of conceptual and theoretical toolkits to approach the increasing datafication of society and its potentially transformative consequences on how news is perceived and practised. Their chapter presents four conceptual lenses on big data and journalism: *epistemology* (the legitimization of new journalistic claims about knowledge and truth based on big data); *expertise* (the negotiation of occupational status, authority, and skill sets as new specializations are developed and deployed in the newsroom); *economics* (the potential for and challenges with new efficiencies, resources, innovations, value creations and revenue opportunities that big data might induce for journalism); and *ethics* (ethical issues raised for the norms and values that guide human decision-making and technological systems design). While each of these addresses relevant questions for news media and helps to guide future research, Westlund and Lewis choose to devote the rest of the chapter to the current research literature on the epistemology of big data and journalism and future directions for such research. Their investigation can never be more timely in this post-truth moment. As the authors argue themselves, exploring the epistemology of journalism and big data is in itself 'an opportunity to understand how information is legitimated as "real" or "fake" to make sense of journalism's co-production of knowledge with publics, to identify dynamics of information production and circulation that lead to various interpretations on the part of audiences'.

All in all, it is our ambition that this book will make a fresh contribution and invite further research to an enquiry area that should have received much more scholarly attention in the literature. In addition, we hope that the book offers the news profession some systematic perspectives and principles to build a pragmatic framework for a more effective, more fruitful interaction between journalism, statistics and society. We have never needed such theoretical and practical work as badly and urgently as we do today.

References

Andreas, P. and Greenhill, K. M. (2010), *Sex, Drugs, and Body Counts: The Politics of Numbers in Global Crime and Conflict*. Ithaca, NY: Cornell University Press.

Best, J. (2001), *Damned Lies and Statistics: Untangling Numbers from the Media, Politicians, and Activists*. London: University of California Press.

Blastland, M. and Dilnot, A. (2008), *The Tiger That Isn't: Seeing through a World of Numbers*. London: Profile Books.

Cushion, S., Lewis, J., Sambrook, R. and Ghallagan, R. (2016), Impartiality reporting of BBC reporting of statistics: A content analysis. Retrieved 30 September 2016 from http://www.bbc.co.uk/corporate2/bbctrust/our_work/editorial_standards/impartiality/statistics.

Davies, W. (2017), 'How statistics lost their power'. *The Guardian*, 19 January. Retrieved 19 January 2017 from https://www.theguardian.com/politics/2017/jan/19/crisis-of-statistics-big-data-democracy.

Dunwoody, S. and Griffin, R. J. (2013), 'Statistical reasoning in journalism education.' *Science Communication*, 35(4), 528–38.

Fioramonti, L. (2014), *How Numbers Rule the World: The Use and Abuse of Statistics in Global Politics*. London: Zed Books.

Harford, T. (2016), 'How politicians poisoned statistics', http://timharford.com/2016/04/how-politicians-poisoned-statistics/ (accessed 15 May 2016).

Howard, A. (2014), *The Art and Science of Data-Driven Journalism*. Tow Centre for Digital Journalism. Retrieved 12 February 2015 from http://tinyurl.com/jmxskoz.

Knight, M. (2015), 'Data journalism in the UK: A preliminary analysis of form and content.' *Journal of Media Practice*, 16(1), 55–72.

Lippmann, W. (1920), *Liberty and the News*. New York: Harcourt, Brace and Howe.

Lippmann, W. (1922), *Public Opinion*. New York: Pearson Education.

Lugo-Ocando, J. and Nguyen, A. (2017), *Developing News: Global Journalism and the Coverage of 'Third World' Development*. London: Routledge.

MacGregor, Phil (2007), 'Tracking the online audience.' *Journalism Studies*, 8(2), 280–98.

Maier, S. R. (2002), 'Numbers in the news: A mathematics audit of a daily newspaper.' *Journalism Studies*, 3(4), 507–19.

Newman, N. (2016), Journalism, Media and Technology Predictions 2016. Oxford: Reuters Institute for the Study of Journalism. Retrieved 30 April 2016 from http://tinyurl.com/h86ak9t.

Nguyen, A. (2013), 'Online news audiences: The challenges of web metrics.' In Karen Fowler-Watt and Stuart Allan (eds), *Journalism: New Challenges*. Bournemouth University Centre for Journalism and Communication Research. Online: http://eprints.bournemouth.ac.uk/20929/1/The%20challenges%20of%20web%20metrics.pdf.

Nguyen, A. and Lugo-Ocando, J. (2015), 'A vaccine against that anti-data journalism brain.' *Data-Driven Journalism*, 4 November. Retrieved 4 November 2015 from http://datadrivenjournalism.net/news_and_analysis/a_vaccine_against_that_anti_data_journalism_brain.

Nguyen, A. and Lugo-Ocando, J. (2016), 'The state of data and statistics in journalism and journalism education: Issues and debates.' *Journalism*, 17(1), 3–17.

Petre, Caitlin (2013), 'A quantitative turn in journalism?' Retrieved 1 March 2015 from http://towcenter.org/a-quantitative-turn-in-journalism/.

Shirky, Clay (2014), Last call: The end of the printed newspaper. *Medium*. Online: https://medium.com/@cshirky/last-call-c682f6471c70 (accessed 1 March 2015).

Tabary, C., Provost, A. M. and Trottier, A. (2016), 'Data journalism's actors, practices and skills: A case study from Quebec.' *Journalism: Theory, Practice, and Criticism*, 17(1), 66–84.

Utts, J. (2002), 'News and numbers: A guide to reporting statistical claims and controversies in health and related fields.' *The American Statistician*, 56(4), 330–31.

Utts, J. (2010), 'Unintentional lies in the media: Don't blame journalists for what we don't teach.' *Invited paper for the International Conference on Teaching Statistics, Slovenia.* Retrieved 30 January 2015 from http://iase-web.org/documents/papers/icots8/ICOTS8_1G2_UTTS.pdf.

Wells, H. G. (1903), *Mankind in the Making*. London: The Echo Library. Retrieved 20 April 2014 from http://www.gutenberg.org/ebooks/7058.

Willnat, L. and Weaver, D. (2014), *The American Journalist in the Digital Age: Key Findings*. Bloomington, IN: School of Journalism, Indiana University. Retrieved 20 April 2016 from http://news.indiana.edu/releases/iu/2014/05/2013-american-journalist-key-findings.pdf.

Section One

Data and Statistics in News Production

1

Common Statistical Errors in the News: The Often-Unfulfilled Roles of Journalists in Statistics–Society Relationship

Fabienne Crettaz von Roten, *University of Lausanne, Switzerland*

Introduction

Journalists hold a crucial function in the proliferation of numerical and statistical elements in the public sphere, as they introduce such information in daily news reports of virtually all areas of society. For example, audiences are frequently exposed to probabilistic estimates of a new disease, confidence intervals of success of a politician for an upcoming election, or results of a survey on a social issue. As one statistician observes, 'I find it hard to think of policy questions, at least in domestic policy, that have no statistical component' (Moore 1998, 1253). Understanding these statistical components is crucial for citizens to participate in public debate and to make political decisions. H. G. Wells's prophecy of a century ago – that statistical thinking will one day be as necessary for citizenship as the ability to read and to write – is becoming a reality (Billard 1998). But, as I argued elsewhere, 'since most citizens lack statistical literacy, it is not surprising that ambivalence is increasing and that the better educated (with more statistical skills) are able to develop more firmly held, more extreme attitudes' (Crettaz von Roten 2006, 247). For modern data-rich society to operate effectively, citizens must be statistically literate to avoid exclusion.

As the news media are a major source of numerical and statistical elements, journalists hold a crucial function in a data-rich society, through their daily work of selecting numerical and statistical information and translating this information into news stories that contribute to the formation of attitudes. As the long tradition of social psychology demonstrates, people form attitudes towards an object by way of information at disposal, the social environment, salience issues and so on (Eagly and Chaiken 1993). As such, the quality of statistical information in the media could exercise a strong influence on the relationship between statistics and society, and more generally between science and society. Mathematical and statistical errors or misunderstandings in the news media, however, are all too common.

As an engaged statistician, that is, a scientist willing to use her expertise for the public good,[1] I often feel the duty to write to editors or journalists when I see a statistical problem in the news. My contact with journalists frequently leaves me surprised and frustrated: journalists either stick to their positions, or reject the related errors with arguments that do not refer to statistics but other aspects of news. Many would argue, for example, that aesthetic considerations can take on the correct processing of information, that it's important to increase the impact of the news even with incorrect charts or that the reader would not understand otherwise. This chapter reflects on that experience to highlight three roles that, in my views, journalists should endorse in dealing with the relationship between statistics and society, namely verifying statistical information, presenting statistics in an accurate manner in the news and providing statistical education for the public. I will start with a personal account on common misuses of statistics in the media with examples from my own country – Switzerland – before moving on to briefly discuss each of the three roles. If these roles call for a rigorous inclusion of statistical thinking in journalism training, this has hardly translated into a reality in journalism education. The final section, therefore, reviews the place of statistical training in journalism education and suggests some ideas for future journalists to shoulder the three roles and to improve the relationship between statistics and society.

Some common misuses of statistics in the media

Statistical literacy can be characterized as 'the ability to understand and critically evaluate statistical results that permeate our daily lives – coupled with the ability to appreciate the contributions that statistical thinking can make in public and private, professional and personal decisions' (Wallman 1993, 1). According to Marriott (2014), the core concept of statistical thinking has evolved in the past century: at the beginning of the twentieth century, it involved averages, minimum and maximum; in the late twentieth century it expanded to expectation, variance, distribution, probability, risk and correlation; and in the twenty-first century it adds data, visualization and cognition. In this context, statistical literacy among journalists should ideally have followed the evolution. But scholars and critics have long documented poor and wrong uses of statistics in the media (Huff 1954; Paulos 1995; Best 2001,Best 2005; Goldin 2009; Nguyen and Lugo-Ocando 2016). Here are a few common misuses that I find from my own experience:

1. Much news involves figures found in releases or documents, but some numbers are so erroneous that they indicate a lack of order of magnitude among journalists. For

---

[1] The Code of Conduct of the Royal Statistical Society states that statisticians 'should always be aware of their overriding responsibility to the public good' (RSS 2014, 2). It means to prevent the occurrence of incorrect statistical results by scientists (Lynn 2016), but it means also to point out errors in the media.

example, one newspaper reported that 77.8 per cent of employees – instead of 7.5 per cent – had been harassed according to a study of the State Secretariat for Economic Affairs (Le Matin 2011); another estimated the remaining costs of tunnelling for the Lötschberg tunnel at 1.3 million Swiss francs instead of at 1.3 billion Swiss francs (Le Matin Dimanche 2010); finally, an article mentioned that Africa's gross domestic product was $2 billion instead of $2,000 billion (Le Temps 2012a).

2. Journalists often fill the news with figures from statistics providers, such as EUROSTAT, but such data series might not be fully comparable (e.g. not the same year, not the same statistics), which is often explained in providers' metadata files. To use the data, caution should be, but is rarely, exercised to read these files fully and critically. For instance, a Swiss newspaper compared gross annual salaries among European countries from EUROSTAT, without noting that gross salaries were summarized by a mean in some countries and by a median in others,[2] causing misleading comparisons and a misleading conclusion. In this example, the problem is not the quality of the data (the information is mentioned by the statistics provider) but lies in the journalist's use of the data. In my experience, journalists fail very often to deliver such notifications or mention them in very small print, making it difficult for readers to realize the difference of the measures at stake.

3. When journalists report the results of surveys, such as of a pre-electoral survey, they usually provide the accurate percentages of intended votes for candidates, but some imperfectly refer to information about variability. They endlessly comment on a 2 per cent or 3 per cent difference between two candidates, or on a 2 per cent or 3 per cent increase of voting intention for one candidate. This is meaningless if the margin of error is greater than 3 per cent – and this piece of technical information is often either unreported or shown in small print at the end of the article.[3] Even worse case is when they compare groups, such as linguistic regions in Switzerland, whose sample sizes would be smaller and the margin of error is bigger (often around 6 per cent). This misinterpretation of statistics is more or less frequent, varying across the types of media and the countries, but in Switzerland it is an ongoing problem.

4. Some technical terms in statistics, such as 'normal', 'significant', 'random' or 'correlation', are frequently used for a non-technical meaning in daily life, creating a common source of confusion. Its consequences can be seen through a recent news story about the price increase of roasted chestnuts (Tribune de Genève 2012). The journalist begins with the presence of a parasite in chestnuts (*dryocosmus kuriphilus*), then talks about droughts in Italy – the main supplier of chestnuts for Switzerland – and continues with the following comment: 'This meteorological phenomenon was correlated with the arrival of a devastating

---

[2] This error was later corrected by the newspaper (Le Temps 2012b).
[3] See the worthwhile visualization of the pre-electoral US surveys in 2016 with the margin of errors realized by the fact-check site Les Décodeurs available at http://www.lemonde.fr/les-decodeurs/article/2016/11/10/data-visualisation-les-26-derniers-sondages-de-l-election-americaine_5028726_4355770.html.

parasite.' What does 'correlate' stand for here? A non-technical term used only as a synonym of 'coincide', or a technical term, that is, a point-biserial correlation or a causal relationship (the more the weather is dry, the more we observe parasites in chestnuts)?[4] Such terms, in my views, should be used with caution (as little as possible), with an indication of its sense (technical or non-technical, and, for the former, the specificities of the technical sense).

5. The concepts of p-value and statistical significance play a central role in reporting quantitative results. For a statistician, these notions have a precise definition in the theory of statistical tests and it is recommended to complete it with effect sizes, whereas for a journalist, they are often only a rhetorical argument. An impressive number of scientific articles list the misuses of the notions and provide their exact meaning – see, for example, Glaser (1999) and Greenland et al. (2016). The media report many significant results, often wrongly as a synonym of practical importance or of causality, even if the design of the research does not allow that. The situation becomes more complicated if one considers that 'the adjective "significant" has a different meaning in the mouth of a scientist and on the ears of the general public' (Duplai 2012, 15). An article about household composition in Geneva stated that 'the structure of the household has experienced two significant changes between 2000 and 2011: the proportion of couples without children decreased by 23% to 21% and lone parent households increased by 7% to 9%7 per cent to 9 per cent' (Tribune de Genève 2015). What is the meaning of 'significant' here? If it is related to a statistical test, the reader needs more information on the context to be able to judge the practical importance of the changes (how many couples without children, how many one-parent families, what effect sizes and so on).

6. Graphs are often used to illustrate a news text with numerical and statistical elements. Data visualization has become more and more professionalized, thanks in a large part to technological advance and to leading figures in academic spheres (e.g. Hans Rosling, professor of international health at Karolinska Institute, Sweden) and media spheres (e.g. Amanda Cox, *New York Times* graphics editor). But misleading visuals can still be found, as was a case in a magazine for a Swiss university community (Wyss and Meyer 2013). The article addressed the distribution of men and women in academic staff of the university and included male/female silhouettes instead of bars (i.e. a bar chart) to illustrate the gender difference. The height of the silhouette was proportional to the data but the width of the silhouette was proportional to the height, leading to a surface that is not proportional to the data. This choice resulted in accentuation of the deficit of women in higher status. When contacted, the journalist acknowledged the error but justified it by citing the need of identification for the reader (more identification to a silhouette than a bar) and for aesthetic purposes. He concluded that the purpose was to raise awareness of the underrepresentation of women in leading positions and his graph achieves this perfectly.

---

[4] In fact, the plant is becoming more sensitive in the presence of the parasite to different types of stress, including drought.

This list is not at all exhaustive, but it is substantive enough for us to be concerned about the state of statistics in the media. This is worrying in light of recent changes to journalism in Western societies. Declining numbers of journalists and increasing time constraints on news production have increased the pressure on journalists to do more in less time, with a shortening of the verification step, meaning more factual errors, among other things. The disappearance of old news providers and the emergence of new ones may have introduced new sources of errors. The great shift to the internet generates some 'net-native' actors (e.g. non-governmental actors, interest groups and so on), blurs the boundaries between science and public communication (see Trench 2007) and allows misinformation to be spread faster and more difficult to correct. The significant challenges of having to validate and assess information at the same time as competing with other actors online for public attention make it hard for journalists to escape overstatement, misreporting or manipulation. Finally, there are challenges posed by the emergence of data journalism that has been fuelled by advanced technologies. New technologies for data collection, data visualization or statistical analysis pose a challenge to journalists, who are generally not trained to handle data and might have to rely heavily on exchange with digital data specialists. Along this exchange, errors may be introduced, later appearing in the news stories. For all these reasons, the new types of statistical errors might appear and on a larger scale than the past.

The three unfulfilled roles of journalists

To more systematically explore the situation depicted in the last section, I will discuss here the three roles that journalists should have in the science–society relationship and how they have done in these roles. These roles fall within fundamental journalistic standards (i.e. core values of accuracy, freedom from bias, independence, integrity and education mission) but have not yet been fulfilled to any substantial extent.

First, before producing any story, journalists must rigorously scrutinize and verify the related statistics, treating them like any other raw material. Journalists' practices imply checking and critically appraising statistical results before they are reported, but this rarely happens. Even in science journalism, Dunwoody (2008) underlined journalists' frequent inability to determine whether a source's assertions are true or not: 'Rather than judging the veracity of a truth claim, the journalist concentrates instead on representing the claim accurately in her story' (ibid., 20). As for statistics, two key aspects need to be scrutinized.

The first is the technical – or the reliability and validity of the data at stake. In the case of survey data, for example, potential bias of sampling methods must be analysed by journalists. But this is far from a standard practice in newsrooms. Due to financial shortage in the Swiss media, some election opinion surveys have been replaced by internet surveys done by the largest private media group, Tamedia. However, due to the non-representativeness of the readers of Tamedia's publications, and despite the use of weighting techniques, there is a risk to repeat the famous mistake of *Literary*

Digest in its 1936 US presidential election prediction,[5] which was based on a huge but unrepresentative sample of 2.4 million of its readers who were much wealthier people than the average American (which represented a sampling error of 19 per cent). Today, as Swiss journalists overlook such problems while reporting the results of Tamedia surveys, they will face similar errors in the future.[6]

The other crucial factor for journalists to detect during statistical data collection for news stories is the trustworthiness of their sources. Statistics sources might sound very reliable, thanks in part to the intrinsic authority of numbers, but they might hide many flaws. Even in the case of peer-reviewed publications, which are a generally trustworthy source of data that can be reported in a news story, journalists should still be cautious. Scientific peer reviews do not systematically apply statistical reviewing for empirical articles. A worrying amount of statistical errors in science publications – see, for example, Button et al. (2013) in neuroscience, or Bakker and Wichert (2011) in psychology – have pushed for progressive change. Voices are raised demanding a more proactive and responsible attitude of editors: 'Journals could recruit qualified statisticians as reviewers, which is done in some medical journals. The statistical review of a manuscript should encompass design, methods of analysis, presentation and interpretation of statistical results' (Crettaz von Roten 2013, 139). If peer review did this job, this would limit the chance for journalists to receive and communicate erroneous results.

The prevalence of statistical misinformation can be reduced by using fact-checking services. These can take the form of specific teams and operations within newsrooms (e.g. Les Décodeurs run by *Le Monde*) or independent non-profit groups. More than fifty fact-checking organizations have been launched across Europe and worldwide (Graves and Cherubini 2016), such as the non-profit Fact Check, which emerged in 2003 in the United States, or Full Fact founded in 2010 in the United Kingdom. Most of them focus on statistics used by journalists and the media. Advanced technology could also offer solutions in as close to real time as possible, with automated systems of fact checking such as Factminder. This field is still in early days, with promising projects having emerged with achievements alongside problems (Greenblatt 2016; Full Fact 2016).

When statistical information pass the verification, the second role of journalists is to translate it into news with a full and correct account of the source data. There is an imperative for journalists to understand and report the original data in an accurate and well-informed manner. If the first role is about taking citizens out of danger by refusing to give credence to dubious sources and avoiding repetition of false data, the second role involves journalists avoiding to introduce their own statistical errors. In other

[5] Before 1936, the very respected magazine had a thirty-year record of accurately predicting the winners of presidential elections, but in 1936 the prediction was that Landon would get 57 per cent of the vote against Roosevelt's 43 per cent, but the actual results of the election were 62 per cent for Roosevelt against 38 per cent for Landon.

[6] Such errors already occurred: the last Tamedia survey before the vote on an initiative 'For the programmed exit of nuclear energy' predicted 57 per cent of voters in favour of the initiative, but 54 per cent of voters rejected the initiative on 27 November 2016.

words, it is the distinction between accurate reporting of inaccurate/flawed data and inaccurate/flawed reporting of accurate research. This involves not only rigorous and creative statistical storytelling but also a thorough check/proofreading mechanism to detect numerical and statistical content in news stories. Journalists, of course, should and could avoid damages by correcting errors quickly but this is not often the case.

During the UEFA Euro 2016, for example, a Swiss journalist wrote an article about the relative age effect in hockey and football, basing himself on a scientific article written by Nolan and Howell (2010) (Le Matin Dimanche 2016). Research into relative age effect in sport analyses whether young players born earlier in the year are more likely to perform well, play professionally and be selected in national teams. The journalist stated that the effect was observed in Switzerland when comparing the distribution of the birth months of players with an equiprobable distribution (i.e. one-twelfth for each month). Yet the scientific article clearly explains that one should not compare with equiprobability but with the distribution of the same age cohort (i.e. the distribution of birth months in the population with the same age as the players).[7] While the journalist had been warned by the article about the common mistake, he was not able to explain it accurately. In response to my letter, however, he refused to publish a corrigendum, arguing that this would not change the result in the end and that he had to simplify because the public would not understand otherwise.

This incident reminds me of Katherine Viner, the editor-in-chief of the *Guardian*, who recently raised questions about the importance of truth for journalists nowadays (The *Guardian* 2016). In an interesting analysis of a print press that is involved in 'a series of confusing battles between truth and falsehood, fact and rumour, ... between an informed public and a misguided mob' (ibid., 3). Viner worries that the values of journalism have shifted to consumerism where the goal is 'chasing down cheap clicks at the expense of accuracy and veracity' (ibid., 8). These concerns joined a long list of reflexions on accuracy in media and communication, as analysed in information on the environment by Hansen (2016).

Finally, the third role of journalists in the statistics–society relationship is an extension of their core mission of education. The inclusion of statistical elements in a piece of news implies the need to explain statistical notions if one wants the reader to understand them. As implicated in one of the mottos of communication, 'Know your public', journalists should ask themselves to which extent their publics understand statistics and to act accordingly. In doing so, journalists play an important part in the improvement of statistical literacy. My informal examination of Swiss press shows some attempts to fulfil this mission, but the task is easier said than done, with errors being found quite often. There are, for example, many opportunities to confuse various measures of central tendency, such as an article about Swiss median salary in 2007 that wrongly defined it as 'the salary received by the greatest number of employees' (i.e. confusing the median with the mode) (L'Hebdo 2007).

---

[7]   In the general population, the distribution of births indicates an increase in the birth rate in the summer months, and a decrease in February. More, the distribution may vary from year to year.

The increase in public statistical literacy can be achieved in more or less formal context. Let's consider one of the key concepts in statistics, uncertainty, which is needed to avoid drawing unfounded conclusions. Sadly, some journalists fail to mention it, arguing that the public does not understand and cope with uncertainty. However, this could be informally explained in the context of sport. Sport journalists never mention information about measurement errors in the Hawk Eye system used in tennis and cricket. Hawk Eye is a virtual reconstruction produced on the principle of triangulation, using the visual images and the synchronization of video data provided by a number of high-speed cameras, which are combined through a mathematical model to create a three-dimensional representation of the trajectory of the ball. Collins and Evans examined this visualization technology to show how designers decided to not incorporate information about measurement errors in the visualization, which leads the graphics to be too seductive and to involve a too rich virtual reality. In other words, viewers 'are not watching something that happened but a picture which has been produced by a calculation based on imperfect data' (Collins and Evans 2008, 301). The authors recommended that 'these graphics should be accompanied by visual error bars and/or numerical statements of confidence' (ibid., 301). Efforts to increase statistical literacy in less formal context can rapidly bear fruit: one study observes a better understanding of quantitative probability statement in a city where people have been exposed to probabilistic weather forecasts for a longer period (Gigerenzer et al. 2005).

Concluding notes

The starting point of this chapter is that the media serve as a main source of knowledge acquisition, attitude formation and agenda setting on issues from science and health to social problems or politics. Studies on the position of statistics within society indicate that people actually hold a favourable attitude to statistics and highlight the need for improving statistical literacy in the population (for a review, see Crettaz von Roten and de Roten 2013). The media should work along this line through scrutinizing and verifying bad statistics, accurately communicating good statistics and educating the public through statistical news stories.

It must be acknowledged that fulfilling these roles is complicated by a quality problem on the supply sides. This includes not only the inclination of some vested-interest actors – such as politicians – but could also come from science, where the various pressures on scientists, like the 'publish or perish' rule, jeopardize the quality of statistical outputs (Crettaz von Roten and de Roten 2013). It would be ideal if some specialists were employed by the media to verify numerical and statistical elements and detect related errors on behalf of journalists.

A less ambitious expectation would be to train journalists to separately check statistical elements of their news. Such training could be done in-house and refreshed throughout their career (i.e. in professional development courses) through the use of good examples[8] or bad examples.[9] Some statistical popularization books – for example, Paulos (1989); Blastland and Dilnot (2008); Blastland and Speigelhalter (2013) and Wheelan (2013) – are excellent resources for such training. In order for journalists to fulfil the above roles over the long term, however, the provision of formal statistical training in journalism schools plays a vital role.

But journalism education has in itself been one source of the problem due to its own negligence of statistics.[10] Research into this is rather sketchy but evidence has emerged. A survey on the place of statistics in British journalism education concluded that 'the level of statistical training provided to journalists in the UK does not match the profile use of statistics in the news' (Kemeny 2014, 34). The author reported no university journalism programme with statistics as a core module or with a course on data journalism, although journalists may benefit from online courses or actions from the statistics community, such as the Royal Statistical Society. A survey of journalism department heads in the United States (Dunwoody and Griffin 2013) found that positive valuation of statistical instruction, stated by the majority of respondents, did not translate into effective actions – mainly due to presumptions about students' inability and/or unwillingness to learn statistics (see Chapter 16 for more data from this study). In Switzerland, my observation is that statistics play a marginal role (in the majority of cases an optional course) in each of the three pathways to becoming a journalist (academic, vocational and professional), but after formal education, various associations of journalism offer continuous training on statistics, visualization or data journalism. Finally, a study on educational strategies in data journalism in six European countries reported that 'data journalism education is still in an experimental stage' and statistical/mathematical skills are more or less developed according to the length of courses, the type of training, teachers' background, and more generally, national media systems (Splendore et al. 2016, 148).

What could be the solutions to improve this situation? According to Dunwoody and Griffin (2013), most heads of journalism departments considered that statistical training should be embedded in a variety of courses – instead of a specific statistical course. The j-chairs, however, anticipated some reservations on the part of students on the belief that they would actively avoid such material and, worse, would be intellectually unable to handle such an instruction, as well as some reservations from faculty members, most of whom would have difficulties in embedding statistical

---

[8] For example, 'International code of practice for the publication of public opinion poll results' available at http://wapor.org/pdf/ESOMAR_Codes&Guidelines_OpinionPolling_v5.pdf or 'Statistical terms used in research studies: a primer for media' available at http://journalistsresource.org/tip-sheets/research/statistics-for-journalists.

[9] To name a few, Tom Lang wrote a series of articles entitled 'Common statistical errors even you can find' published between 2003 and 2006 in American Medical Writers Association Journal (available at http://www.amwa.org/issues_online) or in 2004 in Croatian Medical Journal 45(4), 361–270. Gal (2002) proposed a list of questions that readers and journalists can ask themselves about statistical elements involved in a message that, intentionally or unintentionally, may be misleading, biased, one-sided or incomplete in some way.

[10] See Nguyen and Lugo-Ocando (2016) for a recent overview of the state of statistics in journalism and journalism education.

training. Embedded statistical reasoning would be more accepted by reluctant-to-statistics students, and would have the advantage to be repeated in different contexts. On the negative side, this requires universities to maximize the number of teachers who put in the effort to integrate statistical reasoning in their course and, therefore, to involve teachers who are not necessarily skilled in statistics.

In the broad literature on statistical education at a university level, this proposition is replaced by the distinction between a general statistical course and a subject-oriented statistics course, with a plebiscite for the latter (Meng 2009, 21). Gal (2002) stressed the limits of traditional approaches to teaching statistics and called on educators to 'distinguish between teaching more statistics (or teaching it better) and teaching statistics for a different (or additional) purpose', suggesting them to work together with students on a large range of authentic issues and problems. Many authors have stressed that teachers should teach critical statistical thinking besides statistical knowledge, as 'critical questioning skills are necessary in order to understand that statistical messages and their contexts, as communicated in the media, are shaped by political, commercial, or other agendas, which may not be entirely objective' (Hayward, Pannazzo and Colman 2007, 396). Friedman, Friedman and Amoo (2002) found that humour reduces statistical anxiety of students and improves the relationships between students and teachers. Williams et al. (2008) found a positive relationship between student attitudes towards statistics and student performance, that is, students with positive attitudes are more likely to perform well. A further review of current research into teaching and learning statistics may be found in Garfield and Ben-Zvi (2007). The positive results from such research allow us to be somewhat sanguine about the relationship between the media, statistics and society in a data-rich world.

References

Bakker, M. and Wicherts, J. (2011), 'The (mis)reporting of statistical results in psychology journals.' *Behavioral Research*, 43(3), 666–78.

Best, J. (2001), *Damned Lies and Statistics: Untangling Numbers from the Media, Politicians and Activists*. Berkeley: University of California Press.

Best, J. (2005), 'Lies, calculations and constructions: Beyond how to lie with statistics.' *Statistical Science*, 20(3), 210–4.

Billard, L. (1998), 'The role of statistics and the statistician.' *The American Statistician*, 52(4), 319–24.

Blastland, M. and Dilnot, A. (2008), *The Tiger that isn't: Seeing Through a World of Numbers*. London: Profile books.

Blastland, M. and Spiegelhalter, D. (2013), *The Norm Chronicles: Stories and Numbers About Danger*. London: Profile Books.

Button, K., et al. (2013), 'Power failure: Why small sample size undermines the reliability of neuroscience.' *Nature Reviews Neuroscience*, 14(5), 365–76.

Collins, H. and Evans, R. (2008), 'You cannot be serious! public understanding of technology with special reference to Hawk-Eye.' *Public Understanding of Science*, 17(3), 283–308.

Crettaz von Roten, F. (2006), 'Do we need a public understanding of statistics?' *Public Understanding of Science*, 15(2), 243–9.

Crettaz von Roten, F. and de Roten, Y. (2013), 'Statistics in science and in society: From a state-of-the-art to a new research agenda.' *Public Understanding of Science*, 22(7), 768–84.

Crettaz von Roten, F. (2016), 'Statistics in Public Understanding of Science review: How to achieve high statistical standards?' *Public Understanding of Science*, 25(2), 135–40.

Dunwoody, S. (2008), 'Science journalism.' in Bucchi, M. and Trench, B. (eds), *Handbook of Public Communication of Science and Technology*, 1st edn. London: Routledge, 15–26.

Dunwoody, S. and Griffin, R. J. (2013), 'Statistical reasoning in journalism education.' *Science Communication*, 35(4), 528–38.

Duplain, H. (2012), 'C'est significatif [It is significant].' *Le Temps*, 24 October, 15.

Eagly, A. and Chaiken, S. (1993), *The Psychology of Attitudes*. Fort Worth: Harcourt Brace Jovanovich College Publishers.

Friedman, H., Friedman, L. and Amoo, T. (2002), 'Using humor in the introductory statistics course.' *Journal of Statistics Education*, 10(3). Available at : http://ww2.amstat.org/publications/jse/ (accessed 2 September 2016).

Full Fact (2016). The state of automated Factchecking: How to make factchecking more effective with technology we have now. Available at https://fullfact.org/media/uploads/full_fact-the_state_of_automated_factchecking_aug_2016.pdf (accessed 2 November 2016).

Gal, I. (2002), 'Adults' statistical literacy: Meanings, components, responsibilities.' *International Statistical Review*, 70(1), 1–25.

Garfield, J. and Ben-Zvi D. (2007), 'How students learn statistics revisited: A current review of research on teaching and learning statistics.' *International Statistical Review*, 75(3), 372–96.

Gigerenzer, G., et al. (2005), 'A 30% chance of rain tomorrow: How does the public understand probabilistic weather forecasts?' *Risk Analysis*, 25(3), 623–9.

Glaser, D. (1999), 'The controversy of significance testing: Misconceptions and alternatives.' *American Journal of Critical Care*, 8(5), 291–6.

Goldin, R. (2009), 'Spinning heads and spinning news: How a lack of statistical proficiency affects media coverage.' *Proceedings of the Joint Statistical Meetings*, Section on Statistical Education. Available at: http://math.gmu.edu/%7Ergoldin/Articles/SpinningHeadsSpinningNews.pdf (accessed 18 August 2016).

Greenblat, A. (2016), 'What does the future of automated fact-checking look like?' Available at https://www.poynter.org/2016/whats-does-the-future-of-automated-fact-checking-look-like/404937/ (accessed 2 December 2016).

Greenland, S., et al. (2016), 'Statistical tests, P values, confidence intervals, and power: A guide to misinterpretations.' *European Journal of Epidemiology*, 31(4), 337–50.

Graves, L. and Cherubini, F. (2016), *The Rise of Fact-checking Sites in Europe: Digital News Project 2016*. Reuters Institute for the Study of Journalism. Available at https://reutersinstitute.politics.ox.ac.uk/sites/default/files/The%20Rise%20of%20Fact-Checking%20Sites%20in%20Europe.pdf (accessed 1 December 2016).

Hansen, A. (2016), 'The changing uses of accuracy in science communication.' *Public Understanding of Science*, 25(7), 760–74.

Hayward, K., Pannazzo, L. and Colman, R. (2007), *Developing Indicators for the Educated Populace Domain of the Canadian Index of Wellbeing: Literature Review*. Available at: https://uwaterloo.ca/canadian-index-wellbeing/sites/ca.canadian-index-wellbeing/ files/uploads/files/HistoricalEducated_Populace_Literature_Review__Doc1_ August_2007.sflb_.pdf (accessed 19 October 2015).

Huff, D. (1954), *How to Lie with Statistics*. New York: W. W. Norton & Company.

Kemeny, R. (2014), 'The statistical foundations of the fourth estate'. *Significance*, 11(4), 34–5.

L'Hebdo (2007), '*5845 francs* [5845 Swiss francs, translation by the author]', 23 August, 18.

Le Matin (2011), 'Le Mobbing, c'est du sérieux [Harassment, that is serious]', 4 May, 4–5.

Le Matin Dimanche (2010), 'Le Valais et Berne se mobilisent pour que le deuxième tube du Lötschberg ouvre en 2020 [Wallis and Bern take action to enable the second tunnel tube opens in 2020]', 12 September, 5.

Le Matin Dimanche (2016), '(Dés)avantagé à la naissance [(dis)advantaged at birth]', 8 May, 38.

Le Temps (2012a), 'Les investissements en Afrique [Investments are increasing in Africa]', 9 May, 5.

Le Temps (2012b), 'Les salaires suisses relégués en 6ᵉ position: un abus statistique [Swiss wages confined in sixth position: a statistical misuse]', 26 October, 7.

Lynn, H. (2016), 'Training the next generation of statisticians: From head to heart'. *The American Statistician*, 70(2), 149–51.

Marriott, N. (2014), 'The future of statistical thinking'. *Significance*, 11(5), 78–80.

Meng, X. (2009), 'Desired and feared – What do we do now and over the next 50 years?' *The American Statistician*, 63(3), 202–10.

Moore, D. S. (1998), 'Statistics among the liberals'. *Journal of the American Statistical Association*, 93(444), 1253–59.

Nolan, J. and Howell, G. (2010), 'Hockey success and birth date: The relative age effect revisited'. *International Review for the Sociology of Sport*, 45(4), 507–12.

Nguyen, A. and Jugo-Ocando, J. (2016), 'The state of data and statistics in journalism and journalism education: Issues and debates'. *Journalism*, 17(1), 3–17.

Paulos, J. (1989), *Innumeracy: Mathematical Illiteracy and its Consequences*. London: Penguin books.

Paulos, J. (1995), *A Mathematician Reads the Newspaper*. New York: Basic Books.

RSS (2014), *Code of Conduct*. London: Royal Statistical Society.

Splendore, S., et al. (2016), 'Educational strategies in data journalism: A comparative study of six European countries'. *Journalism*, 17(1), 138–52.

The Guardian (2016), 'How technology disrupted the truth'. Available at: www. theguardian.com (accessed 12 July 2016).

Trench, B. (2007), 'How the Interned changed science journalism'. in Bauer, M. W. and Bucchi, M. (eds), *Journalism, Science and Society: Science Communication Between News and Public Relation*. New York: Routledge, 133–41.

Tribune de Genève (2012), 'Rares, les marrons chauds sont plus chers [Scarce, hot chestnuts are more expensive]', 26 October, 8.

Tribune de Genève (2015) 'Un Genevois sur six vit seul [One inhabitant of Geneva in six lives alone]', 3–4 January, 16.

Wallman, K. (1993), 'Enhancing statistical literacy: Enriching our society'. *Journal of the American Statistical Association*, 88(421), 1–8.

Wheelan, Ch. (2013), *Naked Statistics: Striping the Dread from the Data*. New York: W. W. Norton & Company.

Williams, M., Payne, G., Hodgkinson, L. and Poade, D. (2008), 'Does British sociology count? Sociology student's attitudes toward quantitative methods.' *Sociology*, 42(5), 1003–21.

Wyss, R. and Meyer, F. (2013), 'When careers run up against stereotypes.' *ETH Life*, October 2013, 13.

2

More Light, Less Heat: Rethinking Impartiality in Light of a Review into the Reporting of Statistics in UK News Media

Stephen Cushion, *Cardiff University, UK*
Justin Lewis, *Cardiff University, UK*

Introduction

At no point in history have statistics been more widely used or so freely available. Yet while the production of an ever-expanding supply of data has increased in many Western democracies, it is less clear *how* statistical sources routinely inform debates in the public sphere. This chapter explores how UK news media report statistics, examining which sources routinely inform coverage and how well statistical information is explained and interpreted.

The importance of statistics was recognized by the BBC Trust, the body that regulates BBC content, when it commissioned an impartiality review of statistics in 2015. We informed this review – published in August 2016 – by carrying out a content analysis study of how BBC and commercial broadcasters reported statistics. In this chapter, we draw on both the BBC Trust review (2016) and our study to consider whether the wider information environment has been enhanced by the statistical supply of more facts and figures (Cushion et al. 2016a, b). In doing so, we ask whether statistical sources reproduce the voices and perspectives of sources that scholars have typically found shaping day-to-day news media coverage (Cottle 2000; Wahl-Jorgensen et al. 2016). Or, as statistical information has become more prevalent and accessible, have statistics promoted a more diverse range of actors and views being sourced in routine news reporting? For UK broadcasters legally required to remain impartial, which – according to the BBC's editorial guidelines – is not just about balancing competing perspectives but interrogating competing positions, statistics play a vital role in shaping debates about politics and public affairs. By analysing the range of statistical sources in news coverage and how they are communicated and interpreted by reporters, we can consider the impartiality of statistical information in online and broadcast news media.

In order to do so, we will first offer a general interpretation of the quantitative supply of statistics in news coverage and examine the everyday reporting of statistics across different platforms and broadcasters. We will then explore in detail two qualitative case studies about how well statistics were interpreted and communicated to audiences. Our focus here is on how news media reported the UK government's statistical claims about EU migrants and the tax credits system, which were challenged by independent sources. We focus on the most widely consumed news sources in the UK, including television news on BBC, ITV and Channel Four, as well as an extensive range of BBC radio and online sources. BBC journalism, in particular, reaches more people per day in the UK than any other news service, and along with other broadcasters in a public service system, all its editorial content – including online news – must be impartially reported.

Interpreting the quantitative supply of statistics in news reporting

In recent years, the term 'data journalism' has emerged, prompting debates about its precise meaning and scope (Coddington 2015; Rogers 2011). Generally speaking, it refers to the use of data to enhance journalism, which can range from the use of infographics (a long-standing practice) to the production of statistics drawing on raw data sources. In the digital age, open-source data has become more widely available and accessible, with analytical tools that open up new lines of inquiry for journalists to investigate and report (Lewis and Usher 2013). The rise of data journalism, in this context, creates news gathering techniques that can, potentially at least, enhance the media's watchdog role and democratize the flow of information (Rogers 2011).

However, debates about the role of data journalists often centre on the medium rather than the message, that is, the potential for new technologies to revolutionize different platforms of news attracting more attention rather than the reality of how information sources inform day-to-day journalism or raise editorial standards. As Knight (2015, 58) has pointed out, 'the focus is entirely on how to do it, and how amazingly revolutionary it is, but there is little critique of what data journalism actually is, who is actually doing it and why we should do it'. Moreover, data journalism is primarily viewed as journalists interrogating new found information sources, such as the 2011 WikiLeaks documents, rather than handling routine statistical information. And yet, we would argue, the practice of data journalism is more widespread, making it part of everyday journalistic practice in a data-rich world. Data journalists might be a small and specialist group, but most journalists are obliged to deal with secondary data sets or statistical claims on a daily basis.

Our focus is on these more widespread, everyday encounters with the world of data. While scholars have long studied the use of sources in news coverage (Manning 2001), less sustained empirical attention has been paid to systematically tracking the statistical claims made by sources across different platforms. We know, for example, that institutional actors, in particular political elites, tend to be sourced regularly in news coverage (Cottle 2000; Wahl-Jorgensen et al. 2016), but has the proliferation of statistical information promoted greater access to a wider pool of sources and diversity of views?

In order to track the degree to which statistics informed news coverage and the clarity in which they were expressed and interpreted, we conducted a content analysis of UK television, radio and online news (Cushion et al. 2016a). We examined the source of every type of reference to a statistic and the context in which they typically appeared, such as news about politics, business, crime or health. While our aim was to develop a clear quantitative picture about the volume and nature of how statistics shape news coverage, we did so in the context of assessing the clarity and impartiality of reporting across different media platforms.

Although our study quantifies the extent to which data informs news coverage and the sources regularly drawn upon, our case studies also explore more qualitatively how impartially statistics are dealt with by journalists such as scrutinizing claims by the UK government. Impartiality, of course, is a highly contested term and its conceptual meaning is often difficult to operationalize and apply in news programming (Barkho 2013). For example, is has been argued that a crude adherence of impartiality can lead to relativistic reporting – 'false equivalence' – where competing positions are juxtaposed without interpreting the evidence supporting different perspectives. In other words, impartiality is a construction of journalistic balance, an attempt to avoid being biased or partisan that, if not handled well, could be made at the expense of 'the truth'.

While remaining impartial is a normative aim for many journalists, scholars have long studied how empirically it can be measured in news coverage, such as exploring the type of news agenda pursued by different broadcasters, the sources used in coverage or the degree of contextual information about a particular issue or event (Barkho 2013; Cushion et al. 2012; Wahl-Jorgensen et al. 2016). We contribute to this empirical line of inquiry by understanding the everyday use of statistics in news coverage and interpreting their impartial use across different broadcast and online platforms.

Everyday reporting of statistics in UK news media: An overview

The quantitative part of our BBC Trust review involved examining 6,916 news items in total, including 4,285 references to statistics (with multiple references in some news items) over one month between October and November 2015. We found over one in five items – 22 per cent – featured a statistical reference, a ratio that increased to one in three items in online news. However, most references – approximately 2 out of every 3 mentioned – were fairly vague, providing only limited context. About a third – 35.2 per cent – provided some explanation or included some comparative data. Online news presented generally used statistics more often and with more context and clarity than other news platforms.

Statistics were most likely to be featured in stories about business, the economy, politics, social policy and health coverage. So, for example, three quarters of all economy items featured at least one statistics. Although there is plenty of data about crime or terrorism that may help provide useful context to audiences, these and other issues featured few statistical references. Just 6 per cent of crime news items, for instance, contained a statistic.

The most common source of statistics used in news stories were politicians (20.6 per cent), businesses and government departments or agencies (both 12.3 per cent). Other information sources, such as NGOs (7.3 per cent), academics (6.5 per cent), regulatory bodies (4.4 per cent) and think tanks (3.8 per cent) – were far less prominent. Perhaps most strikingly, nearly three quarters of party political statistical references – 72.6 per cent – came from Conservative politicians, with Labour – the official opposition – making up only 18.4 per cent. We might, however, expect cabinet ministers and government departments to be a dominant source of statistical information – after all, civil servants regularly supply them with data. What is perhaps more concerning is that we identified a lack of clarity in the communication of government statistics, since most appeared in a relatively vague or imprecise form.

This raises important questions about how far the government's statistical claims are routinely challenged and scrutinized. Our quantitative analysis showed that, most of the time, journalists did not contextualize or challenge statistical claims. However, in order to explore the potential for journalists to scrutinize statistics – while maintaining due impartiality – we chose two case studies to where reporters *did* use independent data and expertise to challenge government claims. We examine the reporting of UK Prime Minister David Cameron's use of migration data and the reporting of government claims about changes to the tax credits system. They provide, in this sense, examples that allow us to explore both the potential and limits of those instances when claims were scrutinized.

Case study 1: The UK Prime Minister's claim about EU migrants claiming benefits

In a speech about EU migrants claiming benefits in the UK in November 2015, the Prime Minister stated that 43 per cent of EU migrants claimed benefits in their first four years of being in the UK. In advance of the speech, a journalist at *The Times* raised concerns on Twitter about the methodology and the government's interpretation of the figures. The Code of Practice for Official Statistics asks for the official sources to be released before a speech is made, but in this case the government failed to supply one. Full Fact, a fact-checking organization, formally complained to the UK Statistics Authority, while Channel 4 made it clear that 'until the full figures are published it is impossible to interrogate the government's specific claims'.

The statistic was also ambiguous on a broader level. A number of surveys have suggested widespread ignorance in the UK population about what welfare benefits actually refer to, with many people assuming that the phrase refers mainly to unemployment benefit (rather than in-work tax credits, for example), while significantly overestimating the proportion of fraudulent claims. This is especially germane in this instance, since EU migrants are *more* likely to be working than the rest of the UK population. For many people, then, simply to give the percentage of people 'on benefits' or 'on welfare' is insufficiently clear, signifying, for many, *unemployment* benefits – an impression that is likely to exacerbated by the phrase 'claiming benefits'. In this case study, we therefore explore two issues:

- How far was the statistic accepted, challenged or overlooked?
- To what extent was it – or other, more reliable statistics – contextualized to give an accurate picture of the proportion of EU migrants in work (and who may, if in low paid work, have been in receipt of working tax credits)?

While the figure was questioned and debated, we found several instances in which the figure was presented as either as unproblematic or in isolation. So, for example, a headline on Radio 4's *Today Programme* appeared to accept the government's figures:

> Today in an effort to buttress the case for this last fraught area Number 10 released new figures to underline what it says is the scale of the problem. These suggest some 43% of EU migrants arriving in Britain over the four years receive benefits, two-thirds of them in work benefits such as tax credits at a total cost to the tax payer of about half a billion pounds a year. (Today, BBC Radio 4, 10 November)

During an interview on BBC Radio 5, Conservative MP Michael Fallon quoted the statistic three times, and while the interviewer offered a mild challenge, it was muddled and vague, quickly backing away from questioning the statistic itself:

> Michael Fallon: What we want is to make sure that the benefit system isn't driving migration, around 40% of those who come here from the rest of the Union are claiming benefits straight away and that isn't right.
> Presenter: Well so the Prime Minister says, although those statistics … have been questioned by some this morning – who've said these are in contradiction to what the labour force survey tells us, it all gets complicated.
> Michael Fallon: Yes, these are government figures I think.
> Presenter: Yes, well it's how you interpret them. But the question is, have you got any evidence these in-work benefits are of themselves a draw to migrants and that's what's pulling people into this country?

In the final part of the interview, the cabinet minister is able to re-state his point without any challenge, twice using the somewhat misleading phrase '*claiming* benefits': 'No but what is a fact, is around 40% of those who come here are claiming benefits straight away and what we can't have I think is a system where there is an incentive for people to move to Britain from other European states simply because of the level of benefits that they claim.'

The vague idea that 'some question' the statistic was repeated in other bulletins: however, in these instances the original statistic sets the news framework and is thereby given a degree of authority:

Crucially, he is also demanding benefit curbs for EU migrants, claiming 40% of those coming are on welfare – a figure some question. (BBC News at Six, 10 November)

We did, however, find a number of other instances in which the figure was questioned – both as a statistic *and* in terms of its meaning/interpretation. ITV's evening bulletin adopted a sceptical tone from the outset:

David Cameron's figures today certainly were different. They came from Department for Work and Pensions data from 2013. The government said that around 40% of all EU migrants, more than 220,000 people, were receiving some kind of benefits either for themselves or their dependents. 66% of those EU claimants, around 150,000 people, were on in-work benefits such as tax credits, claiming on average of £6,000 per year per family. Add it all up and the government said in-work benefits for EU migrants cost the UK taxpayers £570 million in 2013. In short a lot higher than most other studies seem to suggest. So why is that? (ITV News at Ten, 10 November)

The report then included interviews with two experts, Madeleine Sumption, Director of the Migration Observatory and Professor Christian Dustmann, Centre for Research & Analysis of Migration at UCL, both of whom pointed to the discrepancy between these other figures. The package ended by accepting these experts' interpretation of EU migrants rather than the government's, with the report adding:

And on the overall impact on the public purse it is worth noting that according to the official independent budget watchdog the OBR, without the economic contribution of migrants, government debt would be 78% higher by 2062. (ITV News at Ten, 10 November)

On the BBC TV early evening news bulletin, sources from academia (including, Christian Dustmann used by ITV), and a research institute both directly challenged David Cameron's statistical claim, and argued migrants benefit the UK economy:

Professor Christian Dustmann, Centre for Research & Analysis of Migration, UCL: Well, they seem to be in contradiction to many of the figures which are conducted by academics, including ourselves, and those figures are far lower. (BBC News at Six, 10 November)

Jonathan Portes, National Institute of Economic and Social Research: The evidence suggests that migration is on the whole a good thing for the British economy. For example, the government's Office for Budget Responsibility says that lower migration would mean over the medium to long term we would have to have higher taxes or lower public spending, because migrants make a net contribution to the public finances over the long term. (BBC News at Six, 10 November)

The BBC's Economics Editor then presented the broader statistical landscape:

> So what do we know about why migrants are coming here? EU migrants represent 6% of the working population but only a bit over 2% of welfare benefits claimants. That suggests that when they arrive from places like Poland and Romania to coach stations like this one in Victoria, they are not coming to sit on their bottoms and claim. But if we look at working tax credits and child tax credits, EU migrants represent around 10% of those. That shouldn't really be a surprise, because the evidence suggests they come here to work and they have to be on relatively low pay. (BBC News at Six, 10 November)

Similarly, while it gave the figure a degree of authority in a news headline, the interviewer on BBC's put the statistic in context:

> You're quoting one set of statistics, there are others. Jonathan Portes, former chief economist at the cabinet office says that if you look at those who are claiming welfare benefits in this country, 2% of them are EU nationals which compares to the fact they make up around 6% of the working age population – so the idea they are the ones who are sucking up the welfare benefits is just not true. (Presenter, Today, BBC Radio 4, 10 November)

Overall, this case study suggests that the PM's statistical claim was, despite its problematic status, allowed to set the agenda for many of the day's news stories. On some BBC bulletins, experts or statistical data *was subsequently* used to question both the veracity and the gist of the figures. ITV's News at Ten were bolder in questioning the government's statistical claim, taking the opportunity to assess the (positive) economic contribution made by EU migrants to the UK.

Channel 4's decision not to cover the story was, perhaps, the boldest assertion of journalistic independence. The broadcaster questioned the figures in an online fact-checking blog.[1] This is in contrast to the BBC who, while often questioning, allowed the government to set the agenda. Nonetheless, ITV's approach, which used the story as a prompt to convey a more accurate statistical picture, may have been more useful in advancing viewers' understanding of the issue.

Case study 2: Reporting changes to tax credits system

This was a story with statistics at its heart. The government, in order to reduce public spending, proposed to cut working tax credits, a mechanism introduced by the Labour government to increase the incomes of the working poor. George Osborne hoped to

---

[1]   https://www.channel4.com/news/factcheck/fact-check-43-eu-migrants-claim-benefits   (accessed 6 January 2017).

compensate those on low incomes by increasing the National Minimum Wage (NMW), shifting the burden of alleviating low pay from taxpayers to employers. A key thrust of the government's case was a statistical claim: those on low incomes would gain more (or as much) from increases to the NMW (and other changes) than they would lose in working tax credits.

Since their proposed changes were based on fairly clear (and widely available) data sets, it was possible to apply detailed scrutiny to this claim, with high-profile independent analyses conducted by the Institute of Fiscal Studies, the Resolution Foundation and the House of Commons Library. In this case, the weight of statistical evidence was fairly clear: all the independent analysis suggested that the government's claims were simply wrong, and that increases to the NMW (and other changes) would *not* compensate for the losses to the working poor, who would be worse off as a result of the propose changes.

The issue came to a head-on on 26 October when the House of Lords threatened to (and indeed, did) reject the proposed changes, a situation that was the lead story on most bulletins. This story is therefore a clear instance where independent data analysis was available to journalists to directly challenge what was, in essence, a highly questionable statistical claim by the government.

While the government continued to argue their case, they did not offer any detailed rebuttal to the IFS or other independent analyses. After defeat in the House of Lords, they decided to delay the implementation of the changes to tax credits. We would argue that the broadcast media played an important watchdog role in this policy change, by using both statistical data and expertise to establish broad truths. Overall, our case study suggests three points, all of which illustrate the broader picture provided by our content analysis:

- It is clearly possible for broadcast journalists to use independent data and expertise to question government claims.
- Despite a number of examples of good journalism calling the government to account, there was still a tendency – especially early on in the story – to resort to a tit-for-tat model of statistical claims.
- We found a tendency to recycle the same statistic(s) – sometimes inaccurately and often without attribution – rather than draw directly from the primary sources of statistical information.

Using independent statistics to challenge claims

We found a number of examples of good practice. So, for example, the BBC's 6 pm radio news broadcast contrasted government claims with the independent IFS research:

Under the proposals, tax credits would start being removed from workers at a lower level of earnings and withdrawn more quickly. The government says that, taken as an overall package, most people will be better off because of a higher minimum wage, to be called the national living wage, and because they'll be able to earn more before paying any income tax. But the independent Institute for Fiscal Studies says more than three million families will still be worse off. (Six O'Clock News, BBC Radio 4, 26 October)

The report then offered a detailed set of statistics, giving the reasons behind the government's proposed cuts, but also suggesting that the government's claims of mitigation did not stand the scrutiny of independent investigation.

> Tax credits were introduced in 2003 but their cost has risen dramatically to £30 billion a year. The government announced in the Budget that it was reducing tax credits; it desperately wants to bring down the deficit and these changes alone would bring in around £6 billion a year by 2020. At the same time it said that the minimum wage would rise sharply to £9 an hour by 2020, compensating people for the cuts. But as Paul Johnson, Director of the Institute for Fiscal Studies, told the Work and Pensions Committee today, most people on tax credits would still be worse off. ... The fear is that these changes will also be a disincentive to people wanting to work longer hours, as some will lose 80% or more of their extra income in tax and lost tax credits. (Six O'Clock News, BBC Radio 4, 26 October)

The BBC's 6 pm television news bulletin also drew upon independent statistical data to question the government's claims in some detail:

> If one parent, typically the mother, works part-time, right now she can earn up to £6,420 a year before her tax credit money gets reduced. Next year, they will start clawing it back if she earns more than £3850. ... And that has a big effect.

The report then looks at a family with 'two working parents, the father working full-time, the mother part-time, both on the minimum wage':

> They will get a rise next year because of the national living wage, but they will lose so much in tax credits that they will end up £1800 a year worse off. It is also hard for a single mother working full-time for the minimum wage. Her pay would go up by £700 a year, but her income overall would drop by £1500.

The report then tests the claim by the chancellor that 'his critics have not taken account of changes like more free childcare' and scrutinizes the claim in relation to a specific example, before moving onto the overall figures which suggest clear shortfalls:

> Looking at all the measures from last July's Budget, the national living wage will boost the incomes of low-paid working families by around £4 billion before tax, but that is set against a £12 billion cut to tax credits and other benefits. If the government funds the extra childcare help and caps social rents, it will help those on low incomes to some extent, but that help is valued at around £1.4 billion. For

most working families on tax credits, that is not nearly enough to make up for cuts. (BBC News at Six, 26 October)

In this instance the reporter uses the statistical evidence available to come to the same conclusion as various independent reports – namely that, under scrutiny, the government's claims (that the working poor would be no worse off) did not add up. It is a good example of journalists using independent evidence to call power to account.

Tit-for-tat statistical claims

This comparatively rigorous statistical journalism was not, however, always a consistent feature of the news coverage of this story. We found a number of examples of a very different style of presentation, in which reporters allowed a fairly crude form of impartiality to trump objectivity. In this quite different narrative, the government's claims are simply presented alongside the counterclaims of 'government critics'.

This was particularly notable during the morning BBC radio news reports. Although all the available data was obtained well in advance of the coverage on 26 October, the morning radio news programmes generally avoided independent statistical assertions or judgements. BBC Radio 4's *Today Programme* covered the story by interviewing MPs for the main political protagonists: they began with Conservative MP Matt Hancock who put forward the government's case, then later in the programme invited Tim Farron for the Liberal Democrats and Owen Smith for Labour to make the opposition case.

During the interview with Matt Hancock, the interviewer made a vague reference to data suggesting people will be worse off, and the Conservative MP responds with more precise (though questionable) figures.

> Presenter: Families are going to be clobbered by this. Now that's a funny kind of support?
> Hancock: No, by next year 8 out of 10 people will be better off as a result, but we are trying to make a bigger change in this country, we are trying to change Britain from country that lives beyond its means to one that lives within its means. … So if you take this as part of the overall package with the new national living wage, the rise in the income tax threshold and the extra child care which makes it easier and more affordable to get out to work, it is overall a package to support people, to support the changes in the country that we need to see. (Today, BBC Radio 4, 26 October)

The presenter does not respond to the statistical claim that '8 out of 10 people will be better off' by next year, but instead anticipates a response later in the programme:

> Well we will hear the opposition view on that from Labour and the Liberal democrats later in the programme. (Today, BBC Radio 4, 26 October)

The programme then repeated the (questionable) '8 out of 10 people will be better off' claim in their next set of headlines:

> We heard earlier from the cabinet office minister Matthew Hancock who said the Lords should not interfere in that way and he argued that 8 out of 10 people affected by the changes would in the end, because of other changes, be better off. (Today, BBC Radio 4, 26 October)

During the subsequent discussion with opposition MPs, both Tim Farron and Owen Smith offered statistical claims to counter the government's position:

> We all know that this is going to hit families who are on low incomes, middle incomes in Britain and all of the things Matthew Hancock was saying would be compensatory measures, increasing the national minimum wage, raising the personal allowance, none of those things in anyway offset the volume of losses people will have, they're just not being truthful about that. (Owen Smith, Labour MP, Today, BBC Radio 4, 26 October)
>
> The thing that matters is the three million families who will lose this money, who'll get a letter just before Christmas telling them how much they'll lose, on average it'll be £1300 per family and these are people for whom that is probably a 10 or 15% cut in their incomes. (Tim Farron, Liberal Democrat MP, Today, BBC Radio 4, 26 October)

In short, the *Today Programme* left it to the politicians to deal with the statistics. The opening headline for the story on 5 live Breakfast took a similar approach, presenting this as a statistical argument between the government and opposition parties, rather than offering any independent verification from the available reports.

> Critics of the changes say millions of vulnerable people on lower incomes will be worse off after the changes, but ministers argue that other changes to benefits and taxes will help to ensure that most people are better off. (5 live Breakfast, BBC Radio 5 live, 26 October)

Claims and counterclaims were then elaborated in subsequent interview with Conservative MP Chris Philp and Baroness Kramer, for the Liberal Democrats, with reporters making only fleeting use of the available statistical data. Unlike the later coverage that clarified the statistical picture, this left the audience to try and make sense of competing claims made by government and opposition politicians.

Exploiting the availability of statistics

While there were no egregious errors in the coverage, there were a number of examples of sloppy reporting. These mainly involved the failure, in some bulletins, to provide (or to misattribute) statistical sources, and the slippage between using households and individual wage-earners as a unit of analysis. While this is understandable in certain circumstances, this particular story was anticipated well in advance with ample time to absorb and process the statistical information before the day began. Even minor mistakes might easily have been avoided by greater use of available expertise, rather than relying on generalist reporters or presenters to quickly master a statistical brief.

What was particularly notable was that despite the abundance of statistical information available from independent sources (much of it presented with both brevity and clarity), many BBC outlets tended to recycle (variations of) the same statistic – one whose source, while extrapolated from credible sources, appeared to be political rather than independent. A claim made by Labour MP Frank Field in the Spectator on 14 September, reads: 'The proposed cuts in tax credits leave 3.2 million strivers in low paid work on average over £1,300 worse off a year.' This is close to the figures provided by a House of Commons Library Briefing Paper (Number CBP7300, 15 October 2015) on Tax Credit changes from April 2016, which drew similar conclusions (although the claim put them together in way that the Paper did not, and referred to individuals rather than families).[2]

Variations of this figure was used across a wide range of BBC news bulletins (sometimes misattributed to the IFS), as in the following examples:

- Plans by the Westminster Government to change tax credits will affect 3 million families, the independent Institute for Fiscal Studies estimates they'll each lose about £1,300 a year. In the House of Lords right now they're debating the pros and cons (PM, BBC Radio 4, 26 October).
- Cuts to tax credits would save the Treasury £4.4 billion a year. It's estimated around three million low income people will be affected. And many may lose up to £1,300 a year (BBC News Channel at 5 pm, 26 October).
- Ministers want to reduce the threshold up to which people can claim the benefit, and once people reach that threshold, to speed up the rate at which tax credits are taken away. But critics, including some on the Government benches, say more than three million families would be an average of £1,300 a year worse off. … The Government has insisted that taken as an overall package, most people will be better off (BBC News at Six, 26 October).
- For more than three million working families of whom most have children, the average estimated loss would be £1,300 (BBC News at Ten, 26 October).

There are some slippages here – between individuals and families, for example. At times, the BBC News channel even muddled the data to suggest an unidentified number of people would lose *up to* (rather than *on average*) £1,300 a year. But our main point is to identify a tendency to latch onto and recycle a particular statistic – despite the wide availability of many other clearer statistics. One in particular, provided in the September 2015 IFS report, included a pithy and germane statistic that captured the

---

[2] Report can be found here: http://blogs.spectator.co.uk/2015 /09/george-osborne-could-revolutionise-welfare -but-does-he-know-what-hes-doing/

thrust of the story – that those 'affected by changes such as tax credit cuts would only be compensated for 27% of their losses by increases in earnings through NMW and allowances' – was ignored entirely. Overall, *we found the statistical picture tended to be presented most clearly by specialist reporters or by experts.*

This story came as close to having clear statistical truths as most political stories can (the weight of independent evidence was very clearly on one side). In short, all the independent analysis showed clearly that the government's claim that tax credit cuts would be mitigated by other measures was not credible. Indeed, the subsequent decision to delay the proposed changes could be seen as an acknowledgement of this. While some of the coverage did convey the general thrust of the independent analysis, *other outlets tended to present the statistics in a party political 'balance' framework, as a set of claims and counterclaims.* This was, for the audience, much more confusing, since it provides no independent touchstone by which to judge the two sets of claims, thereby neglecting one of the key functions of independent journalism.

Rethinking impartiality: Challenging statistical claims

This research shows that statistics regularly inform in news coverage across all broadcast media outlets and particularly online. But in our content analysis we found statistics were routinely invoked in vague and imprecise ways, with limited context and explanation about their meaning or method behind them. Stories about business and economics contained the most statistical references, with regular figures and data sets informing coverage including share prices and growth forecasts. But in other major subjects, notably crime and terrorism, while a great deal of statistical information is available about these topics, the use of statistics was used far less to help contextualize stories and issues. The government of the day was the most widely relied upon source of statistical information, with limited opportunities for alternative sources – such as independent research bodies, think tanks and academics – to counter their perspectives.

Overall, the evidence presented in the study suggests that, far from data supporting a more diverse information environment, the type of sources drawn upon reinforced the institutional perspectives typically found in news coverage (Cottle 2000; Wahl-Jorgensen et al. 2016), with the overwhelming majority of statistical references supplying limited background or context to a story. Since we uncovered little evidence of any major differences in the quantitative flow of statistical information between public service television news and more commercially driven bulletins, it suggests the way statistics are routinely reported is deeply ingrained in the practices and conventions of broadcast and online journalism.

However, our two case studies identified examples of stories where journalists went beyond the government's statistical claims by drawing on independent sources of knowledge. ITV News, for example, sceptically reported the PM's claim about EU citizens claiming benefits, with the final part of its package accepting the experts', rather than government's, interpretation of statistics. There was also a more critical stance taken by news media outlets during coverage of the tax credits story. The IFS, in particular, was regularly used to counter the government's claim that families would not be worse off after changes to the tax credits system. Nonetheless, at times the coverage balanced party political perspectives, with competing statistics traded without independent analysis or journalistic arbitration. This meant audiences would have encountered an abundance of statistical information but limited context or explanation about its veracity and wider significance.

The journalistic need for providing more light than heat when dealing with statistics was revealed in the audience research that informed the BBC Trust's review. Drawing on focus group responses, the study concluded that when 'statistics which are slightly beyond the audience's ability to interpret, and particularly if these are linked to a heated debate or controversy in which there are clearly biased agendas, the audience appear to be asking not for more opinions on the data but for a more definitive and truly illuminating analysis of the data to help them reach a conclusion' (Oxygen Research 2016). The response here resonates with much of the public's engagement with the 2016 EU referendum campaign, since one survey showed more than two-thirds of people did not feel informed ahead of the election (Electoral Reform society 2016). Indeed, evidence suggests many people were also misinformed during the campaign. So, for example, despite the fact that the UK Statistics Authority and other independent sources cast considerable doubt on the veracity of the Leave campaign's claim that the UK government spent £350 million per week on EU membership, IPSOS MORI discovered almost half of the public they surveyed just days before the election accepted this figure (cited in Hawkins and Arnold 2016).

In conclusion, while our cases studies show the ways in which journalists have, at times, successfully used independent data and expertise to challenge statistical claims made by the government, these instances remain the exception rather than the rule. Even in our case studies this was not always done robustly or clearly. In the tax credits story, for example, we found a tendency to recycle the same statistics and, on occasions, to rely on a tit-for-tat claim and counterclaim by political adversaries. In the EU migrants' story, while some broadcasters (ITV and Channel 4) took fairly robust stances, most BBC outlets allowed a questionable statistical claim to set the news agenda. Across the piece, the most coherent and adept presentations of statistics came from those more specialist reporters.

The case studies also show the importance of not only challenging statistical references, but clarifying their context and meaning. In the case of the EU migrants story, a statistic about the proportion of EU migrants claiming benefits is ambiguous, notably because of widespread public misunderstandings of what claiming benefits actually means (many conflating it simply with unemployment benefit). Some broadcasters took the opportunity to clarify the broader statistical picture – an approach that, if used more regularly, might begin to address the significant democratic deficit that exists on a range of issues. All too often, however, issues such as immigration, welfare and the EU have been reported with more heat than light, with the UK's tabloid press often leading the agenda. By rethinking how impartiality is interpreted, broadcasters could apply greater scrutiny to these statistics and, in the process, raise people's understanding of the facts and figures that should shape public debate on these issues.

References

Barkho, Leon (2013), *From Theory to Practice: How to Assess and Apply Impartiality in News and Current Affairs*. Chicago: University of Chicago Press.

BBC Trust (2016), *Making Sense of Statistics*. London: BBC Trust.

Coddington, Mark (2015), Clarifying Journalism's Quantitative Turn: A typology for evaluating data journalism, computational journalism, and computer-assisted reporting. *Digital Journalism*, 3(3), 331–48.

Cottle, Simon (2000), 'Rethinking news access.' in *Journalism Studies*, 1(3), 427–48.

Cushion, Stephen, Lewis, Justin and Ramsay, Gordon (2012), 'The impact of interventionist regulation in reshaping news agendas: A comparative analysis of public and commercially funded television journalism.' *Journalism: Theory, Practice and Criticism*, 13(7), 831–49.

Cushion, Stephen, Lewis, Justin, Sambrook, Richard and Callaghan, Rob (2016a), *Impartiality Review of BBC Reporting of Statistics: A Content Analysis*. London: BBC Trust.

Cushion, Stephen, Lewis, Justin and Callaghan, Rob (2016b), 'Data journalism, impartiality and statistical claims: Towards more independent scrutiny in news reporting.' *Journalism Practice*, Ifirst.

Electoral Reform Society (2016), 'Under a third of voters feel well-informed about EU referendum.' *Electoral Reform Society*, 21 June, http://www.electoral-reform.org.uk/press-releases (accessed 25 November 2016).

Hawkins, Amy and Arnold, Pheobe (2016), 'The BBC must improve how it reports statistics.' *LSE Blog*, http://blogs.lse.ac.uk/politicsandpolicy/the-bbc-must-improve-how-it-reports-statistics/?utm_content=bufferefabf&utm_medium=social&utm_source=twitter.com&utm_campaign=buffer (accessed 25 November 2016).

Lewis, Seth L. and Usher, Nicki (2013), 'Open source and journalism: Toward new frameworks for imagining news innovation.' *Media, Culture and Society*, 35(5), 602–19.

Knight, Megan (2015), 'Data journalism in the UK: A preliminary analysis of form and content.' *Journal of Media Practice*, 16(1), 55–72.

Manning, P. (2001), *News and News Sources: A Critical Introduction*. London: Sage.

Oxygen (2016), *Impartiality Review: BBC Reporting of Statistics Report on Qualitative Research with the BBC Audience*. London: BBC Trust.

Rogers, Simon (2011), *Facts are Sacred: The Power of Data*. London: Guardian Books.

Wahl-Jorgensen, Karin, Berry, Mike, Garcia-Blanco, Iñaki, Bennett, Lucy and Cable, Jonathan (2016), 'Rethinking balance and impartiality in journalism? A case study how the BBC attempted and failed to change the paradigm.' *Journalism: Theory Practice and Criticism*, Ifirst.

3

Numbers that Kill: How Dubious Statistics Shaped News Reporting of the Drone War

Muhammad Idrees Ahmad, *University of Stirling, UK*

Falsehood flies and the Truth comes limping after it; so that when Men come to be undeceived, it is too late; the Jest is over, and the Tale has had its Effect
– Jonathan Swift

Introduction

In July 2001, during the Second Intifada, when several Palestinian militants were assassinated by Israel, the then US ambassador to Tel Aviv Martin Indyk issued a strong rebuke. 'The United States government is very clearly on record as against targeted assassinations,' he said. 'They are extrajudicial killings, and we do not support that.' The then CIA director George Tenet declared that for someone in his position it would be a 'terrible mistake' to 'fire a weapon like this' (Mayer 2009). By 2009, however, when US president Barack Obama took office, such concerns had been set aside. The administration embraced extrajudicial killings as a less messy alternative to capturing terrorism suspects. On several occasions, it avoided the political complications of detaining terrorism suspects by having them killed instead (Woods 2014; Mazzetti 2013; Klaidman 2012). The new CIA director Leon Panetta declared unmanned aerial vehicles (UAV) – or drones – as 'the only game in town'. Remote warfare quickly became the administration's signature approach to military engagement overseas.

This U-turn in the drone war was made possible thanks in a large part to the persistent public image of armed drones as a technological panacea for security problems – problems that were separated from the intractable political challenges that underlie them. Unlike many of the military excesses that were licensed by September 11 and subsequently reversed, the drone war was dramatically escalated, even when America's other military engagements were scaled back. The tactic has retained strong public support over the past fifteen years. Unlike boots on the ground, using armed drones as a weapon of war appeared less onerous. It lifted political constraints by deferring costs onto the targeted population and it circumvented moral constraints by cultivating the myth of precision. But where drones have proved politically liberating –

making it possible to wage war without putting personnel at risk, circumventing public and congressional scrutiny – they haven't eliminated risk altogether. Despite their much-touted precision, drones have not always discriminated between combatants and civilians. In the longer term their unseen, and often unreported, human cost creates the conditions for blowback – a CIA term for 'the unintended consequences of covert operations' (Johnson, 2000). While they have been used successfully to eliminate high-profile terrorists, they have killed enough innocents to create pools of new recruits for terrorist organizations (Pilkington and McAskill 2015).

The question, then, is why such a wrong image of the drone persists for so long, with so much impact. The answer to this is complicated but one critical factor is a long-running political and military attempt to decouple drones from the predictable consequences of their use, in which the media play a central role. This chapter aims to demonstrate how credulous news coverage in a docile press, especially with respect to the way dubious drone-related statistics were employed, helped to legitimize the drone war. Given the secrecy surrounding the conflict, reporters have struggled to find accurate information on the war's casualties. Whatever data is collected has been aggregated into seemingly precise numbers that are then reported as fact. But the statistics are neither objective nor precise; and the validity and rationale for these numbers remains by and large unexamined. Dubious claims have as a consequence been uncritically relayed to the public, the public remains passive, and the passivity feeds political inertia. In the following pages, I will present a brief overview of the drone war, examine the underlying causes for its poor media coverage, and, through a case study, illustrate how journalists, in the debate over blowback, have disregarded questions about the validity and rationale of the statistics. This kind of uncritical reporting has led to the dramatic shift in political and public attitudes towards extrajudicial killing with dangerous consequences.

A history of violence

In 1976, following the Watergate Scandal, when the church committee hearings revealed that the CIA had made several attempts on the life of the Cuban president Fidel Castro, US president Gerald Ford issued Executive Order 11905. 'No employee of the United States Government shall engage in, or conspire to engage in, political assassination,' it said. Though successive US governments found ways to circumvent it, the proscription survived until 2001. Shortly after 9/11, President George W. Bush signed an intelligence 'finding' that authorized the CIA to engage in 'lethal covert operations' in a so-called 'War on Terror' (CNN 2012). Under Bush, however, drones were used sparingly. Between November 2001, when the first drone struck Afghanistan, and January 2009, when Bush left office, the technology had been used on fifty-one occasions. According to the Bureau of Investigative Journalism, there were at least 530 uses of drones beyond active combat zones under Obama.

Though Barack Obama had asserted his preference for remote warfare on the campaign trail, his embrace of the tactic was controversial and initially resisted. But eager to distinguish himself from Bush, Obama rejected the then dominant counterinsurgency doctrine, with its large-scale troop commitments, in favour of counterterrorism, with its light footprint and reliance on Special Forces and precision weapons (Woodward 2010). By 2013, the administration's use of drones was so extensive that in a speech, Obama had to confess that the administration had come 'to view drone strikes as a cure-all for terrorism' (Obama 2013). In Pakistan alone the United States has launched 424 drone strikes since 2002, 373 of them under Obama. According to the Bureau of Investigative Journalism (BIJ), these strikes have killed between 2,499 and 4,001 people, at least 424–966 of them civilians, including 172–207 children. Fewer than 4 per cent of these were identified as al-Qaeda (Serle 2014); only 2 per cent as 'high value targets' (Kaag and Kreps 2014). The identity of the majority remains unknown.

The drone war is consequential for the international order (Boyle 2013; Kaag and Kreps 2014). The technology is cheap to acquire and easy to replicate. By November 2013, eighty-seven states were in possession of drone technology (Taylor 2013). Since then even non-state actors have deployed them. Their offensive use across borders will not remain exclusive for long. Though the practice has been considerably rolled back since 2012, this was mainly in response to anger overseas. At home the propaganda was successful. Polls in the United States have consistently shown high support for the drone policy (Fuller 2014), although the results may have been skewed by the way pollsters framed questions, failing to acknowledge the disputed status of the targets (Kaag and Kreps 2014).[1]

By accepting official claims about the status of those killed, the media has lowballed civilian deaths thereby denying the public an opportunity to fully assess the policy's humanitarian and moral implications. The moral and legal questions have in turn been obfuscated with the claim that the strikes are welcomed by the targeted populations. All of this has contributed to the muted public response and the political inertia. A reckoning is necessary.

How rituals of objectivity bias statistics in the media[2]

Statistics are an authoritative way of describing the scope of a social or political problem. Numbers convey a sense of precision. We live in 'a hyper-numeric world preoccupied with quantification,' write Andreas and Greenhill (2010). 'In practical political terms', they argue 'if something is not measured it does not exist … if there are no "data", an issue or problem will not be recognized, defined, prioritized, put on the agenda, and debated' (1). Statistics, writes Best (2001, 10), can 'become weapons in political struggles over social problems and social policy'. 'The creation, selection,

---

[1] For example, Yougov/Economist poll: 'Do you approve or disapprove of the Obama Administration using drones to kill high-level terrorism suspects overseas?'; Gallup: 'Do you think the US government should or should not use drones to … launch airstrikes in other countries against suspected terrorists?'; NBC/Wall Street Journal: 'Do you favor or oppose the use of unmanned aircraft, also known as drones, to kill suspected members of Al Qaeda and other terrorists?' (Kaag and Kreps 2014).

[2] This section is adapted from Ahmad (2016).

promotion, and proliferation of numbers are thus the stuff of politics' (Andreas and Greenhill 2010, 2).

Statistics are not always neutral: they are inscribed with the interests of their producers. Supporters and opponents of policies inflate or minimize numbers based on their serviceability (Best 2001, 10). This matters 'because quantification is politically consequential', write Andreas and Greenhill (2010, 2). 'Both proponents and opponents of any given policy will marshal reams of data to bolster their position and to weaken support for rival positions.' Because they 'shape both public and closed-door policy debates', they 'serve to legitimize some positions and undercut others' (Andreas and Greenhill 2010, 135).

This becomes particularly significant in the case of conflicts as 'prevailing estimates of the scale of violence, its complexion, and its measurable consequences undeniably play a role in shaping policy priorities and objectives' (Andreas and Greenhill 2010). Statistics, therefore, can potentially prolong or end a conflict. If politicians or their publics have a distorted view of a war's progress – an inevitability since inflating success and downplaying setbacks becomes necessary for maintaining military morale and public support – it pre-empts reappraisal.

The drone war is an object study in such distortions. In 2011, President Obama's chief counterterrorism adviser John Brennan insisted that 'nearly for the past year [in the drone war] there hasn't been a single collateral death'. Obama concurred: 'Drones have not caused a huge number of civilian casualties. … This is a targeted, focused effort at people who are on a list of active terrorists trying to go in and harm Americans.' The administration's spokesman Jay Carney added: 'Drone strikes are legal, they are ethical, and they are wise' (quoted in Kaag and Kreps 2014). These claims are self-serving. But the media for the most part overlooks the conflict of interests.

For Tuchman (1972), this dispensation has to do with a certain notion of 'objectivity' that privileges process over outputs. Commonly understood, objectivity means a judgement that is free from prejudice and bias (Gaukroger 2012). But in the practice of journalism, writes Tuchman (1972), 'objectivity' now comprises strategic rituals that are geared not so much for the pursuit of truth as for protecting journalists from flak or libel. All journalists are pressed by deadlines, and when someone makes a claim, a journalist might not have the time to verify its validity. But 'newspapermen regard the statement "X said A" as a "fact"' – Tuchman notes – 'even if "A" is false'. And by virtue of carrying an institution's credibility with them, official claims are deemed more credible even if the officials belong to institutions themselves in need of scrutiny, such as the military and intelligence agencies. When unverifiable claims from these institutions become the source for statistics, then regardless of methodological rigour and journalistic objectivity the output will be misleading. Garbage in, garbage out, as the old truism has it.

A further dilemma is presented when access becomes a barrier to objectivity. When an individual or an institution has privileged information, this grants them power by allowing them to restrict access only to pliant conduits. This dilemma has been recognized and debated since at least the sixteenth century when, in his *Dialogues on History*, Francesco Patrizzi (1560) argued that a historian can be either impartial or informed, but he can't be both. Rulers grant access to information only to partisan observers who are sympathetic to them rather than to objective observers who will provide an unbiased account of this information (Gaukroger 2012). Limiting the supply of information thus inflates its value, regardless of quality.

Transposed to the world of journalism, this means that objective observes with a reputation for probity are less likely to be granted access by the powerful. In the case of the secretive drone war, this has allowed its prosecutors to shape narratives and elude scrutiny. Their claims of success face no immediate challenge, and subsequent refutations get relegated to back pages because the news cycle has moved on. Consequently, the drone war for the most part has failed in the United States to rise to the level of a public issue. It only entered the national debate after a leaked Justice Department memo revealed that the Obama administration has legalized killing Americans abroad if they 'present an imminent threat to national security' (Isikoff 2013). The drone war has also benefited from another feature of reporting. As Galtung and Ruge (1965) noted, when it comes to foreign conflicts, an incident is considered newsworthy only if it crosses a certain fatality threshold. Drones kill on a small enough scale to escape scrutiny and strike frequently enough to have lost the element of the unexpected (a factor that earlier brought them some attention).

Reality is thus distorted when this defective reportage passes through the legitimizing process of 'calculative practices'. Bad primary data are laundered into ostensibly credible numbers and, regardless of a statistician's good faith, the output misleads. For the media and the public, statistics represent 'hard facts' – regardless of their provenance. The semblance is aided, notes Best, by 'widespread confusion about basic mathematical ideas' (2001, 19–20). A consequence of this innumeracy is that 'many statistical claims about social problems don't get the critical attention they deserve':

> We use statistics to convert complicated social problems into more easily understood estimates, percentages, and rates. Statistics direct our concern; they show us what we ought to worry about how much we ought to worry. In a sense, the social problem becomes the statistic and, because we treat statistics as true and incontrovertible, they achieve a fetishlike, magical control over how we view social problems. We think of statistics as facts that we discover, not as numbers we create. (Best 2001, 160)

Even when produced by experts – as statistics often are – they involve choices that shape the outcome. A necessary simplification, they can sometimes contribute to a simplified, even superficial, understanding of complex realities (Best 2001, 161). This, as we shall see, is particularly true of the drone war.

Drones and statistical misfires in news coverage: A case study

Official denials and continuing attempts to minimize the scale of the atrocities notwithstanding, by 2013 even President Obama had to acknowledge that the drones weren't always hitting their intended targets. It was clear that whatever counterterrorism success the drone war had had – and it did help eliminate many high-profile terrorists – it had come at a high human cost. Beyond the quotidian terror of living under drones, the attacks also brought immiseration to the larger population of Pakistan as militants took their revenge on a country that they deemed complicit in the policy. Since the beginning of the war, Pakistan has lost over 30,000 civilians to terrorist attacks, many of them killed in retaliation for US drone strikes. These consequences weren't lost on US diplomats stationed in the country. In a leaked cable to Washington from 2009 the then ambassador Anne Patterson warned:

> Increased unilateral operations in these areas risk destabilising the Pakistani state, alienating both the civilian government and military leadership, and provoking a broader governance crisis in Pakistan without finally achieving the goal (quoted in Ahmad 2015).

But by 'shielding US citizens, politicians, and soldiers from the risks associated with targeted killings', Kaag and Kreps (2014) argue, the drone war has created a 'moral hazard'. Ethical concerns have gone out the window as there are no immediate risks or legislative constraints. Success is rewarded but failure carries no sanction. Most drone strikes barely make it into the media (Ahmad 2016).

In this context when the *Washington Post*'s respected Monkey Cage blog – a site curated by leading political scientists to host significant new studies in journalistic form – posted a study concluding that drone blowback was a myth (Shah 2016), it predictably garnered much attention. If drones were precise, causing few civilian casualties, and if blowback was a myth, then the CIA was in possession of an infallible counterterrorism tool. The researcher, Professor Aqil Shah of the University of Oklahoma, presented his findings in solid quantitative form. Based on a survey, he asserted that drone strikes had had no adverse effect on public opinion in the region and hence there was no correlation with the terrorist backlash. He rejected what he called 'the blowback thesis' and insisted that

> the data contradict the presumed local radicalization effects of drones. In fact, 79 percent of the respondents endorsed drones. In sharp contrast to claims about the significant civilian death toll from drone strikes, 64 percent, including several living in villages close to strike locations, believed that drone strikes accurately targeted militants.

The study received wide coverage and Shah made appearances on various media, including the BBC, to elaborate on his findings. On social media, Sam Gad Jones, the defence and security editor of *Financial Times* called it a 'an important data-rich read'; Professor Michael C. Horowitz of the University of Pennsylvania exhorted his followers to learn from Aqil Shah; Dhruva Jaishankar of Brookings India called it 'superb research'; Shashank Joshi of Royal United Services Institute devoted a whole Tweet storm to it. This article and the reactions to it are symptomatic of the ways journalism has failed in the case of the drone war. A respectable publication chose to publish a survey whose rationale and validity were both suspected. Neither publisher nor the audience asked if opinion polling was an appropriate method for determining a historical reality; nor was the question raised if the opinion data were representative since they were based on a small non-random sample of individuals made available to the researcher by the Pakistani military. But because the findings were presented in confident, statistical form, they were not only accepted but praised for their 'data-richness'. In deconstructing this survey therefore we'll also be diagnosing common problems with reporting on the drone war.

Let us first examine the rationale for the stats. Since blowback is a fringe reaction, a public opinion survey can reveal little about it. Severe blowback can happen even when there is little public support (e.g. 9/11). Blowback is a matter of fact, not perception. To the extent that it happens, its reality cannot be wished away on the strength of feeling any more than climate change can be wished away just because Middle America denies its existence. There is a more straightforward method for measuring blowback: simply looking at instances of terroristic activity where the drone war was cited as motivation. Consider the following:

- After a Jordanian double agent blew up a CIA base in Khost on 30 December 2009, a Taliban commander told the *Wall Street Journal*: 'We attacked this base because the team there was organizing drone strikes in Loya Paktia and surrounding area' (Gopal, Gorman and Dreazen 2010).
- When Faisal Shahzad appeared in court after his failed 1 May 2010 attempt to detonate a bomb at Time Square, he cited 'drone strikes in Somalia and Yemen and in Pakistan' as one of his motivations for the attack.
- When Umar Farouk Abdulmutallab pleaded guilty for his December 2009 attempt to bomb a passenger airliner over Detroit, he cited as one of his motivations the 'killing of innocent and civilian Muslim populations in Yemen, Iraq, Somalia, Afghanistan and beyond' – the main arenas for drone ops (Wheeler 2013).
- Before he and his brother bombed the Boston Marathon in April 2013, Dzhokhar Tsarnaev told a high school friend that drone strikes were one of the reasons why terrorist attacks against the United States could be justified (Reitman 2013).
- At the Nanga Parbat base camp when eleven mountaineers were killed in June 2013, the Taliban claimed it as retaliation for the drone strike a month earlier that had killed the movement's second-in-command Waliur Rehman (Mir 2013).
- When in December 2013 the Pakistani Taliban attacked a Pakistani military checkpoint, killing five and wounding thirty-four, they cited a drone attack that had killed their leader as the reason (Sherazi 2013).

Unmanned technology makes it possible to wage war without putting personnel at risk, but it is not free of consequences. The reactions are indirect, but they are real. They might be anecdotal but are substantive enough to prove that the consequences of drone attacks range from passive resentment, military action to terroristic retaliation. There are also ambient effects such as the general radicalization of a targeted society that heighten the likelihood of blowback (Robertson 2013). Drone technology may be new, but there is nothing novel about being killed in an explosion. The way a laser-guided missile is experienced on the ground is as violence, regardless of how it is delivered – and violence has consequences. Focus on the technology in the coverage of the drone war draws attention away from the mundane realities of violence and pushback.

Let us now examine the validity of the data. Advocates for the drone war have made three related arguments to deny a link between drone strikes and the militant backlash: (1) that drones are accurate hence discriminating; (2) that they are effective for countering terrorism; and (3) that they are welcomed by the targeted populations. But as the following discussion shows each of these claims merits further interrogation since the statistics have little validity when the underlying data are so contradictory. If statistics is the science of 'extracting meaning from data' (Hand 2008), then it can't reveal much if the data itself is meaningless.

Accurate and discriminative?

There is little doubt that targeting technologies have improved over the past century. But in practice the gap between the ability to bomb precisely and the capacity to identify targets accurately has yet to be fully bridged. The precision of a weapon gives little advantage when it is deployed on fallible information. On 13 February 1991, during the Gulf War two US F-117 stealth bombers dropped 2,000-pound GBU-27 laser-guided bombs precisely through the air ducts of a compound in Baghdad's Amiriyah neighbourhood. The CIA had identified the building as 'command-and-control' facility. The target was in fact a civilian air raid shelter and 408 women and children were incinerated.

In January 2015, after a drone strike accidentally killed western aid workers Warren Weinstein and Giovanni Lo Porto, the Obama administration admitted that it did not know whom it was targeting (Ackerman, Siddiqui and Lewis 2015). 'It has become clear', said the *New York Times*, 'that when operators in Nevada fire missiles into remote tribal territories on the other side of the world, they often do not know who they are killing, but are making an imperfect best guess' (Shane 2015). In an analysis of classified data from US airstrikes carried out in Afghanistan from mid-2010 to mid-2011, Lawrence Lewis of the Center for Naval Analyses found that drone attacks were ten times more likely to result in civilian casualties than attacks by non-remotely controlled aircraft (Briggs 2013).

The technology may be flawless, but based on flawed intelligence, it can only lead to flawed outcomes. In its first successful use in the 'war on terror', on 16 November 2001, a drone strike on a small hotel in Afghanistan killed al-Qaeda deputy Mohammed Atef and his six companions. But the attack also killed 'close to 100 people' who happened to be in the vicinity. In Pakistan, on 13 January 2006, a drone strike on Bajaur killed what the media insisted were 'four al-Qaeda terrorists'; in fact the strike had killed 18 villagers, mainly women and children. On 30 October 2006 another drone struck Bajaur reportedly killing 'between two and five senior al Qaeda militants'; the target was in fact a seminary and all but one of eighty-one dead were children (Woods 2015: 94–6). All of the dead were initially reported as 'militants' or 'al-Qaeda' or 'terrorists'.

The criteria for drone strikes became more permissive under Barack Obama. But reporting remained just as credulous. In data produced by the New America Foundation (NAF), the most cited source for drone fatalities, many of these civilian deaths went unacknowledged until 2011. The record was corrected only after this author, followed by the Bureau of Investigative Journalism, started raising the issue. In a previous article, I have shown that the basis for much of this seemingly precise data is dubious (Ahmad 2016). The data is culled from early news reports that in turn rely on statements from unnamed Pakistani and American officials. Predictably, all those killed are pronounced 'militants', 'terrorists' or 'al-Qaeda'. Many of these claims are contradicted by subsequent revelations. But there are also more mundane reasons for doubting the claims. Because of the insecurity, FATA is an information void. US and Pakistani governments don't have a mechanism for verifying the identity of those killed in these strikes; nor do they make an effort (ibid.). But even in the rare cases where the innocence of the victims is established, the data bases are rarely revised (the BIJ is an exception since it makes a determined effort to establish the identities of those killed).

If the administration often doesn't know whom it is killing, then for the media to accept that those being killed are all combatants is merely an act of faith. The administration has in fact taken measures to give itself legal cover, mainly by defining 'militant' expansively, thereby shrinking the space for a target to be counted as a civilian. Definitions, writes Best (2001), are important, because they 'specify what will be counted' (44–5) – or, conversely, what isn't counted. As the *New York Times* revealed, the administration had 'embraced a disputed method for counting civilian casualties' that – according to several administration officials – 'in effect counts all military-age males in a strike zone as combatants … unless there is explicit intelligence posthumously proving them innocent' (Becker and Shane 2012). Dehumanizing labels such as MAM ('military-aged males') have further eased moral constraints by turning a whole category of humans into acceptable targets.

The media rarely interrogates the legal basis for these categories. International Humanitarian Law (IHL) discriminates even among combatants, drawing a distinction between 'direct and indirect participation in hostilities and between legal and illegal targets'. However, as Kaag and Kreps (2014) note, most of the 'militants' killed in Pakistan are 'lower-level foot soldiers … who are "neither presently aggressing nor temporally about to aggress"'. The administration has also tried to circumvent legal barriers by adopting an expansive definition of 'imminent' that 'does not require the United States to have clear evidence that a specific attack on US persons and interests will take place in the immediate future' (Isikoff 2013).

Consequently, Kaag and Kreps (2014) argue, the drone war satisfies neither the *jus ad bellum* (right to war) nor *jus in bello* (conduct in war) principles of a just war. Unsurprisingly the UN's special rapporteur on extrajudicial killings Philip Alston, his successor Christoph Heyns, and the UN's special rapporteur on human rights and counterterrorism Ben Emmerson QC have all questioned the administration's legal justifications for the drone war (Alston 2010; Bowcott 2013).

Effective?

In May 2016, the leader of the Afghan Taliban Mullah Akhtar Mansoor was killed in a drone strike in Pakistan. Drones have previously eliminated two former leaders of the Pakistani Taliban. They have also killed top al-Qaeda leaders. If eliminating the leadership of an insurgency is proof of their effectiveness then the drone war is a success. But the undiminished force of the insurgency seems to suggest otherwise. Part of the reason is related to the first question. If drones have been successful in eliminating the leadership of the Taliban, it has to do less with accuracy and discrimination than with perseverance and ruthlessness. A study conducted by the human rights group Reprieve has shown that in targeting the twenty-four individuals in Pakistan, drones caused a total of 874 deaths, with many of the targets reported dead on multiple occasions (Ackerman 2014).

Before the Obama administration scaled back the drone war in 2013, under expanded authority granted to the CIA, drone strikes weren't confined only to 'high value targets'; anyone suspected of being a combatant could be killed in a controversial policy known as 'signature strikes'. To circumvent the time and resources needed for gathering human intelligence, the CIA was authorized to use dubious tools such as social network analysis (SNA), a social science method that focuses on the relations between actors rather than their individual attributes, and 'pattern of life' intelligence that picks targets based on remotely observed behavioural patterns instead of actual evidence. In an infamous incident in Uruzgan, a US drone blew up twenty-three members of a family based on such 'patterns of life' factors as the ablutions that all practicing Muslims carry out ahead of prayers. In other instances, even modes of urinating have been used to determine guilt or innocence (Cockburn 2015).

Likewise, the drone war has shown the limits of SNA. In Afghanistan, the US National Security Agency (NSA) records every telephone conversation, and the CIA and the Joint Special Operations Command (JSOC) have sometimes used International Mobile Subscriber Identity (IMSI) numbers to track targets. In one instance after associating an IMSI with insurgent leader Mohammed Amin, prolonged drone surveillance helped track him down along with five of his companions, leading to a targeted strike. Except the man they killed wasn't Amin, but the *anti-Taliban* leader Zabet Amanullah, along with five of his relatives, one of them seventy-seven years old. The initial intelligence was defective (Cockburn 2015).

Popular?

Since the early days of the drone war there has been an attempt to deflect attention from the moral, legal and political questions with the claim that drone attacks are popular among the targeted populations. The claim originated in the form of a survey carried out 'on the ground in FATA' by an entity named the Aryana Institute for Regional Research and Advocacy. It claimed that 55 per cent of respondents did not think that the attacks caused 'fear and terror in the common people'; 52 per cent found them 'accurate in their strikes'; and 58 per cent did not think they increased anti-Americanism (Taj 2009).

The 'survey' was uncritically reported in the US press and enthusiastically embraced by drone advocates [even though a contemporary Gallup/Al Jazeera (2010) poll showed only 9 per cent support for drone attacks in Pakistan]. The 'institute' that produced the study was in fact a letterhead organization with a brief virtual existence and the 'survey' was little more than an op-ed by the controversial pro-drone activist Farhat Taj. No data was published to back it up. Taj claimed that the survey was carried out in 'parts of FATA that are often hit by American drone'. But besides North and South Waziristan, the only region she listed was Parachinar (Kurram Agency), which had never been struck by a drone. Parachinar, the capital of Kurram Agency, is predominantly Shia and its population is hostile to the virulently anti-Shia Taliban. [To understand the significance of this, consider that according to a 2010 survey of FATA, 99.3 per cent of respondents in North Waziristan considered drone attacks to be 'never justified' as opposed to 12.9 per cent in the Kurram Agency (Shinwari 2010)].

The credibility of Taj's claims was further undercut when a 2010 professional survey by the NAF and Terror Free Tomorrow (TFT) showed that 76 per cent of respondents in FATA opposed the drone attacks; 40 per cent held the United States primarily responsible for the violence; 16 per cent thought the drones were only targeting combatants. The 2011 surveys by the local NGO Community Appraisal & Motivation Programme (CAMP) reinforced these findings. It revealed that 63 per cent of FATA residents thought drone attacks were 'never justified' (Shinwari 2012).

It is therefore only with alarm that one could read the august *Washington Post* resurrect this long debunked claim in 2016. What is most troubling about this, however, is the shoddiness of the research. The researcher, Aqil Shah, had used a selected sample of 147 people displaced from one of FATA's seven agencies to draw statistical conclusions about the whole region. These findings weren't representative of North Waziristan, let alone of FATA. There is no methodological justification for drawing statistical conclusions from a non-random sample. Yet they were published and reported widely without demurral.

But methodological shortcomings are the least of Shah's problems. Consider how he deals with past surveys. Shah does not mention the NAF/TFT poll at all; and to conceal the degree to which his claims are at odds with past research, he interprets the 2011 CAMP survey in a peculiar way. He writes that 'when the results are disaggregated, support for drone strikes is the highest in North Waziristan, the FATA agency (district) where the CIA has carried out most of its lethal drone operations'. This is an example of a statement being factually correct without being truthful. The CAMP survey does show that 16 per cent of North Waziristan residents believe that drones strikes are 'sometimes justified *if properly targeted and excessive civilian casualties are avoided*'. But it also reveals that while only 0.2 per cent in North Waziristan support drones unconditionally, 58 per cent believe that drone attacks are '*never* justified'.

That the Monkey Cage survey was misconceived is obvious. But it also demonstrates that innumeracy can cloud what Mills called the sociological imagination. In its eagerness to decouple a policy from its predictable consequences, the research also makes the absurd assumption that blowback is a geographically delimited phenomenon. As noted earlier, and as Shah acknowledges, multiple surveys have shown consistent opposition to drones across Pakistan. So even if Shah's findings were true, would a retaliatory act originating from the rest of Pakistan not count as blowback?

Limits of contrarianism

When the Monkey Cage research was published, some in the media praised it for going 'against the current' or revealing a 'counterintuitive' truth. A counter-thesis carries a certain cache. It suggests a willingness to swim against the tide and risk obloquy. The attempt to dissociate drone attacks from their consequences however is not a dissent, but an affirmation of a superseded orthodoxy. It is not a useful contribution to the debate but a dilution of truth. Inconvenient realities cannot be wished away by soliciting convenient opinions. There are some facts about the drone war that are incontrovertible: drones are accurate but the intelligence they operate on is fallible; drones have eliminated some terrorists but most of the people they kill are low-level insurgents; drones have caused many civilian casualties; and drones are resented by the targeted populations. To say that all of this is free of consequences, one will have to disprove human nature.

It took many years of effort to bring to light the impact and consequences of the drone war. The aggravating consequences have even been acknowledged by a victim of Taliban terror like Malala Yusufzai (Kumar 2014). In response, the Obama administration finally changed its policies. And though the administration became more sparing in its use of drones and more careful in its targeting practices, the legal, political and moral questions were left unresolved. This oversight may prove costly with Donald Trump in power. Journalism simply cannot afford to be as credulous and complacent as it has been during the Obama years. It is therefore imperative that journalists exercise greater caution in reporting without letting skepticism shade over into cynicism, especially when the reporting involves numbers and statistics.

In his aforementioned investigation into the conflict between objectivity and access, Patrizzi had arrived at the grim conclusion that it is 'utterly and totally impossible for human actions to be known as they were actually done' (quoted in Gaukroger 2012). But as his contemporaries responded, questions of evidence and reliability can be addressed the same way that courtrooms establish the credibility of witnesses and the plausibility of their claims – through verification and corroboration.

Finally, it should be made clear that the chapter is not a case against statistics. To be sure, though statistics can sometimes mislead, in most cases they remain a powerful tool for understanding the scope and intensity of social and political issues. To quote Frederick Mosteller, 'It is easy to lie with statistics, but easier to lie without them.'

What I propose is three simple considerations that can have a salutary effect on the use of statistics in journalism. In the era of fake news, it will be doubly important for journalists to check sources, verify evidence and corroborate claims. A journalist must always ask: (a) Is the issue amenable to statistical analysis? (b) Are the data credible? and (c) Is the method rigorous? In short, journalists will have to accept statistics only if their rationale and validity aren't in doubt. For numbers can kill.

References

Ackerman, S. (2014). '41 men targeted but 1,147 people killed: US drone strikes – The facts on the ground.' *The Guardian*, 24 November. Retrieved 14 March 2016 from https://www.theguardian.com/us-news/2014/nov/24/-sp-us-drone-strikes-kill-1147.

Ackerman, S., Siddiqui, S. and Lewis, P. (2015), 'White House admits: we didn't know who drone strike was aiming to kill.' *The Guardian*, April. Retrieved 14 March 2016 from https://www.theguardian.com/world/2015/apr/23/drone-strike-al-qaida-targets-white-house.

Ahmad. M. I. (2016), 'The magical realism of body counts: How media credulity and flawed statistics sustain a controversial policy.' *Journalism*, 17(1), 18–34.

Ahmad, M. I. (2015), 'Death from Above, Remotely Controlled: Obama's Drone Wars.' *In These Times*, 9 July. Retrieved 9 July 2015 from http://inthesetimes.com/article/18163/drones-andrew-cockburn-kill-chain-chris-woods-sudden-justice.

Alston, P. (2010), Study on targeted killings: Report of the Special Rapporteur on extrajudicial, summary or arbitrary executions. UN Human Rights Council, 13 May. Retrieved 30 July 2012 from http://www2.ohchr.org/english/bodies/hrcouncil/docs/14session/A.HRC.14.24.Add6.pdf.

Andreas P. and Greenhill K. M. (2010), *Sex, Drugs, and Body Counts: The Politics of Numbers in Global Crime and Conflict*. Ithaca, NY: Cornell University Press.

Becker, J. and Shane, S. (2012), 'Secret "Kill List" proves a test of Obama's principles and will.' *The New York Times*, 29 May. Retrieved 9 July 2015 from http://www.nytimes.com/2012/05/29/world/obamas- leadership-in-war-on-al-qaeda.html.

Best, J. (2001), *Damned Lies and Statistics: Untangling Numbers from the Media, Politicians, and Activists*. London: University of California Press.

Boyle, M. J. (2013), 'The costs and consequences of drone warfare.' *International Affairs* 89(1), 1–29.

Bowcott, O. (2013), 'Drone strikes by US may violate international law, says UN.' *The Guardian*, 18 October. Retrieved 9 July 2015 from https://www.theguardian.com/world/2013/oct/18/drone-strikes-us-violate-law-un.

Briggs, B. (2013), 'Study: US drone strikes more likely to kill civilians than US jet fire.' NBC News, 2 July. Retrieved 9 July 2013 from http://www.nbcnews.com/news/investigations/study-us-drone-strikes-more-likely-kill-civilians-us-jet-v19254842.

CNN (2012), U.S. policy on assassinations. 4 November.

Cockburn (2015) reference is: Cockburn, A. (2015), *Kill Chain: Drones and the Rise of High-Tech Assassins*. London: Verso.

Fuller, J. (2014), 'Americans are fine with drone strikes. Everyone else in the world? Not so much.' *Washington Post*, 15 July. Retrieved 23 July 2014 from http://www.

washingtonpost.com/blogs/the-fix/wp/2014/07/15/americans-are-fine-with-drone-strikes-everyone-else-in-the-world-not-so-much/.

Gallup/Al Jazeera (2010), 'Pakistan: State of the nation.' 13 August. Retrieved 20 July 2011 from http://www.aljazeera.com/focus/2009/08/2009888238994769.html.

Galtung J. and Ruge M. H. (1965), 'The structure of foreign news: The presentation of the Congo, Cuba and Cyprus crises in four Norwegian newspapers.' *Journal of Peace Research*, 2, 64–91.

Gaukroger, S. (2012), *Objectivity: A Very Short Introduction*. Oxford: Oxford University Press.

Gopal, A., Gorman, S. and Dreazen, Y. (2010), 'Taliban Member Claims Retaliation.' *Wall Street Journal*, 4 January. https://www.wsj.com/articles/SB126246258911313617.

Hand, D. (2008), *Statistics: A Very Short Introduction*. Oxford: Oxford University Press.

Isikoff, M. (2013), 'Justice department memo reveals legal case for drone strikes on Americans.' *NBC News*, 4 February. Retrieved 20 July 2014 from http://investigations.nbcnews.com/_news/2013/02/04/16843014-justice-department-memo-reveals-legal-case-for-drone-strikes-on-americans.

Kaag, J. and Kreps, S. (2014), *Drone Warfare*. London: Polity.

Klaidman, D. (2012), *Kill or Capture: The War on Terror and the Soul of the Obama Presidency*. New York: Houghton Mifflin.

Kreps, S. (2014), 'Flying under the radar: A study of public attitudes towards unmanned aerial vehicles.' *Research & Politics*, 1(1), 1–7.

Kumar, A. (2014), 'Nobel Prize winner Malala told Obama U.S. drone attacks fuel terrorism.' 10 October 10. Retrieved 9 July 2015 from http://www.mcclatchydc.com/news/politics-government/article24774460.html.

Mayer, J. (2009), 'The predator war: What are the risks of the C.I.A.'s covert drone program?' *The New Yorker*, 26 October. Retrieved 9 July 2015 from http://www.newyorker.com/magazine/2009/10/26/the-predator-war

Mazzetti, M. (2013), *The Way of the Knife: The CIA, a Secret Army, and a War at the Ends of the Earth*. London: Penguin Press.

Mir, S. (2013), 'Nanga Parbat assault: Tragedy in Himalayas.' *Express Tribune*, 24 June. Retrieved 9 February 2014 from https://tribune.com.pk/story/567527/nanga-parbat-assault-tragedy-in-himalayas/.

Obama B. (2013), Text: Obama's speech on drone policy. *National Defence University*, 23 May. Retrieved 9 July 2016 from http://www.nytimes.com/2013/05/24/us/politics/transcript-of-obamas-speech-on-drone-policy.html.

Pilkington, E. and McAskill, E. (2015), 'Obama's drone war a "recruitment tool" for Isis, say US air force whistleblowers.' *The Guardian*, 18 November. Retrieved 9 July 2016 https://www.theguardian.com/world/2015/nov/18/obama-drone-war-isis-recruitment-tool-air-force-whistleblowers.

Reitman, J. (2013), 'Five revelations from Rolling Stone's Boston bomber cover story.' *Rolling Stone*, 16 June. Retrieved 9 July 2013 from http://www.rollingstone.com/culture/news/five-revelations-from-rolling-stones-boston-bomber-cover-story-20130716.

Robertson, N. (2013), 'In Swat Valley, U.S. drone strikes radicalizing a new generation.' CNN, 15 April. Retrieved 15 April 2013 from http://edition.cnn.com/2013/04/14/world/asia/pakistan-swat-valley-school/.

Serle, J. (2014), 'Only 4% of drone victims in Pakistan named as al Qaeda members.' *The Bureau of Investigative Journalism*, 16 October. Available at: https://www.

thebureauinvestigates.com/2014/10/16/only-4-of-drone-victims-in-pakistan-named-as-al-qaeda-members/.

Shah, A. (2016), 'Drone blowback in Pakistan is a myth. Here's why.' *Washington Post*, 17 May. Retrieved 9 January 2017 from https://www.washingtonpost.com/news/monkey-cage/wp/2016/05/17/drone-blow-back-in-pakistan-is-a-myth-heres-why/?utm_term=.46b354f828b7.

Shane, S. (2015), 'Drone strikes reveal uncomfortable truth: U.S. Is often unsure about who will die.' *New York Times*, 23 April. Retrieved 9 July 2015 from https://www.nytimes.com/2015/04/24/world/asia/drone-strikes-reveal-uncomfortable-truth-us-is-often-unsure-about-who-will-die.html?_r=2.

Sherazi, Z. (2013), 'At least 23 suspected militants killed in North Waziristan.' *Dawn*, 19 December 19. Retrieved 10 August 2015 from https://www.dawn.com/news/1075069/at-least-23-suspected-militants-killed-in-north-waziristan.

Shinwari, N. (2010), Understanding FATA – Vol. 4. Retrieved 9 July 2015 from http://www.understandingfata.org/files/Understanding-FATA-Vol-IV.pdf.

Shinwari, N. (2012), Understanding FATA – Vol. 5. Retrieved 9 July 2015 from http://www.understandingfata.org/uf-volume-v/Understanding_FATA_Vol-V-11.pdf.

Taj, F. (2009). 'Drone attacks—a survey'. *The News* (Pakistan). 5 March. Retrieved 12 July 2017 from http://web.archive.org/web/20090312040744/http://www.thenews.com.pk/daily_detail.asp?id=165781.

Taylor, G. (2013), 'U.S. intelligence warily watches for threats to U.S. now that 87 nations possess drones.' *The Washington Times*, 10 November. Retrieved 10 November 2013 from http://www.washingtontimes.com/news/2013/nov/10/skys-the-limit-for-wide-wild-world-of-drones/.

Tuchman, G. (1972), 'Objectivity as strategic ritual: An examination of newsmen's notions of objectivity.' *American Journal of Sociology*, 77(4), 660–79.

Wheeler, M. (2013). 'In guilty please, Abdulmutallab named Awlaki as inspiration, not as co-conspirator'. *Empty Wheel*. 22 May. Retrieved 12 July 2017 from http://www.emptywheel.net/2013/05/22/in-guilty-plea-abdulmutallab-named-awlaki-as-inspiration-not-as-co-conspirator/.

Woodward, B. (2010), *Obama's Wars*. London: Simon & Schuster.

Woods, C. (2015), *Sudden Justice: America's Secret Drone Wars*. London: Hurst.

4

Poor Numbers, Poor News: The Ideology of Poverty Statistics in the Media

Jairo Lugo-Ocando, *University of Leeds, UK*
Brendan Lawson, *University of Leeds, UK*

Introduction

In 2016, the government of Mexico seemed to have achieved an amazing and completely unexpected task: it managed to dramatically cut poverty levels at a simple stroke. This went against the fact that poverty and social exclusion levels had been rising for the previous ten years in that country despite increasing average incomes. Could this be the most successful and well-orchestrated government intervention into poverty of all times? Well, not quite so. What the Mexican government, under Enrique Peña Nieto, did was much simpler; they changed the methodology used by their National Statistics Institute to show that Mexico's poor people were better off in terms of income by a third compared with the previous year.[1] The news media did not miss the opportunity to expose the attempt of the already embattled Peña Nieto – often accused of corruption, incompetence and close links with drug cartels – to meddle with numbers to create a misleading positive spin on what by all means has been a very flawed administration (Temkin Yedwab and Salazar-Elena 2012; Vilchis 2013).

Mexico is not the only country to have done such things. Many others have gone unnoticed when altering the basis of their national statistics and key measures of poverty and social exclusion in order for politicians and bureaucrats to show 'improvements' in poverty reduction, employment and people's life during 'their turn in office'. In Napoleonic France, officials manipulated statistics to highlight the French welfare programmes (Desrosières and Naish 2002; Perrot and Woolf 1984). In Italy, under Benito Mussolini's rule, the Fascist produced the numbers as to show that they had managed to reduce inequality, creating the now widely used GINI coefficient (Ipsen 2003; Prévost 2009). In the 1980s in the United Kingdom, the late Margaret Thatcher started to count people in fractional and fix-termed short contracts as

---

[1] David Agren (18 July 2016), 'CityMexico cuts poverty at a stroke – by changing the way it measures earnings'. *The Guardian*. https://www.theguardian.com/world/2016/jul/18/mexico-cuts-poverty-national-statistics-changes-earnings-measurements (accessed 20 October 2016).

being fully employed (Webster 2002). More recently, in Venezuela, the then president Hugo Chavez authorized his Office for National Statistics to include the use of public services and other government benefits as part of people's income, although this did not preclude the fact that there was a significant reduction of poverty between 1999 and 2005 (Weisbrot, Sandoval and Rosnick 2006, 4). So many others, in the present and the past, have followed suit to adopt these tactics to give a better picture of poverty in their own countries. However, what is unique in the particular case of Mexico is that the unhealthy play with numbers was caught red-handed and thereafter fiercely exposed by the media. Few have been done so in history.

The question, then, is why the news media have failed consistently to critically report and expose these PR stunts around statistics in many places. Why have mainstream journalists effectively failed to bring into account those who have unscrupulously used statistics on poverty to advance their own political agenda while neglecting to address the structural issues? This chapter attempts to provide a possible answer for this by looking at the political communication of statistics in the context of news coverage of poverty. It does so by examining the relationship between politicians, journalists and statistics in relation to the news coverage of poverty. The central thesis is that while the dissemination of poverty statistics by government officials, multilateral organizations, NGOs and other actors continues to be highly mediatized and journalists are able to tap upon almost unlimited resources to do so, their ability to critically analyse numbers and counter hegemonic discourses continues to be diminished by decreasing resources in the newsroom and the lack of statistical preparation among newspeople.

To be sure, over the years, officials have allocated a substantial amount of resources and efforts into public relations efforts towards spinning poverty statistics to the audiences, therefore allowing mediatization practices to shape the way these numbers go out into the public. This has happened in parallel to a growing deficit in the ability of journalists to deal properly with numbers, given their lack of knowledge, time and other resources. The consequence of this news deficit is an over-reliance upon the explanatory frameworks offered by official sources who not only provide the interpretation of the statistics but also define their meaning in the public imagination. This, as we argue in this piece, is not only fundamentally problematic for the purported role of journalism as a watchdog of democracy but also profoundly detrimental to the process of public policy formulation, allowing politicians to get away with flaw and ideologically driven social policy.

By the mediatization of statistics, we do not mean a policy-making process directed by the media. Rather, we refer to a policy-process in which publicity-seeking activities and political decision-making become closely interlinked (Cater 1965; Cook 2005), to the point that the production and analysis of statistics become inextricably tied to the active management of its mediated representation. Recognized today as a process concurrent with modernity (Thompson 1995), in which the logic of the media starts to permeate multiple social subsystems (Schrott 2009, 42), the process of mediatization of poverty statistics implies tailoring the presentation of these numbers to fit the media requirements, in the search for public support and legitimation of the policies and actions that they aim at underpinning (Esser 2013; Marcinkowski and Steiner 2014). The enactment of authority in a policy-making process therefore will require treating statistics as both a communicative achievement as much as a political one (Crozier 2008; Hajer 2009).

The circular logic

Scholars of poverty and social exclusion know that multilateral organizations and institutions have an intimate and long history of using numbers to establish, maintain and expand their authoritative power over the 'management of poverty' (Fioramonti 2013, 2014; Masood 2016; Philipsen 2016) as these numbers underpin their role of 'expert voices'. This role emerged from a specific process in which relative experiences from the individuals are transformed into objective ones – hence, they are institutionalized, through the singularization of their definition. Thereafter, these numbers and statistics can be used to measure and determine these seemingly 'objective' experiences, enabling actors and institutions to appear capable of addressing problems by increasing or decreasing relevant measurements. Thus, a circular logic develops: the problem is defined, measured and solved by the same set of people who 'establish' the problem in the first place as an issue in the public imagination. As Arturo Escobar (1995) reminds us, before 1945 we did not have a 'Third World' to develop, it only came into existence when the elites realized that it was possible to 'manage' poverty to underpin power.

Indeed, a collection of literature that can be broadly categorized as the post-Development school of thought provides an interesting theoretical framework to approach the use of statistics in the articulation of poverty. This school emerged in the 1990s with the works of Arturo Escobar (1995), James Ferguson and Larry Lohmann (1994), Gustavo Esteva (1992), Claude Alvares and Wolfgang Sachs (1992), among others. They challenged and critiqued the relatively modern conception of development – the implementation of so-called objective economic models across a plurality of different cultures, nations and communities. This type of development systematically rejects local knowledge, interest and relativism in favour of modernization (Rapley 2004, 352). The economic rules imposed by 'development institutions' and which is often presented as objective, emerged from Western European and Northern America. Its modern origins are placed firmly in the post-Second World War era, encapsulated by Truman's Point IV Programme and central to the Structural Adjustments Programmes (SAPs) in the 1980s (Lugo-Ocando and Nguyen 2017). This specific form of development is the one that the post-Development school so strongly criticizes.

In this sense, the ability for a geographically specific conception of development economics to be legitimized as universal depends on, among other things, statistics. Quantifiable information, such as gross domestic product (GDP), has the power to validate the underlying ideology of the imposition of external economic models on a range of different cultures, histories and contexts (Fioramonti 2013; Masood 2016). However, in order to use numbers in this way, it is necessary for 'experts' to be involved in a process where the definitions of certain problems are singularized. Poverty is an excellent example of this. If there is a multiplicity of definitions for poverty, then how can statistics lay claim to the idea that they represent the reality of poverty? This is especially prevalent when the definition of poverty is not quantifiable or if it is positioned as relative to specific circumstances or people.

In *Global Poverty: A Pauperising Myth*, Majid Rahnema (2001) explores how the current Western discourse of poverty was constructed. He argues that the current discourse of poverty that has come to dominate the world is a binary one in which poverty is the opposite of rich. He explains, nevertheless, that there exists, and has existed, a plurality of concepts of poverty that conflict with this dominant discourse (Rahnema 2001, 2). For him, there was a rejection of the old concepts of poverty in favour of a new economic-centric discourse that was forced upon the developing world in the post-Second World War Development era. In addition, Michel Mollat (1986, 3) argues that in the Middle Ages, the words covering the range of conditions under the concept [of poverty] were well over forty. For example, in those times, being poor could mean falling from one's estate, being deprived of one's instrument of labour and the loss of one's status (1986, 5).

Rahnema argues that this plurality of the poverty concept was singularized through a pauperization process from the Middle Ages to the modern period. This occurred through a changing of 'needs' of the poor. Traditionally, this was defined within the local communal relief system. However, through the relatively recent process of 'economization' of society – by which it is meant trying to rationalize society's dynamics in economic terms, ascribing economic value to all things (Kurunmäki, Mennicken and Miller 2016; Madra and Adaman 2014) – the poor came to be defined through their inability to acquire the new commodities and services that the news industrial and market society had to offer (2001, 19). Thus came the singular definition of what it is to be 'poor' and what can be called 'poverty'. For Rahnema, this historical process had four themes: universal notions of poverty, creation of objective economic programmes, focus upon a healthy world economy to save the poor and the post-Second World War concept that poverty could be conquered through increased productivity (25–6).

While the plurality of definitions has all but disappeared in the English language, there still exist alternative discourses in other areas of the world. Rahnema draws heavily from probably the most well-known African linguist, John Iliffe, whose arguments draw upon wide-spanning ethnographic studies and existing literature. He explains that 'in most African languages, at least three to five words have been identified for poverty' (2001, 6). Furthermore, Iliffe (Iliffe 1987, 28) draws upon colloquialisms and proverbs from Nigeria and Ethiopia to highlight the pluralistic nature of poverty. The relativity of these concepts is rejected by the modern discourse in place of dominant notions of 'lack' or 'deficiency' (Rahnema 2001, 8). Consequently, having a singular definition allows statistics to perform their role as *the* indicators of poverty.

In this sense, GDP provides a good example of this. This statistic has been used by multilateral development organizations to evaluate national economic performance and by implication poverty reduction. The assumption for years was that increased GDP leads to a more economically prosperous, and therefore successful, country. This, of course, has been challenged both by economists (Stiglitz, Sen and Fitoussi 2010) and other academics such as Fioramonti (2013), Jerven (2013) and Philipsen (2016), among others, who have exposed the highly contingent nature of a supposed universal fact. At a fundamental level, there are flaws in its simplicity. For example, this statistic does not account for the depreciation of assets, for the fact that natural resources are not free (as the equation presumes) and does not include the cost of pollution or the informal market (Fioramonti 2014, 4). Nevertheless, despite these gaps, it possesses great political power to singularize and define 'advances' in addressing poverty. As Philipsen asks: how did something describing a crisis – pointing to the emergence of GDP during Second World War – subsequently turn into something prescribing what we do, serving as a substitute for democratic deliberation and political ideals (2016, 6)? Indeed, GDP often operates at a macro level yet is linked to the objective poverty statistics, such as the World Bank's X number of dollars per day as a measurement of individual poverty, explored later in this chapter.

Statistics are also the key feature in defining the so-called symptoms of poverty such as famines. In relation to this, de Waal explains how the medieval discourse of famine was starkly different to the modern discourse. It is hard to tell precisely how the English term 'famine' was used in the Medieval Period, but it seems that until the eighteenth century 'famine' and 'hunger' were interchangeable (2005, 14). Interestingly, de Waal points to dearth as being the root of the word famine. Dearth did not imply starvation or dying due to starvation. This is most clearly demonstrated by the severe 1315–17 famine. On the basis of records from Winchester, Kershaw (1973, 11) argued that the famine years saw a crude death rate of near 10 per cent. de Waal (2005, 14) explains, however, that in reading the accounts of the famine there was a 'primary focus upon economic or agrarian crisis, and only secondary (if at all) a crisis of starvation and population fall'.

This leads de Waal to conclude that the notion of famine 'was closer to the notion of famine as hunger and dearth than the modern notion of famine as mass starvation and death' (de Waal 2005, 15). He explains that the modern discourse of famine, centred on death and starvation, emerged in the late eighteenth century with the works of Malthus. The consequence of which was a rejection of the traditional, pluralist definitions. Malthus's *An Essay on the Principles of Population* (1890), published in six editions between 1798 and 1826, analysed the relationship between demography and resources. He outlined the five checks on populations, the fourth and fifth being famine (Malthus 1890, 106). According to de Waal (2005, 17), Malthus's work on famine caught the imagination of the nation and led to a re-conceptualization of famine as mass starvation leading to death. The importance of Malthus's contribution was not the validity of his work, which has been brought into question (de Waal 2005, 19), but the mortality and starvation-centric framework that the work provided, which has become central to the modern discourses on famine. In doing so, the Malthus Factor (Ross 1998) introduced the 'mathematization' of famine where the 'arithmetical' production of resources would never catch up with the 'geometrical' expansion of the population, therefore provoking famine at the end. This conception of famine as mathematical notion created a vital role for statistics and numbers in the articulation of poverty as a news issue in the works of journalists such as Henry Mayhew (1812–87),

Charles Dickens (1812–70) and Jack London (1876–1916). Fast forward today, and we find the United Nations saying that

> a famine can be declared only when certain measures of mortality, malnutrition and hunger are met. They are: at least 20 per cent of households in an area face extreme food shortages with a limited ability to cope; acute malnutrition rates exceed 30 per cent; and the death rate exceeds two persons per day per 10,000 persons. (UN 2011)

By terming famine in numerical terms, those who do not experience the famine could then nevertheless quantify it, declare its severity and seem to provide solutions. More importantly, it conferred power to construct social reality around famine in ways it could be politically 'managed'. This is because the phenomenon then passed from being an individual and collective experience into a set of statistics that determined who and who did not receive aid and support. Moreover, through the process of developing a single definition for both poverty and famine, certain statistical models can be used by development or humanitarian agencies to measure the phenomena and allocate resources to address it. Then, in the same statistical sense, the solutions provided by these organizations can be judged against a certain set of numbers to determine its success. It is within this context that journalists covering famine and poverty are positioned. They, of course, find them useful because it offers the 'objective' assessment that allows them to provide 'detached' reporting of these events. More important, it facilitates a definition of poverty that otherwise would be far too problematic for 'objective reporting' within commercial journalism practice.

The definitional morass of who is poor

When covering any news story, journalists have to approach the issue in three phases. First, they conceptualize the issue as a news item, which means creating broad conventions around particular topics so they fit a relevant news beat. For example, the break of cholera in Haiti can be conceptualized as a foreign affair issue, which means the foreign desk will deal with the story, or as a health issue, which means that the health editor will look after that particular piece. This process of 'thematic allocation of a particular issue or subject also known as the "news beat" has, in addition, the function of setting the network of "expert voices" who are considered to have enough legitimacy and authority to speak about that topic'. Secondly, journalists have to define the issues in terms of common codes that their audiences can understand and relate to. Hence, a journalist covering general elections in Tanzania speaks about the party in power and the party in opposition as well as about the 'centre-right party', the 'centre-left opposition' and the 'far left', to cite some examples. This frames the story within specific boundaries of understanding, allowing journalists to perform an 'interpretative' function (Berkowitz and TerKeurst 1999; Zelizer 1993). Thirdly, journalists have to operationalize the issue; that is to make it tangible and measurable in order to present it in 'objective terms'. This process of making things 'measurable' is not only limited to assigning it a number – for example the number of doctors sent by the UK to Sierra Leona to help the locals to deal with the outbreak of Ebola – but is also about offering evidence that can be corroborated – for example making reference to the Panama Papers to show how the then prime minister of the UK, David Cameron, had benefited from tax avoidance schemes set by his late father. All in all, taking a news item from conceptualization to some measurement is always an intrinsic part of journalism work and it is far from being detached or neutral.

Hence, among the challenges that journalists face in attempting to cover statistics on poverty is to find broad agreement around what it actually means and how can it be observed. Indeed, journalists find it hard to deal with poverty because of the difficulties they find in conceptualizing, defining and measuring poverty in objective terms (Harkins and Lugo-Ocando 2016; Lugo-Ocando 2014). This problem is present in both the coverage of national levels of poverty as well as in the international terrain. Poverty is in fact an elusive term that is highly contested, problematized and politicalized in public discourse in general and in media representations in particular. This is not, of course, only a problem of journalism as the output of journalists' work reflects in part much broader disagreements from experts and officials. To be sure, journalists working collaboratively in Europe and the United States would find it almost impossible to make effective and credible comparisons in relation to poverty at both sides of the Atlantic. This is because, while the US Bureau of National Statistics establishes a threshold for poverty in absolute terms (a minimum income per family) across the whole of that country, the equivalent offices in Europe has set that threshold instead in relative terms to the average income of each nation (Olson and Lanjouw 2001; Ravallion 2003). This is not to say that journalists do not carry out these comparisons all the time. On the contrary, stories about poverty are full of cross-national statistical comparisons. However, statistically speaking, most of them are flawed in terms of validity and reliability and, consequently they tend to present a distorted picture, which does not take into account significant differences among countries in the way poverty is measured.

It is important to highlight that this is not just an abstract discussion. Instead, the definition of poverty plays a pivotal role not only in the selection of the type of statistics on poverty that journalists use but it also reflects the ideological stance of journalists in relation to issues such as globalization. As Martin Ravallion (2003) points out:

> The measurement choice does matter. Roughly speaking, the more 'relative' your poverty measure, the less impact economic growth will have on its value. Those who say globalization is good for the world's poor tend to be undisguised 'absolutists'. By contrast many critics of globalization appear to think of poverty in more relative terms. At one extreme, if the poverty line is proportional to mean income, then it behaves a lot like a measurement of inequality. Fixing poverty line relative to mean income has actually been popular in poverty measurement in Western Europe. This method can show rising poverty even when the standards of living of the poor has in fact risen. While we can agree that relative depravation

matters, it appears to be very unlikely that individual welfare depends only on one's relative position, and not at all on absolute levels of living, as determined incomes. (2003, 741)

In fact, there are numerous examples to be found in the daily beat of news media outlets that corroborate this alignment between the selection of statistics and the ideological framework of a particular news item. Take, for example, a story from the *Guardian* produced in October 2016, titled 'World Bank renews drive against inequality'. The piece relays the fundamental message in the World Bank's annual publication – *Poverty and Shared Prosperity* – through the voice of its then president, Jim Yong Kim:

> 'It's remarkable that countries have continued to reduce poverty and boost shared prosperity at a time when the global economy is underperforming – but still far too many people live with far too little,' said Kim. 'Unless we can resume faster global growth and reduce inequality, we risk missing our World Bank target of ending extreme poverty by 2030. The message is clear: to end poverty, we must make growth work for the poorest, and one of the surest ways to do that is to reduce high inequality, especially in those countries where many poor people live.'[2]

The alignment here is clear: the poverty statistics are used to support the humanitarian-neoliberal ideology of a global financial institution that predicates solving poverty by pushing for constant economic growth. The basis of Kim's commentary is the measure of global poverty by those who are living on less than $1.90 a day – a rather arbitrary threshold produced by the World Bank itself (Lugo-Ocando and Nguyen 2017). This validates the claim that poverty has been declining since 1990 due to increased globalization and also that certain remedies need to be adopted to arrest a potential future decline. Given this backdrop, the author ends the article by outlining six of the bank's aims, ranging from universal health coverage to better roads and electrification. As displayed in this news piece, these 'objective' poverty statistics are not only used to 'demonstrate' – deliberately or inadvertently – that globalization is reducing global inequality but also then to suggest that yet inequality may increase if the World Bank's 'objective' remedies are not adopted.

There have been some attempts to produce a critique to prevalent statistics on poverty used by the media. In an article in *the Independent* titled 'Can we solve UK poverty', Felicity Hannah explores the recent work of the *Joseph Rowntree Foundation* (JRF), a think tank that aims to have a poverty-free UK, on defining and solving poverty. Hannah begins with an explanation of the objective poverty statistics used by the government to define poverty.

---

[2] Elliot, L. (2016), 'World Bank renews drive against inequality'. *The Guardian*, 2 October 2016.

The government typically uses median income as its definition of poverty; a household with an income below 60 per cent of the median income is considered to be in poverty. Under this measure, there are 13 million people in the UK living in poverty, of which 3.7 million are children. That's a quarter of all the children in the country.[3]

This approach is then contrasted with the suggested adoption of the 'standard of living' statistical model by the JFR,

The JRF annually calculates the Minimum Income Standard, the amount the public believe is essential for an adequate standard of living, including essentials but also transport, cultural and social costs. It measures poverty as having 75 per cent or less than that standard. In 2016, a couple with two children (one pre-school and one primary school age) would need £422 per week to achieve the Minimum Income Standard, while a single working-age person would need £178 per week.[4]

In her comparison, the inadequacy of the state in tackling poverty was clearly highlihgted. This is reinforced by references to the scrapping of the Child Poverty Act, the lack of headway on the policy to end child poverty by 2020 and the alternative remedy to poverty as suggested by JRF.[5] However, these types of comprehensive and critical discussions in the news media around the meaning of numbers and what they tell us about poverty are far and in between.

Over-reliance on 'official sources'

Another very important problem that journalists face in reporting poverty statistics is the over-reliance upon the use of official sources, which remain unchallenged and unquestioned at large and as such, what they say tend to be reported uncritically. Indeed, journalists' overdependence on official sources is a long-standing issue that has been well established by a variety of scholars (Brown et al. 1987; Lewis, Williams and Franklin 2008; Manning 2001). This problem is not only limited to the predominant and systematic selection of sources among the elites (i.e. government officials, bankers and other members of the establishment) but also refers to the disproportionate airtime and space these sources receive against other voices. In the case of statistics, the issue is further problematized by the way these numbers are used by policymakers and journalists as factual and objective evidence to underpin 'truth'.

---

[3] Hannah, F. (2016), 'Can we solve UK poverty? The UK could solve poverty by 2030, according to one think-tank, but how?'. *The Independent*, 2 November, 2016.

[4] Ibid.

[5] Hannah, F. (2016), 'New definitions of poverty are re-defining financial hardship'. *The Independent*, 14 April 2016.

More importantly, as many journalists have neither the ability nor training to conduct statistical work of their own, they have to rely almost entirely upon the public relations departments of the institutions dealing with poverty. Their statistics are not dissected and journalists do not produce alternative statistical models or research themselves. Thus, this circularity is never interrogated nor contested. The frameworks within which the current statistics on poverty have developed are crucial in understanding how poverty statistics are used to articulate and frame news on statistics. Moreover, journalists' over-reliance and uncritical acceptance of official statistics explains in great part their central role in producing public knowledge regarding poverty in our times.

This problem is made more acute as journalists covering poverty see and treat statistics as both a news source, awarding these numbers the same level of authority and credibility than the one they confer to elites, and – at the same time – as 'objective facts' upon which interpretative claims can be made (Lugo-Ocando and Brandão 2015). This duality in the consideration of the nature of the 'object' to be reported – the statistics on poverty – problematizes further the issue as it renders these numbers unquestionable to external scrutiny. Journalists report the numbers as an 'objective fact' in spite of being a mathematical representation that summarizes and interprets a given phenomenon. So, when reporters write that the per capita income of a country is improving and assume that people are better off because of that, their readers assume that this is a matter of fact, even though it is simply an estimation followed by an interpretation.

Take the example of an article published in *The Telegraph* by Matthew Lynn in which he explains how the World Bank has decided to get rid of the term 'developing countries'. The evidence used to support this claim is from the World Bank's official statistics on poverty,

> The World Bank cites the example of Mexico, which now has a gross national income (GNI) per capita of $9,860. It is ridiculous to bundle it in with a genuinely poor country such as Malawi, with a GNI of $250.[6]

The collapse of this difference ends up underpinning certain views on global poverty that reflect more neoliberal and pro-globalization approaches. This is further corroborated by the use in the same news item of supporting evidence from the CATO Institute (2016), a US-based think tank 'dedicated to the principles of individual liberty, limited government, free markets and peace'. The author explains,

> It is about time that the Left, and indeed a lot of mainstream liberal opinion, caught up to the way that the global economy has changed – instead of constantly ramping up the rhetoric about how evil the West is.[7]

---

[6] Lynn, M. (2016), 'Why the title of "developing country" no longer exists'. *The Telegraph*, 23 May 2016.

[7] Ibid.

In the article, the author fails to recognize that the same logic of objective poverty statistics initially used by the World Bank to delineate developing and developed countries to assert that this distinction, no longer exists. Instead of identifying the potential inadequacy of these measures to understand poverty, the author uses this spurious evidence to validate a particular ideological position.

This issue is not confined to objective poverty statistics. In June 2016, Gabriella Bennett in an article for *The Times* relies almost entirely on quotes or references to official sources to create the content. For statistical validation, she uses a report from the Scottish Government. The report states,

> The proportion of people in poverty in working households increased in the latest year. The move into employment was largely into part-time work, especially for women, meaning that while people were in employment, they remained in poverty. In 2014–15 income inequality increased. The top 10 per cent of households saw the largest increases in income while the bottom 10 per cent saw no real change.[8]

The subsequent commentary is provided entirely from two sources. The director of *Poverty Alliance* calls on the readers to work together for social justice while the Scottish social security secretary critiques the Conservative government's welfare cuts. Bennett does not provide either a critical perspective on the statistics used or an individual commentary on the report itself. In a similar case, *The Daily Mirror* published an article titled 'Ireland poverty crisis; 1.3 million struggle without basics'. Trevor Quinn, the author, relies entirely upon Social Justice Ireland (SJI) for the evidence and commentary to back up the claim in the headline.

> ALMOST a third of the population – 1.3 million people – are in poverty and going without the basics, Social Justice Ireland said yesterday. The advocacy body released the figures as it called on the Government to address the inequality and deprivation in our society in next week's Budget. Director Dr Sean Healy said hundreds of thousands are 'going without basic necessities such as heating their home or replacing worn-out furniture'. He added: 'Why are these people not a priority for Government as it prepares for Budget 2017?'[9]

Regardless of the intentions of these organizations, these extracts exemplify journalist's over-reliance on the sources who provide the statistics when telling the story of poverty. Indeed, even though Quinn acknowledges that the release of the statistics was a tactical move by the SJI to pressure the government, the link between these 'facts' and the ideology it looked to serve was treated uncritically. In fact, connection between the statistic and the ideology formed the very basis of the article itself.

---

[8] Bennett, G. (2016), 'Increase in Number of Working Households in Poverty'. *The Times*, 29 June 2016.

[9] Quinn, T. (2016), 'Ireland Poverty Crisis; 1.3 million struggle without basics'. *The Daily Mirror*, 5 October 2016.

Overall, most journalists lack the resources, time, training or ability to either challenge official statistics or produce their own numbers to explain poverty. This results in a lack of critical engagement with the ideology that underpins the use of certain measurements elements of poverty. This has developed a specific phenomenon within the production of news articles: statistics are used as a facade of objectivity that hides the highly contingent and relative ideologies that either created them or that they serve. This is not to say that journalists do this wittingly. While some journalists do use certain measurements of poverty to push their own conceptions of globalization or anti-free trade movements, many, if not most, others just operate as a statistical mouthpiece for governments, the third sector or corporations.

Towards a conclusion

The fact remains that poverty is not a natural phenomenon and there is nothing unavoidable or unsolvable about it. However, by 'economizing' poverty as a news subject, journalists tend to reduce the issue to a phenomenon that seemingly 'just happens' and that inevitably 'will continue to happen'. In this sense, issues such as inequality are rarely brought into news stories and for a long time explanatory frameworks that take into consideration structural causes for poverty are ignored or dismissed altogether. The problem is made more acute by the way statistics on poverty are used as they are often awarded the same level of authority as those used in natural sciences. In so doing, journalists 'naturalize' these numbers and, consequently, articulate poverty as a seemingly natural phenomenon. This is partly to do with the 'special status' that statistics as a discipline has acquired in news reporting, where it is seen as a natural science very close to mathematics, therefore considered to be infallible in its premises.

Indeed, statistics over the years establish itself as a reliable scientific discipline (Desrosières and Naish 2002) even when they are produced by ideological conventions like any other thing in social science. To be sure, the social facts that are encapsulated in poverty statistics have become somehow 'objective things' for everyone who uses statistical techniques, and 'these techniques are intended to back political arguments in 'scientific' terms. Thus, once the necessity of being recognized an objective discipline – as outlined previously – was conceived to the discipline of statistics, it meant that 'we can begin to decree famine, judge schools and allocate foreign aid resources. In this context, it is important to remind ourselves that all mathematical language facilitates and reinforces the myths of transparency, neutrality and independence and that this fits very well with the journalistic discourse, primarily because 'mathematics is a system of statements that are accepted as truth' (Zuberi 2001, xvi).

To be sure, the journalistic discourse seeks a mythical objectivity that is professionally ideological and impossible to achieve fully (Koch 1990; Maras 2013; Schudson and Anderson 2009). Given their aim to appear objective and unbiased, media outlets tend to exclude opinions in their hard news reporting or ascribe those opinions to 'others' (Gans 2014; Tuchman 1978). In reality, however, news items are gathered, selected and constructed on the basis of journalists' individual and collective preconceptions, values and worldviews. Journalists themselves are the vehicles for these opinions and as subjective actors are not neutral. Nevertheless, given the deontological requirements of their profession they need to be neutral and try constantly to make a clear-cut distinction between facts and opinions. In this sense, statistical information on poverty help them construct a narrative that ensures creditability and that reinforces current worldviews on why 'it happens'. After all, these numbers are seen as neutral entities that underpin journalistic 'truth' and not as contested tools that help to construct social reality in specific ideological terms.

However, the quantification of reality in the media has alienated many individuals on the following ground: those who had come to trust almost blindly these mathematical discourses are the ones who now seem to distrust them. This is because in many cases, when they hear some statistics in the news that crime, unemployment and poverty are down, their personal experience tells them a distinctively different story. Despite the barrage of statistics confirming that globalization has improved the life of many, the benefits seem to have been felt only by the very few. The use of statistics seems instead to further de-humanize the experience of poverty by turning it into an abstract numerical notion: that not only defines who and who is not in a state of poverty, but also determines the allocation of resources while defining policy and action.

Does this all mean that the use of statistics to report poverty should be avoided? Not at all, statistics brings about a sense of dimension and perspective that very few news items can convey. As Krisztina Kis-Katos (2014) from the Universität Freiburg puts it, 'It is important to take statistics seriously in order to come to meaningful conclusions'. For Kis-Katos, one of the key elements that statistics allow us to understand is how difficult and complex it is to try to disentangle the different effects that processes such as globalization have on those living in poverty. They help us to step back and provide more rational action towards addressing poverty through policy. As such, statistics should serve as a very important source of information, one that allows journalists to contextualize and evaluate poverty in societal dimension. The key issue is that journalists should recognize that there is an ideological aspect to statistics and, as such, that problems could arise when statistics are used to define reality. Poverty is a distinctive human phenomenon that no statistic can encapsulate or explain. As most sociologists and economists would recognize, to live in poverty is far more than just dealing with low income or the lack of access to basic services and goods. Charles Dickens made his best depictions of poverty not when he worked as a journalist but when as a writer – and there is a reason for that. Sometimes it is our imagination that allows us to connect to our fellow human beings rather than the facts summarized in numbers. That is the true balance that journalism needs to achieve when reporting poverty.

References

Alvares, C. and Sachs, W. (1992), *The Development Dictionary: A Guide to Knowledge as Power*. London: Zed Books.

Berkowitz, D. and TerKeurst, J. V. (1999), 'Community as interpretive community: rethinking the journalist-source relationship.' *Journal of Communication*, 49(3), 125–36.

Brown, J. D., Bybee, C. R., Wearden, S. T. and Straughan, D. M. (1987), 'Invisible power: Newspaper news sources and the limits of diversity.' *Journalism and Mass Communication Quarterly*, 64(1), 45.

Cater, D. (1965), *The Fourth Branch of Government*. New York: Vintage Books.

Cook, T. (2005), *Governing with the News*. 2nd edn. Chicago: The University of Chicago Press.

Crozier, M. (2008), 'Listening, learning, steering: New governance, communication and interactive policy formation.' *Policy & Politics*, 36(1), 3–19.

de Waal, A. (2005), *Famine that Kills*. New York: Oxford University Press.

Desrosières, A. and Naish, C. (2002), *The Politics of Large Numbers: A History of Statistical Reasoning*. Cambridge, MA: Harvard University Press.

Escobar, A. (1995), *Encountering Development: The Making and Unmaking of the Third World*. Princeton, NJ: Princeton University Press.

Esser, F. (2013), 'Mediatization as a challenge: Media logic versus political logic.' In H. Kriesi (ed.), *Democracy in the Age of Globalization and Mediatization*, 155–76. Basingstoke: Palgrave Macmillan.

Esteva, G. (1992). 'Development' In W. Sachs (ed.), *The Development Dictionary*, 6–25. London: Zed.

Ferguson, J. and Lohmann, L. (1994), 'The anti-politics machine: "Development" and bureaucratic power in Lesotho.' *The Ecologist*, 24(5), 176–81.

Gans, H. J. (2014), 'The American news media in an increasingly unequal society.' *International Journal of Communication*, 8, 12.

Hajer, M. A. (2009), *Authoritative Governance: Policy Making in the Age of Mediatization*. Oxford: Oxford University Press.

Harkins, S. and Jairo, L.-O. (2016). All people are equal, but some eople are more equal than others: How and why inequality became invisible in the British press. In J. Servaes and T. Oyedemi (eds), *The Praxis of Social Inequality in Media: A Global Perspective. Communication, Globalization, and Cultural Identity*. London: Rowman & Littlefield.

Harkins, S. and Lugo-ocando, J. (2016), 'All people are equal, but some people are more equal than others: How and why inequality became invisible in the British press.' In J. Servaes and T. Oyedemi (eds), *The Praxis of Social Inequality in Media: A Global Perspective. Development Statistics and What to do About It*. New York: Cornell University Press.

Kershaw, I. (1973), 'The great Famine and Agrarian crisis in England, 1315-1322.' *Past and Present*, 59(1), 3–50.

Kis-Katos, K. (2014), Globalization and the poor, a look at the evidence. TED Talk. Retrieved 23 July 2015 from https://www.youtube.com/watch?v=dIldvz0jygE&t=135s.

Koch, T. (1990), *The News as Myth: Fact and Context in Journalism*. Westport, CT: Greenwood Publishing Group.

Kurunmäki, L., Mennicken, A. and Miller, P. (2016), 'Quantifying, economising, and marketising: Democratising the social sphere?' *Sociologie du Travail*, 58(4), 390–402.

Lewis, J., Williams, A. and Franklin, B. (2008), 'A compromised fourth estate? UK news journalism, public relations and news sources.' *Journalism Studies*, 9(1), 1–20.

Lugo-Ocando, J. (2014), *Blaming the Victim: How Global Journalism Fails Those in Poverty*. London: Pluto Press.

Lugo-Ocando, J. and Brandão, R. (2015), 'Stabbing news: Articulating crime statistics in the newsroom.' *Journalism Practice*, 10(6), 715–29.

Lugo-Ocando, J. and Nguyen, A. (2017), *Developing News: Global Journalism and the Coverage of 'Third World' Development Hardcover*. Abingdon, OX: Routledge.

Madra, Y. M. and Adaman, F. (2014), Neoliberal reason and its forms: De-politicisation through economisation. *Antipode*, 46(3), 691–716.

Malthus, T. R. 1888. *An essay on the principle of population*. 9th edn. London: Reeces & Turner.

Manning, P. (2001), *News and News Sources: A Critical Introduction*. London: Sage.

Fioramonti, L. (2013), *Gross Domestic Problem: The Politics behind the World's Most Powerful Number*. London: Zed Books.

Fioramonti, L. (2014), *How Numbers Rule the World*. London: Zed Books.

Harkins, S. and Jairo, L.-O. (2016). All people are equal, but some people are more equal than others: How and why inequality became invisible in the British press. In J. Servaes and T. Oyedemi (eds), *The Praxis of Social Inequality in Media: A Global Perspective. Communication, Globalization, and Cultural Identity*. London: Rowman & Littlefield.

Iliffe, J. (1987), *The African Poor: A History*. Cambridge: Cambridge University Press.

Cato Institute (2016). Cato's Mission. Retrieved 11 November 2016 from https://www.cato.org/mission.

Ipsen, C. (2003), 'Under the stats of fascism: The Italian population projection of 1929-31.' In J. Fleischhacker, H. A. Gans and T. K. Burch (eds), *Populations, Projections, and Politics: Critical and Historical Essays on Early Twentieth Century Population Forecasting*, 205–24. West Lafayette, IN: Purdue University Press.

Jerven, M. (2013), *Poor Numbers: How We Are Misled by African Development Statistics and What to Do About it*. Ithaca: Cornell University Press.

Maras, S. (2013), *Objectivity in Journalism*. Hoboken, NJ: John Wiley & Sons.

Marcinkowski, F. and Steiner, A. (2014), 'Mediatization and political autonomy: A systems approach.' In H. Kriesi (ed.), *Mediatization of Politics: Understanding the Transformation of Western Democracies*, 74–89. Basingstoke: Palgrave MacMillan.

Masood, E. (2016), *The Great Invention: The Story of GDP and the Making and Unmaking of the Modern World*. New York: Pegasus Books.

Mollat, M. (1986), *The Poor in the Middle Ages: an Essay in Social History*. London: Yale University Press.

Olson, J. and Lanjouw, P. (2001), 'How to compare apples and oranges: Poverty measurement based on different definitions of consumption.' *Review of Income and Wealth*, 47(1), 25–42.

Perrot, J.-C. and Woolf, S. J. (1984), *State and Statistics in France, 1789-1815*. London: Taylor & Francis.

Philipsen, D. (2016). *The Little Big Number. How GDP Came to Rule the World and What We Can Do about It*. Princeton, NJ: Princeton University Press.

Prévost, J.-G. (2009), *Total Science: Statistics in Liberal and Fascist Italy*. McGill-Queen's Press-MQUP.

Rahnema, M. (2001). Poverty. Retrieved 19 November 2016 from http://www.pudel.uni-bremen.de/pdf/majid2.pdf.

Rapley, J. (2004), 'Development studies and the post-development critique.' *Progress in Development Studies*, 4(4), 350–54.

Ravallion, M. (2003), 'The debate on globalization, poverty and inequality: Why measurement matters.' *International Affairs*, 79(4), 739–53.

Ross, E. (1998), *The Malthus Factor: Poverty, Politics and Population in Capitalist Development*. London: Zed Books.

Schrott, A. (2009), 'Dimensions: Catch-all label or technical term?' In K. Lundby (ed.), *Mediatization: Concept, Changes, Consequences*, 41–61. Bern: Peter Lang.

Schudson, M. and Anderson, C. (2009). 'Objectivity, professionalism, and truth seeking in journalism.' *The Handbook of Journalism Studies*, 88-101.

Stiglitz, J., Sen, A. and Fitoussi, J. P. (2010), *Mis-measuring Our Lives: Why GDP Doesn't Add Up*. New York: The New Press.

Temkin Yedwab, B. and Salazar-Elena, R. (2012), 'México 2010-2011: los últimos años de una gestión cuestionada.' *Revista de ciencia política (Santiago)*, 32(1), 193–210.

Thompson, J. B. (1995), *The Media and Modernity: A Social Theory of the Media*. Cambridge: Polity Press.

Tuchman, G. (1978), *Making news: A Study in the Construction of Reality*. New York: Free Press.

United Nations (2011), 'When a food security crisis becomes a famine.' Retrieved 09 November 2016 from http://www.un.org/apps/news/story.asp?NewsID=39113#. WCMDnvmLTIU.

Vilchis, R. R. G. (2013), El regreso del dinosaurio: un debate sobre la reciente victoria del PRI en la elección presidencial de 2012. *Estudios Políticos*, 28, 145–61.

Webster, D. (2002), 'Unemployment: How official statistics distort analysis and policy, and why.' *Radical Statistics*, 79, 96–127.

Weisbrot, M., Sandoval, L. and Rosnick, D. (2006), 'Índices de pobreza en Venezuela: En búsqueda de las cifras correctas.' *Informe Temático, Washington DC*. Washington: DC Center for Economic and Policy Research.

Zelizer, B. (1993), 'Journalists as interpretive communities.' *Critical Studies in Media Communication*, 10 (3), 219–37.

Zuberi, T. (2001), *Thicker Than Blood: How Racial Statistics Lie*. Minneapolis, MN: University of Minnesota Press.

5

Statistics in Science Journalism: An Exploratory Study of Four Leading British and Brazilian Newspapers

Renata Faria Brandão, *King's College London, UK*
An Nguyen, *Bournemouth University, UK*

Introduction

Science journalism is one of the news beats that, by its own label, signifies the inevitable use of statistical data in storytelling. Statistics are inseparable from science: they are understood as part of the positivist paradigm of natural science, which assumes the existence of a world of facts that is independent of human subjectivity and can be empirically traced, measured, quantified and numerically represented. Every scientist would agree that science is impossible today without statistics. As such, an effective representation of the statistics that underline science discoveries and their social, economic and political implications is a precondition for science journalism to get its messages across the lay public. The question: can journalists writing about science do that well – and if yes, how well?

This chapter offers some preliminary insights into this issue, using data from an exploratory content analysis of science coverage in leading British and Brazilian newspapers. We begin with a discussion of journalism's traditional deference to statistics from cultural and epistemological perspectives. Although there are good reasons to hope that science journalism might be an exception to that traditional relationship, our ensuing historical review of science journalism suggests that it might not be the case. For most of its history, science journalism has existed primarily to promote science as a monolithic institution that produces universal and unquestionable knowledge, thus failing to bring science developments into their socio-cultural context and to equip lay people with the ability to critically engage with and question science when needed. Science journalism has been too faithful to science to remember that the latter is, at the end of the day, a social construction, rather than a human-free mirror, of the world and therefore needs to be socioculturally contextualized and critically understood by lay people. In that context, we suspect that science statistics, as the language of science, would be represented as unquestionable and undisputed facts that are not only far from people's interests and concerns but also fail to enlighten and empower them to become critical and judgemental of science developments. Our content analysis of 1,089 science articles in four leading broadsheets in the UK (The *Guardian* and *The Times*) and Brazil (*Folha de S. Paulo* and *O Globo*) finds strong evidence to support the hypothesis about the lack of critical reporting of science statistics, although there was evidence that journalists no longer deprive people of the opportunities to understand science statistics in non-science contexts.

Journalistic deference to statistics

Statistics have an extensive role in facilitating public understanding of social and natural issues (Simpson and Dorling 1999; Porter 1996). To Gödel (1986) and Frege (1977), only mathematical language could explain the universe objectively and such language implies allusion to and measurement over theoretical objects. This long-time mathematization of public discourse has created a general compulsion to understanding the world out there through statistical goggles. Such understanding comprises an elaborate view of reality that started to prevail when society began using numbers in the same way as science did to logically and progressively construct and produce scientific truths (Frege 1960, 21). Today numerical perspectives govern most aspects of daily lives, shape our perception, establish conditions and provide directions for societal developments. As such, statistics are featured in almost every area of news coverage, with headlines across media outlets making use of some sort of statistical information and/or validation.

This creates a mathematical social filter in many decision-making processes that, unfortunately, only a few can fully understand. Without some intellectual support, most lay citizens do not have the necessary skills and knowledge to critically peruse or challenge statistical information (Utts 2003, 74). Over time, this results in a generally inadequate understanding of statistics and, hence, less informed decisions (as seen in recent seismic political events such as the UK's Brexit vote or the US election in 2016). Zuberi (2001) believes there is an existing need to present quantitative information in a more critical and honest fashion, by means of a better understanding of numbers and their methodological approaches (176). This is the gap where the purported truth-seeking and truth-telling profession of journalism must fill if it wants to help people make sense of the world and to make rational, informed decisions about it.

The problem, as the chapters in this book and other research (Nguyen and Lugo-Ocando 2016; Howard 2015; Utts 2003, 2010) show, is that journalism has never performed that role well and, indeed, it has rarely assumed it as a professional mission. On the one hand, there is a traditional stubborn news culture in which newspeople, despite their intensive and extensive use of statistics as news material, see themselves as wordsmiths, not 'number crunchers', and are generally hostile to numerical information. This has often been worsened by the chronic lack of statistical training and education in newsrooms as well as journalism schools (see also the Introduction).

On the other, and at a more latent level, is an epistemological deference to statistics in news reporting that comes directly from journalism's traditional deference to science. Journalists hold a genuine belief and an almost unconditional faith in science as an exact, precise and authoritative knowledge about the world and therefore aspire to gain its objectivity standards and methods for their own truth-seeking mission (Kovach and Rosenstiel 2001). In their mind, scientific knowledge, including statistics, has been successful in its explanation of the world thanks to its rationality and objectivity; therefore, its strict objectivity-driven procedures should be something that journalism should use to compensate for its inherent weaknesses (Streckfuss 1990, 974). Walter Lippmann (1920, 1922) and others (Myers 2003; Schudson 2001; Stigler 1986) have long advocated the adoption of scientific objectivity as a working method for journalism to measure systematic errors and to reduce the influence of subjectivity, including personal judgement and bias, in verifying the truth and socially constructing events and issues.

As journalism aims to create news that resembles the supposedly universal, unbiased and trusted scientific knowledge, statistics are intimately associated with its core idea and 'strategic ritual' of objectivity. As such, their persuasive power is often used as an ultimate form of evidence to support and validate news and current affairs. In attempting to exercise scientific principles, journalists habitually rely on quantitative data – more exactly the surface credibility that such data confer on argumentation – to make rhetorical appeals in news storytelling. Over time, they develop the habit of taking statistics for granted without further scrutiny and assessment. In most cases, they depend excessively on some authoritative sources to provide data and data-based claims, which are then integrated into the shallow but safe 'he said/she said' formula of news.

Can science journalism be an exception?

That said, there are a number of important reasons to believe that science journalism could be an exception to, or at least a more advanced area than, this aforementioned general journalism–statistics relationship. Since science stories cannot be reported without statistics, it is reasonable to expect that those who write about science are more statistically adept and experienced, thus more judgemental of numbers, than most of their newsroom colleagues. As Holger Wormer puts it in Chapter 14, if a journalist is to avoid dealing with numbers, then he/she is unlikely to be in science reporting as a career. Indeed, a global survey of science journalists reported that 68 per cent see numeracy/grasp of statistics as essential to make a good science journalist (Bauer et al. 2013) – a proportion that is in quite stark contrast to the general hostility to numbers among journalists. Moreover, it might be reasonable to expect that science journalists are more likely to be trained and/or educated about statistics, formally or informally, and therefore more aware of the need to help the public to deal with uncertainty concepts and to be cautious and critical over the validity and conclusiveness of statistical data. Indeed, the above survey found that science journalists have a very high education level, with one-third in Europe and North American holding a PhD degree, which suggests that science journalists might be more trained and equipped to understand and handle data and statistics in the interest of an informed lay public.

Yet, is that what we would observe in reality? If science journalism performs better than other news beats in handling data and statistics, why do we still see so many misleading and damaging science articles that are based on flawed and questionable research data – for example, the influx of news about Andrew Wakefield's false but deadly single-study link between the MMR jab and autism in the UK in the 1990s, or the rather chaotic picture of climate data that continues in the global news arena today? Are these cases only the exception to the norm? Our initial inclination is that they are not, not least because science journalism is by and large still an enterprise that sees itself as a follower and promoter, rather than a monitor and watchdog, of science and its institutions. This has a long history in which journalism regards itself, and is regarded, as not much more than a branch of science communication.

That history starts from the very early days of science, with seventeenth-century 'natural philosophers' such as Francis Bacon in England and Galileo Galilei in Italy, and continues into today. In this history, the science establishment – including scientists and science policy-making institutions – has always been conscious of the critical need to get science across to the lay public (Dick 1955; Shamos 1995). As science evolves and expands alongside the time, so does the perceived demand for lay people to understand the world through scientific lenses. The specific reasons for that demand has changed along the ebb and flow of history – from legitimizing and popularizing science in the face of religion to encouraging people into science to winning the public heart and ensuring public funding – but the ultimate message has remained the same: science is a wonderful world with a noble calling that the lay public should learn, support and even aspire to. This mindset is exemplified in the inspiring work by Carl Sagan (1995) who argues that, despite numerous opportunities for abuse, scientific knowledge may be the 'golden road' away from ignorance and poverty (37). It informs society of the dangers of its world-altering advances, educates us about the most profound issues of origin, life and prospects, and shares the same values as democracy itself (38). To him, science and civilized democracy are concordant and 'in many cases indistinguishable' (Sagan 1995, 38).

Following this logic, the assumed function of science communication is merely to diffuse science facts so as to promote and improve science literacy and to garner and maintain public admiration and support for science (Bueno 1985; Weigold 2001). The dominant mode of science communication has been that of the 'deficit model', despite recent calls for alternative paradigms. In this model, people are seen as 'empty vessels' needing to be filled with facts and figures about science and its discoveries, and once filled, they will learn to appreciate science. In other words, according to this line of reasoning, the more they know about science and its processes, the more they love it. Emerging and growing within this framework, science journalism has been assumed as just another channel to promote public understanding of science (a more recent term for science literacy). For the science establishment, it seems, journalism would do a good job as long as it (a) educates lay people about science facts and events, especially new science discoveries and their benefits, in an accurate manner and (b) provides scientists with ample space to use their expertise and authoritative power to intervene in debates about science-related events and issues.

And science journalism has done exactly that in much of its history – from its emergence in the seventeenth century to its professionalization in the first decades of the twentieth century to its rise to the status of a notable news beat in the 1960s and 1970s till today (Bauer and Bucchi 2007; Bucchi and Mazzolini 2003; Bucchi and Trench 2014; Feyerabend 2011; Fjæstad 2007). With the exception of a rather short-lived surge of critical science reporting that came out of the shadow of Cold War politics of the 1970s and 1980s (Miller 2001; Rensberger 2009), science journalists – growing out of the aforementioned tradition of journalistic deference to science – have treated science as unquestionable knowledge and scientists as authorities to pass on that knowledge to the public, who is perceived to have a fundamental lack of science proficiency and sometimes an ill-informed hostility to science (Durant 1996; Gregory and Miller 1998; Levy-Leblond 1992; Myers 2003). As a result, journalism acts primarily as a mouthpiece and cheerleader for science, representing it as the salvation of society and its practitioners as dedicated, disinterested and unified people who use rigorous methods to make miraculous discoveries for humanity sake. In the words of Ben Goldacre (2008), 'science is portrayed (in the news) as a series of groundless, incomprehensible, didactic truth statements from scientists, who themselves are socially powerful, arbitrary, unelected, authority figures' (225).

The problem with such accounts is the ignorance of the obvious fact that science and scientists, as knowledge producers, are not exempt from inherent human problems: they are inscribed with their own systemic and non-systemic ideologies, with their own inevitable biases, flaws, dilemmas, politics, controversies and struggles (Allan 2002; Bucchi and Trench 2008; Gregory and Miller 1998; Holliman et al. 2009). In representing scientific knowledge as above, according to Goldacre (2008), journalism 'distorts science in its own idiosyncratic way' (225) and obstructs rather than facilitates science communication as a whole. Koch (1990) argues that at the core of contemporary science news lays innate, regulated inclinations, biases and fundamental flaws. He criticizes the institutional myth touted as science journalism, which 'brands' its practitioners as spokesmen and women of the general public in disseminating information without any interest or influence. Nelkin (1987) and Durant (2005) argue that one reason for the knowledge gap about science lies in the way a non-realistic image of its activities and an almost sanctification of its methods and professionals has been created.

Further, the notion of a passive news audience who complies to scientific knowledge as unquestionable and undisputed facts underestimates the complexity of public perception and understanding of science and technology (Gregory and Miller 1998). For one thing, it does not give an appropriate weight to dynamic aspects of informational sense construction, trading messages, motivation and emotional connotations that lead citizens to building their own social representation of science. For another, such understanding does not address scientific culture as a dynamic, collective and social attribute, ignoring that science understanding depends on the social environment in which knowledge becomes operative (Irwin, Wynne and Jasanoff 1996). Also, seeing news as a vehicle for science messages to flow undisrupted from the science establishment to society does not consider the broad and dynamic exchanges between contemporary science and other social institutions (Ziman 2000).

In that light, while it is important that science journalism equips citizens with the necessary knowledge to understand and appreciate science, it should not be its ultimate cause, much less its only cause. The ultimate job of science journalism should be to provide lay people with the cognitive resources they need to understand, question and, when necessary, stand up to science and its huge sociopolitical and ethical implications. Nelkin (1987) and Durant (2005), among others, argue in favour of a representation of science as it is, that is, a form of knowledge that is often controversial, sometimes wrong and seldom certain. Science journalism must not forget that science, as Collins and Pinch (1993) point out, is clumsy, dangerous and 'does not mean it understands the truth' (2). For these authors, it is the uncertainties and disagreements that matter most to the public at large, and it is only through an open explanation of how science really works that one can truly understand such complexity.

Presenting science in that way would better serve the public's understanding of science and create a more reasonable imagination of science, providing citizens with science competencies that they need in contemporary society (Shapin 1992). Such competencies include, according to Sturgis and Allum (2004), the ability for lay publics to participate in scientific debates and often hold governments accountable for their science decisions and policies. Similarly, Nguyen and McIlwaine (2011, 2) argue that science developments must be a potent component of public debate because science has never confronted humanity as it does in the twenty-first century. According to them, the lay public, as the paymaster and ultimate recipient of science outcomes, must have a strong voice over many science developments that impinge on physical and moral values – such as genetic modification, global warming, animal cloning, embryonic stem-cell research, artificial intelligence, and technological disasters like Chernobyl, BSE and, more recently, Fukushima.

In other words, science must be deeply embedded in the intersection between the media, the public and policymakers, on both local and global scales (Bauer and Bucchi 2007; Bucchi and Mazzolini 2003; Bucchi and Trench 2014). In this context, Perreira, Serra and Peiriço (2003) contend that the purpose of science journalism is – aside from entertaining and enlightening the public – the exposition of future discoveries that may change society and the way it is constructed. In broader terms, news communication of science should act as an arena for the public to have a voice over the assumptions, values, attitudes, language and general implementation of science knowledge (Valerio and Bazzo 2006). In doing so, it also positions itself as a vehicle for social inclusion (Bueno 1985).

To accomplish these, science journalism needs to break from its traditional top-down model, accepting that most people are not that eager to learn about science facts to appreciate the wonder of science. Instead, as Nguyen and McIlwaine (2011, 14) observe from their EU survey data: 'Lay publics are generally and keenly interested in science as they perceive it affects them. They want to learn about science issues with a proximity to their life and to have a voice in the debate surrounding these issues.' This interest in turn requires journalism to stand on the side of the people, not the science establishment, stop bombarding the former with classroom-like science facts, and start helping them to not only understand science but also understand it critically.

It is in this framework that the reporting of statistical data in science journalism should be explored. We propose that science statistics, as the language of science, must be reported in ways that at least help citizens to (a) connect science with their daily private and public affairs and (b) maintain a healthy caution of science benefits and risks and a balance between hopes and fears about possible science outcomes. But, as said above, the above culture of science journalism does not allow us to hope for a prevalence of such good reporting of science statistics in the news. In order to gain some preliminary data to test this belief, we conducted a secondary analysis of a data set on science statistics in four respected UK and Brazil newspapers.

Our exploratory content analysis

The content analysis data set at stake was derived from a sample of 1,089 articles that used statistical information to report or discuss science events and issues in 2013 in four broadsheet dailies – the *Guardian* and *The Times* in the UK and *O Globo* and *Folha de S. Paulo* in Brazil. The British sample includes 371 relevant science articles of all types (straight news, features, editorials and others), which were collected through a search in the index for *Science and Technology* in Lexis Nexis (but not Humanities and Social Sciences, Computer Science, Maths and Education, Science Funding, Science Policy). For the Brazilian newspapers, a search for all articles with the search term *ciência* (meaning 'science' in Portuguese) in 2013 was conducted on their corporate archives, with each then being individually examined for the presence of quantitative data. The resulting sample includes 718 articles. The four newspapers have long been known for their generally serious investment in and commendable quality of science and environment journalism (Carvalho 2007; Guedes 2000; Lacey and Longman 1993). As such, they are good case studies for this exploratory study on the quality of statistical reporting in their science news output because the findings would shed some preliminary light on the state of journalism's relationship with science statistics.

This data set was originally collected for Brandao's PhD thesis, but has been recoded and re-analysed for this chapter with a very different focus and approach that are beyond the original work. This secondary analysis of the data set was performed in a limited scope, on only six of the original variables that happened to be more or less relevant to this chapter's purpose. These variables represent six different aspects of statistical reporting that we considered important and helpful for lay readers to relate to and evaluate the data being reported in a science story. We grouped them under two key themes along the key criteria for good science journalism identified above: (a) attempts to make science statistics relevant to the lay public and (b) attempts to help the public to critically understand science statistics. Four of the original variables were recoded to serve the purpose of this chapter's enquiry.

Attempts to make science statistics relevant to lay publics

In order to explore this issue, we studied three variables, shown in Table 5.1. The first is the extent to which science stories reported statistics in some visual form. Although visualization is not the only – and not always the best – way for journalism to make science statistics accessible to the lay public, we believe that it is a major technique whose use is compatible with normal news standards. The result, however, shows that the vast majority of reported statistics in the sample are presented as text-only within the story. Visualization was only found in 13 per cent of the sample, with the Brazil newspapers being much more likely to provide it (17 per cent) than their British counterpart (less than 5 per cent). This low proportion of science statistics being visualized in the news might be disappointing for some observers. It might be worth noting that the stories were sampled from 2013, a year in which data journalism, as a new form of storytelling, was becoming a remarkable trend, especially in more advanced news environments such as the UK.

Second, we considered what the key statistic in each of the sampled stories was used for. The main statistic was defined as the most crucial number on which the story or its main claim is based. Where there was only one statistic in the story (37 per cent of the sample), it was taken as the story's key statistic. As seen in Table 5.1, in only 19 per cent of the cases (12 per cent of UK stories and 23 per cent of Brazil ones) did we find the main statistic as the main thrust of the story, that is, the essential fact without which the related event/issue does not make it to the story at stake. The majority of the sampled stories used the key statistic either as essential explanatory material (34 per cent) or background information (also 34 per cent), and this spreads quite evenly

Table 5.1 To what extent does science journalism make science statistics accessible and relevant to people's life (% of sample/sub-sample)

	UK (n = 371)	Brazil (n = 718)	All stories (n = 1089)
Visual presentation of data in the story			
Yes	4.6	17.0	12.8
No	95.4	83.0	87.2
Use of the key statistic in the story			
Main thrust of story	12.1	23.1	19.4
Essential explanation for story	36.1	33.1	34.2
Background information for story	34.0	33.7	33.8
Others	17.8	10.0	12.7
Main perspectives on statistics in the story			
Government	6.7	9.6	8.6
Think tanks/NGOs	15.6	21.6	19.6
Universities	18.9	24.4	22.5
Other stakeholders	26.9	30.2	29.1
Unidentified sources	31.8	14.2	20.2

across the UK and Brazil sub-samples. This suggests that science journalism at the four newspapers do not treat statistics merely as dry numerical facts but tend to nest them within some sort of socio-textual context.

Third, in relation to that, we explored whether journalists at the four newspapers attempted to bring science statistics beyond the scientific context and into daily life. We did this by examining the major presenters of statistics in a news story. The main presenter, it should be noted, is not necessarily the producer or owner of the statistics at stake, but the organization that cites/uses them to make some claims in the story. As can be seen, universities, the primary source of science research, accounted for less than a quarter (22.5 per cent) of the key statistics reported in the entire sample, less than the combined total of think tanks/NGOs and governmental sources (28.2 per cent). The miscellaneous category of other sources, which include corporate science sources, individual professional experts (e.g. doctors) and so on, only accounted for more key presenters of statistics in the news than universities (29 per cent). In 20 per cent of the sample (14 per cent of Brazil stories and 32 per cent of UK stories), there was no third-party presenter of the reported statistics – that is, the reporter/writer just cites a number to serve some narration purpose without telling readers where it is from. Overall, the four news outlets seemed to maintain a rather balanced mix of different stakeholders of science statistics, including those that are less involved in the production of science knowledge and more in science activism and policy making. In other words, when reported in these newspapers, science statistics do escape from the ivory tower of science to come closer to the interests and concerns of the lay people that governmental and non-governmental organizations represent.

In short, although the above data do not allow a deep insight the overall quality of news reporting of science statistics, they do indicate an attempt by the four news outlets to place those numbers in social context and to help ordinary citizens understand and interpret them in ways beyond their pure scientific meanings. But would such understanding be critical enough?

Attempts to help lay publics to critically understand science statistics

To explore this, we looked at three key technical aspects that readers would need to judge the nature of the statistics being reported to them (see Table 5.2). First, we explored whether it was common for science stories to identify the research design of the reported statistics (observational, experimental or self-reported data). This information is an important aspect that journalists should include and explain in the story to help readers to consider the validity of the claims that are based on the reported data. For example, a research conclusion about a cause–effect association between a lifestyle and a medical condition should be cautioned to readers if it is merely based

Table 5.2 News reporting of key methodological information about reported science statistics (% of sample)

	UK (n = 371)	Brazil (n = 718)	All stories (n = 1089)
Research design of data			
Reported	34.2	24.6	27.9
Not reported	65.8	75.4	72.1
Sample size associated with data			
Reported	24.0	19.4	20.9
Not reported	76.0	80.6	79.1
Indication of data conclusiveness			
Reported	39.6	27.3	30.9
Not reported	60.4	73.7	69.1

on observational rather than experimental data. The result, as shown in Table 5.2, was not particularly positive: over 72 per cent of the sampled stories do not mention the research design of reported data. The UK sample, with 66 per cent, did a little better than the Brazilian one (75 per cent). Thirteen per cent of the whole sample reported data from observational research, 8 per cent from experiments and 7 per cent from self-reported studies.

Second, we looked for information about the sample size from which the reported data were derived. This is not to discredit small samples or to promote large ones – the sufficiency of a sample size depends on the type of research design and various other factors – but again to explore whether science journalists were conscious of reporting crucial issues around sampling for readers to judge the associated data. The analysis (Table 5.2) shows that about eight in ten stories in the whole sample do not report the size of the sample of the presented data. Again, the UK was in a slightly better performance, with 76 per cent (cf. 81 per cent of the Brazil sample).

Third, we examined whether science articles in the four newspapers indicated the conclusiveness of a statistic by highlighting whether it is derived from a single or multiple study. Again, this is an important piece of information for readers to be aware of the scope of the data and their conclusiveness. If a number comes from a meta-analysis of multiple studies, it is more conclusive than one based on a single study. The result (Table 5.2) shows that almost 70 per cent of the sampled stories did not indicate whether reported data were based on single or multiple studies. This was worse in the case of the Brazilian sample, 73 per cent of which did not mention this aspect of the data (cf. 60 per cent of the UK sample). Twenty-seven per cent of the whole sample were based on single studies and only 4 per cent on multiple studies.

Concluding notes

Before a conclusion can be drawn, it should be acknowledged that the above findings are from comprehensive or conclusive. As the data are based on a limited secondary content analysis, we could only produce a small amount of indicative data to serve the central argument of this essay. Although this is not ideal, it still contributes some initial insights into an issue that has never been studied in the literature.

With that in mind, we tentatively conclude that the four analysed Brazil and UK broadsheets treat science statistics in a way that seems to go beyond the traditional habit of reporting science merely for science literacy's sake. The science statistics in the sampled stories were not only used primarily for explanation and background purposes but also for a balanced mix of science and non-science perspectives. As mathematical language varies with respect to the values and rationale of the stakeholders who promote it (Jablonka 2002), these findings suggest that many science statistics would be projected into the news within some social context to address the interests and concerns of the ordinary people that governmental and non-governmental organizations represent. This, of course, can be for better or worse, depending on whom journalists allow to present numbers in reporting science, and the data above do not allow us to check this. But the days when science journalism 'bombarded' people with one decontextualized science fact to the next, primarily for science's own sake, seem to have bygone. The framing of science statistics in non-science contexts would continue to exercise a strong influence on how scientific knowledge is received and constructed among lay people.

Along with that 'good news', however, is the clear fact that most reporting of science statistics in the four flagship newspapers remains rather indifferent to the provision of the key methodological information and explanation that lay people need to understand science critically and judgementally – such as the research design, sample size and conclusiveness level of the reported statistics. These findings add a fresh perspective to the existing literature on news communication of scientific uncertainty, which argues that contemporary society lives in an age of uncertainty precisely because journalists omit certain background information about how the reported research was conducted (Friedman, Dunwoody and Rogers 1999; Nowotny, Scott and Gibbons 2001). This is a significant problem that journalists need to realize if they want to meet the new demands for science democracy in the contemporary world, one in which lay citizens are no longer passive recipients of intelligence from the science world through the news media but increasingly want to have a voice over science developments and to hold science accountable for its own enterprise (Nguyen and McIlwaine 2011).

Towards that aim, we would like to outline a few potential reasons for the omission of key methodological information in science news reporting. The first might be that many science reporters do not have enough skills, knowledge and perhaps courage to critically assess or challenge statistical information from science institutions. While the data set write which stories, we suspect a substantial proportion of them are written by the increasing number of general reporters assigned to cover science stories, rather than specialist science journalists. The lack of statistical reasoning training among journalists is often worsened by the deadline-driven intense newsroom, where relevant resources, such as time, are in an increasingly critical shortage state. Apparently, the inevitable product of all these is the overwhelming amount of insufficient or oversimplified news accounts of science events and issues that are offered to the public. Indeed, this content analysis did attempt to look for cases in which the writers challenged or raised critical points about some science statistics, but found none.

Another potential reason for the relative absence of key methodological details behind science statistics is the common assumption among journalists that ordinary people do not have sufficient knowledge to understand such information. This needs to be revisited. First, if most ordinary citizens do not have the ability to understand methodological information, then it is the job of the journalist to report, explain and educate them about it, and not ignore or dismiss it. Secondly, there has been evidence that people are smarter and more sophisticated than journalists might think: when given enough explanation and the chance to assess numerical information at their own will, they will be ready to do it in an informed and critical manner. This is especially so if people find the event/issue behind the data close to their heart and mind and if they feel valued as an active participant in an open social dialogue about science and its issues (Allan 2002, Durant 1996). And let us be straight: if a journalist cannot find an accessible way to explain to a layperson, for example, the difference between an observational and experimental data set, he/she is not that great a journalist, much less a science journalist.

All this leads us to a final point: the critical need for a better, more systematic training for journalists in reasoning about and with statistics. If newsrooms and journalism schools want to be at the helm of both science communication and science democratization, they need to address the current statistical skill shortages wholeheartedly and quite urgently. Such training should entail not only the ability to make science data and knowledge accessible, comprehensive, transparent and accountable but also to present statistics in a way that motivates lay audiences to think judgementally about them. Without such news reporting of science statistics, citizens might have just enough information to think they understand science but not enough to realize that this understanding is seldom adequate for them to make effective decisions about science, both for their daily life operation and their participation in public science issues. In other words, while, statistics remain an important rhetorical instrument in news discourses of science, beause they work in accord with the rituals of journalistic objectivity, the lack of critical news reporting about them may lead journalism to being no more than the mouthpiece of science and those who have a stake in it. The common result, as we have seen in the discussion above, is the misconstruction, even unwitting distortion, of science, or the relentless spin doctoring of scientific information in the news. The cost of such reporting could be hefty. For example, the MMR scandal, which was due in a large part to an excessive reliance of journalists on an odd data set from a single study, resulted in huge damages to both individual and collective welfare due to outbreaks of otherwise controllable diseases. This and other problematic incidents and areas of science statistics (e.g. global warming, genetic modification) tell us that the stake is rather high and we must learn dear lessons from the past for a better future of statistical reporting in science journalism.

References

Allan, S. (2002), *Media, Risk and Science*. Buckingham: Open University Press.

Bauer, M. W. and Bucchi, M. (2007), *Journalism, Science and Society: Science Communication between News and Public Relations*. London: Routledge.

Bauer, M., Howard, S, Yulye, R., Luisa, M. and Luis, A. (2013), *Global Science Journalism Report*. Retrieved 24 March 2016 from http://eprints.lse.ac.uk/48051/1/Bauer_Global_science_journalism_2013.pdf.

Bucchi, M. and Mazzolini, R. G. (2003a), Big science, little news: Science c. (1986). In S. Feferman (ed.), *Collected Works*. New York: Oxford University Press.

Bucchi, M. and Mazzolini, R. G. (2003b), Big science, little news: Science coverage in the Italian daily press, 1946–1997. *Public Understanding of Science*, 7–24.

Bucchi, M. and Trench, B. (2014), *Routledge Handbook of Public Communication of Science and Technology,* 2nd edn. New York: Routledge.

Bueno, W. C. (1985), *Jornalismo científico no Brasil: os compromissos de uma prática dependente*. São Paulo: Escola de Comunicação, Universidade de São Paulo (Tese de Doutorado).

Carvalho, A. (2007), Ideological cultures and media discourses on scientific knowledge: Re-reading news on climate change. *Public Understanding of Science*, 18(2), 223–43.

Collins, H. and Pinch, T. (1993), *The Golem: What Everyone Should Know about Science*. Cambridge: Cambridge University Press.

Dick, Hugh G. (1955), *Selected Writings of Francis Bacon*. New York: Modern Library.

Durant, J. (1996), Science Museums or Just Museums of Science? In S. Pearce (ed.), *Exploring Science in Museums*. London: The Athlon Press.

Durant, J. (2005), O que é Alfabetização Científica? In L. Massarani, J. Turney and I. C. Moreira (eds), *Terra Incognita. A Interface entre Ciência e Público*, 13–26. Rio de Janeiro: Casa da Ciência/UFRJ, Museu da Vida/Fiocruz, Vieira & Lent.

Feyerabend, P. (2011), *The Tyranny of Science*. Cambridge: Polity Press.

Fjæstad, B. (2007), Why Journalists Report Science as They Do. In M. W. Bauer and M. Bucchi (eds), *Journalism, Science and Society: Science Communication Between News and Public Relations*. London: Routledge.

Frege, G. (1960), *The Foundations of Arithmetic – A logico-Mathematical Enquiry into the Concept of Number*, 2nd edn. Trans. J. L. Austin. New York: Harper Torchbooks.

Frege, G. (1977), *Logical Investigations*. Oxford: Blackwell.

Friedman, S., Dunwoody, S. and Rogers, C. (1999). *Communicating Uncertainty: Media Coverage of New and Controversial Science*. London: Lawrance Erlbaum Associates, Publishers.

Gödel, K. (1986), *Collected Works*. New York: Oxford University Press.

Goldacre, B. (2008), *Bad Science*. London: Fourth Estate.

Gregory, J. and Miller, S. (1998), *Science in Public: Communication, Culture and Credibility*. Cambridge, MA: Basic Books.

Guedes, O. (2000), Environmental Issues in the Brazilian Press. *International Communication Gazette*, 62(6), 537–54.

Howard, A. (2014), *The Art and Science of Data-Driven Journalism*. Tow Centre for Digital Journalism. Retrieved 12 February 2015 from http://tinyurl.com/jmxskoz.

Holliman, R., Whitelegg, L., Scanlon, E., Smidt, S. and Thomas, J. (2009). *Investigating Science Communication in the Information Age: Implications for Public Engagement and Popular Media*. Oxford, UK: Oxford University Press.

Irwin, A., Wynne, B. and Jasanoff, S. (1996), Misunderstanding Science? *Social Studies of Science*, 27(2), 350–5.

Jablonka, E. (2002), Information: Its interpretation, its inheritance, and its sharing. *Philosophy of Science*, 578–605.

Koch, T. (1990), *The News as a Myth: Fact and Context in Journalism*. Westport: Greenwood Press.

Kovach, B. and Rosenstiel, T. (2001), *The Elements of Journalism*. London: Guardian Books.

Levy-Leblond, J. M. (1992), About misunderstandings about misunderstandings. *Public Understanding of Science*, 1(1), 17–21.

Lippmann, W. (1920), *Liberty and the News*. New York: Harcourt, Brace and Howe.

Lippmann, W. (1922), *Public Opinion*. New York: Pearson Education.

Miller, S. (2001). Public understanding of science at the crossroads.. *Public Understanding of Science*, 10 (1), 115–20.

Myers, G. (2003), Discourse studies of scientific popularization: Questioning the boundaries. *Discourse Studies*, 5(2), 265–79.

Nelkin, D. (1987), *Selling Science. How the Press Covers Science and Technology*. New York: W. H. Freemand and Company.

Nguyen A. and Lugo-Ocando J. (2016). The state of data and statistics in journalism and journalism education: Issues and debates. *Journalism*. 17(1), 3–17.

Nguyen, A. and McIlwaine, S. (2011), 'Who want to have a voice in science issues – and why? A European survey and its implications for science journalism.' *Journalism Practice*, 5(2), 1–17.

Nowotny, H., Scott, P. and Gibbons, M. (2001), *Re-Thinking Science, Knowledge and The Public in an Age of Uncertainty*. Cambridge: Polity Press.

Lacey, C., and Longman, D. (1993). 'The press and public access to the environment and development debate.' *The Sociological Review*, 41(2), 207–43.

Pereira, A., Serra, I. and Peiriço, N. M. (2003), Valor da Ciência na Divulgação Científica. In C. M. Sousa, N. P. Marques and T. S. Silveira (eds), *A Comunicação Pública da Ciência*, 59–63. Taubate: Cabral.

Porter, T. M. (1996). *Trust in Numbers. The Pursuit of Objectivity in Science and Public Life*. Princeton, NJ: Princeton University Press.

Rensberger, B. (2009), Science journalism: Too close for comfort. *Nature*, 459, 1055–6.

Sagan, C. (1995), *The Demon-Haunted World: Science as a Candle in the Dark*. New York: Ballentine Books.

Schudson, M. (2001). The objectivity norm in American journalism. *Journalism*, 2(2), 149–70.

Shamos, Morris H. (1995), *The Myth of Scientific Literacy*. New Brunswick: Rutgers University Press.

Shapin, S. (1992), Why the public ought to understand science-in-the-making. *Public Understanding of Science*, 1(1), 27–30.

Simpson, S. and Dorling, D. (1999), 'Statistics and "the truth"'. In D. Dorling and S. Simpson (eds), *Statistics in Society: The Arithmetic of Politics*, 414–23. London: Arnold.

Stigler, S. M. (1986), *The History of Statistics – The Measurement of Uncertainty before 1990*. Cambridge: The Belknap Press of Harvard University Press.

Streckfuss, R. (1990), Objectivity in journalism: A search and a reassessment. *Journalism and Mass Communication Quarterly*, 67(4), 973–83.

Sturgis, P. and Allum, N. (2004), Science in society: Re-evaluating the deficit model of public attitudes. *Public Understanding of Science*, 13(1), 55–74.

Utts, J. (2003). 'What educated citizens should know about statistics and probability.' *The American Statistician*, 57 (2), 74–9.

Utts, J. (2010). *Unintentional Lies in the Media: Don't Blame Journalists for What we Don't Teach*. Accessed 2 May 2015. http://citeseerx.ist.psu.edu/viewdoc/summary?doi=10.1.1.205.227.

Valerio, M. and Bazzo, W. (2006), O papel da divulgação científica em nossa sociedade de risco: em prol de uma nova ordem de relações entre ciência, tecnologia e sociedade. *Revista de Ensino de Engenharia*, 25(1), 31–9.

Weigold, M. F. (2001), 'Communicating science: A review of the literature.' *Science Communication*, 23(2), 164–93.

Ziman, J. (2000), *Real Science: What it is and What it Means*. Cambridge: Cambridge University Press.

Zuberi, T. (2001), *Thicker than Blood: How Racial Statistics Lie*. Minneapolis: University of Minnesota Press.

6

Data Journalism at its Finest: A Longitudinal Analysis of the Characteristics of Award-Nominated Data Journalism Projects

Julius Reimer, *Hans Bredow Institute for Media Research, Germany*
Wiebke Loosen, *Hans Bredow Institute for Media Research, Germany*

Introduction: Journalism's response to the datafication of society

'Datafication' has become a grand narrative used to describe how numerical data have come to represent and influence social reality. Journalism as a field is deeply concerned with covering the consequences of datafication while at the same time is itself profoundly affected by them (e.g. Nguyen and Lugo-Ocando 2016). One observable outcome is the emergence of data-driven journalism (DDJ) – a new style of reporting based on ever more available data sets (Borges-Rey 2016).

The extensive attention that practitioners pay to DDJ has also fuelled 'an explosion in data journalism-oriented scholarship' (Fink and Anderson 2015, 476). Given the relative novelty of the phenomenon, however, this research body is based primarily on case studies, cursory observations, and/or samples that are limited in spatial and temporal terms. As such, research so far has not investigated the international breadth and 'rapidly changing nature' (Royal and Blasingame 2015, 41) of the field in any depth. Building on existing work, we wanted to take a first step to fill this gap by conducting a longitudinal study of what may be considered the gold standard of the field: the international projects nominated for the Data Journalism Awards (DJA) from 2013 to 2015. Using a content analysis of nominated projects during 2013–15 (n = 179), we examine if and how, among other things, data sources/types, visualization strategies, interactive features, topics, and types of nominated media outlets have changed over the years. Results suggest that the set of structural elements and presentation forms on which data-driven pieces are built remain rather stable, but data journalism is increasingly personnel-intensive and progressively spreading around the globe. Finally, we find that journalists, while still concentrating on data from official institutions, are increasingly looking to unofficial data sources for their stories.

What we (don't) know about data journalism

The scholarship on DDJ has been dominated by three particular areas of study. First, researchers have tended to focus on the actors involved in the production of data journalism. Data journalists in Belgium (De Maeyer et al. 2015), Germany (Weinacht and Spiller 2014), Norway (Karlsen and Stavelin 2014), Sweden (Appelgren and Nygren 2014), the United Kingdom (Borges-Rey 2016) and the United States (Boyles and Meyer 2016; Fink and Anderson 2015; Parasie 2015; Parasie and Dagiral 2013) have been interviewed about and/or observed for their journalistic self-understanding and the organization of their work in newsrooms.

Secondly, scholars have tried to clarify what data journalism is and how it is similar to and different from investigative journalism, computer-assisted reporting, computational journalism and the like. Such exercises in classification (e.g. Coddington 2015; Fink and Anderson 2015; Royal and Blasingame 2015) often contradict each other. While Anderson (2013, 1005), for instance, places data journalism within the broader field of 'computational journalism', Coddington (2015) explicitly distinguishes between these two concepts. Consequently, Fink and Anderson (2015, 478) lament the 'lack of a shared definition of data journalism'. However, scholars have more or less agreed on the following key characteristics of DDJ:

1. it usually builds on large sets of digitally stored and distributed quantitative data as 'raw material' that is subjected to some form of (statistical) analysis in order to identify and tell stories (Coddington 2015; Royal and Blasingame 2015);
2. its results 'often need visualization' (Gray, Bounegru and Chambers 2012), that is, they are presented in the form of maps, bar charts and other graphics (Royal and Blasingame 2015);
3. it is 'characterised by its participatory openness' (Coddington 2015, 337) and 'so-called crowdsourcing' (Appelgren and Nygren 2014, 394) in that users help with collecting, analysing or interpreting the data; and
4. it often adopts an open-data and open-source approach, that is, it is regarded as a quality criterion of DDJ that journalists also publish the raw data a story is built upon (Gray, Bounegru and Chambers 2012).

A third strand of research analyses the actual data-driven news content. These studies focus on the above-mentioned elements and affirm their status as key characteristics of a data-journalistic reporting style. However, in spatial or temporal terms, their samples are rather limited: Parasie and Dagiral's (2013) study comprises pieces from one Chicago outlet, which were published before March 2011; Knight (2015) analyses articles in fifteen UK newspapers over a two-week period in 2013; Tandoc and Oh (2015) turn to 260 stories published in the *Guardian's Datablog* between 2009 and 2015; and Tabary, Provost and Trottier (2016) examine projects produced between 2011 and 2013 by six Quebec media outlets. Such research reflects (a) certain knowledge about DDJ actors, their self-image as journalists and their integration inside and outside established newsrooms and (b) various attempts to define DDJ as a distinguishable reporting style that revolves around some apparent core characteristics. Initial studies that empirically analyse DDJ products, however, are restricted in their scope. One key problem is the lack of a longitudinal analysis on a broader geographical scale that advances our understanding of how DDJ as 'an *emerging* form of storytelling' (Appelgren and Nygren 2014, 394; *emphasis added*) is currently evolving over time and around the globe. In what follows, we take a first step to fill this gap, using a content analysis to answer the following research questions:

RQ1: What structural elements and forms of presentation are data-driven pieces composed of and how is this composition evolving?

RQ2: How have the topics covered in data-driven projects changed over time?

RQ3: How is the field of actors producing data journalism (media organizations, in-house teams, external partners) developing?

Our content analysis of DJA-nominated/ awarded projects

We seek to answer our research questions by conducting a standardized content analysis (e.g. Krippendorff 2013) of data journalism pieces from a three-year period and on a broad geographical scale.

Since data journalism is such a 'diffuse term' (Fink and Anderson 2015, 470), it is difficult, or rather *preconditional*, to identify respective pieces for a content analysis. We, therefore, decided to take an inductive and pragmatic approach that avoids starting with either too narrow or too broad a definition of what counts as data journalism: our sample (Table 6.1) consists of pieces nominated for the annual *Data Journalism Awards* (DJA) of the Global Editors Network[1] in 2013, 2014 and 2015. In other words, *the field itself* regards these projects as data journalism and as significant examples of this new reporting style. Similar approaches to sampling have already proven useful for analysing particular reporting styles such as investigative reporting (Lanosga 2014) and aspects of storytelling like the emotionality of reports (Wahl-Jorgensen 2013). We must, however, take into account that our sample is doubly biased. First, the analysed pieces are based on self-selection as any data journalist can submit her/his work to be considered for nomination by the organizing committee. Second, nominees for a data journalism award are not

---

[1] Cf. http://www.globaleditorsnetwork.org/about-us/ (accessed 20 December 2016).

Table 6.1 Data set overview[I]

		2013	2014	2015	Total
Submissions	Freq	>300.0[II]	520.0	482.0	>1,302.0
Nominated projects	Freq	72.0	75.0	78.0	225.0
	% of submissions	<24.0	14.4	16.2	<17.3
Projects suited for analysis	Freq	56.0	64.0	59.0	179.0
	% of nominees	77.8	85.3	75.6	79.6
	% of projects analysed	31.3	35.8	33.0	100.0
Award-winning projects	Freq	6.0	9.0	13.0	28.0
	% of projects analysed	10.7	14.1	22.0	15.6

[I] If a nomination referred to a media outlet as a whole and not to a specific project, the case was excluded from the analysis as our unit of analysis is a single data-driven piece. A list of (and links to) all projects nominated for a DJA in 2013, 2014 and 2015 is available on: http://community.globaleditorsnetwork.org/projects_by_global_event/744 (accessed 14 October 2015).

[II] The GEN does not specify the number of submissions for 2013, but only states that 'more than 300 entries' had been submitted (http://www.globaleditorsnetwork.org/programmes/dja; accessed 17 February 2014).

likely to represent 'everyday' data journalism. In this already diversified field, our sample is likely to consist of 'an extensive, thoroughly researched, investigative form of data journalism' which, as Borges-Rey (2016, 841) found in the UK, can be distinguished from 'a daily, quick turnaround, generally visualised, brief form of data journalism'.

With respect to the research objectives stated above, this sample allows us to track developments over several years as well as to identify differences between those data journalism pieces that were only nominated and those that actually won an award. Our codebook comprises, among others, the presumed key characteristics of DDJ listed in the literature review (see above). Most variables and their assigned values were developed inductively in 2013, based on an explorative analysis of a subsample from that year. Some categories were inspired by Parasie and Dagiral's (2013) study, the only content analysis of data-driven pieces available at the time; others were suggested by fellow researcher Julian Ausserhofer and data journalist Lorenz Matzat. A pre-test was conducted with two coders and a subsample of 10 per cent of cases. All variables reached an intercoder reliability coefficient (Holsti and Krippendorff's Alpha) equivalent to or higher than 0.7 which is generally considered sufficient for exploratory research (Krippendorff 2013).

The final codebook contains twenty-eight categories which can be grouped roughly into five dimensions (Table 6.2). Over the three years we did not have to change or add variables or values to capture new kinds of data, visualizations or other elements, the only exemption being audio files which were used in projects from 2015.

Table 6.2 Dimensions and variables of the codebook[I]

Dimensions	Variables
Authorship	Medium; type of medium; external partners; number of people involved mentioned by name
Story properties	Headline; topic; reference to a specific event[II]; question(s) posed to data; number of related articles[III]; length of article; language; winner of DJA
Data	Data source(s); type(s) of data source(s); access to data; kind of data; additional information on data[II]; geographical reference; changeability of data set[IV]; time period covered; unit of analysis
Analysis and journalistic editing of content	Personalized case example[V]; call for public intervention or criticism[III]; purpose of data analysis[VI]; visualization
Interactive features	Interactive functions; online access to the database[III]; opportunities for communication

[I] We will provide the complete codebook on request.
[II] Suggested by data journalist Lorenz Matzat.
[III] Adopted from Parasie and Dagiral (2013: 5–14).
[IV] Suggested by (data) journalism researcher Julian Ausserhofer.
[V] Inspired by Holtermann (2011).
[VI] Inspired by Gray, Bounegru and Chambers (2012: n.p.).

Findings

This section will answer the research questions in reverse order. It starts by looking at the actors producing DDJ (RQ3) and the topics they cover (RQ2) as well as some formal story elements. These provide the background against which we will then present the results regarding the 'key characteristics' of DDJ: data-drivenness, visualization, and interactivity (RQ1). To put our findings into perspective, we will place them in the context of results from previous content analyses.

The actors producing DJA-nominated pieces

In our sample, newspapers represent by far the largest – and continuously growing – group among the nominees (2013: 41.1 per cent; 2015: 47.5 per cent; 2013–15 on average: 44.1 per cent) as well as among the award winners (2013: 50.0 per cent; 2015: 38.5 per cent; 2013–15 on average: 46.4 per cent). Another important group is investigative journalism organizations such as *Pro Publica* and *The International Consortium of Investigative Journalists* (ICIJ) (19.0 per cent). Print magazines (8.9 per cent), public and private broadcasters (5.6 per cent and 2.8 per cent), native online media, news agencies and non-journalistic organizations (5 per cent, respectively), university media (3.4 per cent) and other types of authors (1.1 per cent) are represented to much lesser levels. This does not mean that nearly three quarters of all data journalism output is produced by print media and investigative organizations. Instead, the composition of the sample carries the inherent bias of awards towards best practices and established, high-profile actors (Jenkins and Volz 2016).

Our results also show that data journalism is usually a collaborative effort. Of all projects with a byline (n = 154), on average, over five individuals are named as authors or contributors (M = 5.16). This is probably due to the division of labour into data analysis, visualization and writing which several studies have found to be common in the field (Tabary, Provost and Trottier 2016; Weinacht and Spiller 2014). Furthermore, data journalism seems to have become ever more personnel-intensive, with the average number of people involved in production increasing from 4.13 in 2013 to 5.67 in 2015.

Nearly a third (31.3 per cent) of all projects were realized in association with external partners, who either contributed to the analysis or designed visual elements. The proportion of projects that were realized with external partners, however, decreased substantially to 23.7 per cent in 2015, compared with 33.9 per cent in 2013 and 35.9 per cent in 2014. This could be a consequence of intensified training within news organizations and/or an increase in the recruitment of in-house data journalism personnel.

Nearly half of the nominees come from the United States (48.6 per cent), followed at a distance by Great Britain (12.8 per cent) and Germany (7.3 per cent). This probably reflects that data journalism has a longer history in English speaking countries. However, the number of countries represented by the nominees grew with each year, reaching a total of twenty-seven countries from all five continents over the three years. This suggests that data journalism is increasingly spreading around the globe. Additionally, it appears that data journalists increasingly wish to appeal to an international audience as bi- or multi-lingual projects (15.6 per cent) are the second most frequent (after English-language projects with 67.0 per cent) and their share has grown from 14.3 per cent to 18.6 per cent over the three-year period. In most of these cases, the projects are published in English and in the producer's native language.

The topics and formal story elements of DJA-nominees

Almost half of the pieces (48.6 per cent) analysed cover a political topic; about one-third (34.6 per cent) deal with societal issues (e.g. census results, crime reports); nearly a quarter (23.5 per cent) focus on business and the economy; and more than a fifth (21.2 per cent) are concerned with health and science (multiple coding possible). Education as well as sports and culture projects accounted for very small portions of the nominees (2.2 per cent to 6.1 per cent). Above that, data-driven stories appear to have a clear thematic focus: nearly two-thirds of the projects deal with only one category of topic (65.9 per cent) while only one-third spreads two or more different topical areas (e.g. economical and educational matters by looking at how much education pays off in the salary people receive in Mexico; or political decisions and their societal impact by investigating how weapon laws influence the number of mass shootings).

Many pieces in the politics section deal with elections, which tend to generate vast amounts of quantitative data. In more than half of the cases (51.9 per cent), political data-driven stories distinctly assume a watchdog role, since we found elements of criticism or even calls for public intervention. Examples include stories that fact-checked claims that Brazilian politicians made during their campaigns; showed who gave how much money to political players in the United States and if that shaped political decisions; revealed how district boundaries in New York get manipulated to benefit a political party; researched what the Argentine senate, the US military or the Egyptian government spent tax money for; and exposed how the Republicans in the United States tried to remove minority voters from the election rolls with the Crosscheck program. The share of projects with such a critical position is only slightly smaller among the nominees in general (49.2 per cent) and remains more or less stable over the three years. Among the award winners, three-fifths of projects assume such a critical watchdog role (60.7 per cent).

Data sets, data sources and data analysis in DJA-nominated stories

The data journalism works we analysed rely to a large extent on geo-data (44.4 per cent), financial data (43.3 per cent), and data gathered by sensors or with measuring tools (42.1 per cent), such as aircraft noise, train speeds and carbon emissions (Table 6.3). While the last category has gained prominence over the years, only 28.6 per cent of award-winning projects are based on this kind of data. Another type of frequently analysed data is socio-demographics (32.6 per cent of the nominated projects). Two types of data are used above the average level in award-winning pieces: financial and personal data – that is, information that can be attributed to individual persons – and this is the only significant difference between award winners and non-winners. Metadata (i.e. 'data about data', such as information about individual instances of application use) and data from polls and surveys are the least frequently used in the sample.

As expected, however, some kinds of data are used significantly more often in pieces dealing with particular topics. For instance, information from public opinion polls is included significantly more often in political stories than in non-political ones (23.0 per cent vs. 7.6 per cent; Fisher's exact test: $p < 0.01$). Economic and business pieces draw on financial data more often than other stories (83.3 per cent vs. 30.7 per cent; Fisher's exact test: $p < 0.001$). In turn, works on societal topics are significantly more likely than non-societal news to contain socio-demographic information (56.5 per cent vs. 19.7 per cent; Fisher's exact test: $p < 0.001$) while measured values appear significantly more often in pieces that deal with health or science (78.9 per cent vs. 31.9 per cent; Fisher's exact test: $p < 0.001$).

The type of data also appears to affect the assumption of a watchdog role since pieces based on measured or financial data are significantly more likely to contain criticism or a call for public intervention than the rest (60 per cent vs. 41.4 per cent; Fisher's exact test: $p < 0.05$; and 62.3 per cent vs. 39.2 per cent; Fisher's exact test: $p < 0.01$).

Table 6.3 Kind of data (multiple coding possible)

%	2013 (n = 55)	2014 (n = 64)	2015 (n = 59)	Not awarded (2013–2015) (n = 151)	Awarded (2013-2015) (n = 28)	Total (n = 179)
Geo data	47.3	39.1	47.5	46.0	35.7	44.4
Financial data	45.5	45.3	39.0	41.3	53.6	43.3
Measured values	34.5	43.8	47.5	44.7	28.6	42.1
Sociodemogr. data	38.2	25.0	35.6	32.0	35.7	32.6
Personal data	21.8	32.8	32.2	26.0[1]	46.4[1]	29.2
Metadata	12.7	20.3	13.6	16.0	14.3	15.7
Poll ratings/survey data	14.5	10.9	20.3	17.3	3.6	15.2
Other data	-	-	-	0.7	3.6	1.1

[1] Fisher's exact test: $p < 0.05$.

Only about a quarter of the pieces rely on only one type of data (24.6 per cent) while most refer to two (40.8 per cent) or three (24.0 per cent) different types. Furthermore, the average number of data types used in a project has grown slightly over the years (2013: $M = 2.14$, $SD = 0.96$; 2014: $M = 2.17$, $SD = 0.99$; 2015: $M = 2.36$, $SD = 1.05$). The most frequent combinations are geo-data with measured values (e.g. radiation levels or noise exposure) (21.2 per cent), geo-data with socio-demographic information (17.9 per cent), and geo-data with financial data (16.8 per cent).

It is considered a quality criterion in data journalism that data sources should be cited (Gray, Bounegru and Chambers 2012, 6). Yet 6.1, not 0.1 per cent of these purportedly best-practice cases did not indicate where their data were from (see Table 6.4). However, this is not the case for any of the award-winning pieces.[2]

Of those pieces that cited sources, most use data from official institutions like Eurostat, other statistical offices and ministries (68.2 per cent). This reflects Tabary, Provost and Trottier's (2016, 75) finding that data journalism exhibits a 'dependency on pre-processed public data'. The second largest group consists of pieces that use data from other, non-commercial organizations, including universities, research institutes and NGOs (44.1 per cent). Roughly 20 per cent of the pieces analyse data that the respective media organization collected itself, for example through its own survey or by searching its own archives ('own source'). A comparison between years shows that basing stories on one's own data, after a drop in 2014, is on the rise again. In addition, the share of pieces that report data from private companies has grown consistently. This rise over the years suggests that data journalists are increasingly looking to unofficial data sources for their stories. This assumption is supported by the fact that the average number of *different* types of sources referred to in a data-driven piece has risen from 1.40 in 2013 ($SD = 0.66$) to 1.68 in 2014 ($SD = 0.63$) and 1.67 in 2015 ($SD = 0.80$).[3]

As far as the access to data is concerned, most of the analysed pieces that provide the respective information rely on data that is publicly available. This cannot be explained entirely by the fact that most data originate from official institutions because with a share of, stories that draw on an official source are significantly more likely than those without a stated official source to be based on information belonging to the category of requested data, that is, data that was not publicly available but had to be inquired from the institutions that produce them. Data obtained through Freedom of Information requests belong to this category and are sometimes explicitly included in additional information about the data. Notwithstanding a drop in 2014, the use of such requested data seems to be rising again. So does the number of stories based on data collected by journalists themselves (Table 6.5).

The share of leaked, requested and collected data is considerable. Yet, it does not appear to be as strong as the link that scholars and practitioners often establish

---

[2] This is a much smaller portion than the 40 per cent share that Knight (2015, 65) found in data-driven stories from UK national newspapers.

[3] Kruskal–Wallis test because of heteroscedasticity: $\chi2 = 6.992$, $df = 2$, $p < 0.05$; pairwise Games–Howell tests revealed only one significant difference ($p < 0.10$) between 2013 and 2014.

Table 6.4 Type of data source (multiple coding possible)

%	2013 (n = 56)	2014 (n = 64)	2015 (n = 59)	Not awarded (2013-2015) (n = 151)	Awarded (2013-2015) (n = 28)	Total (n = 179)
Official institution	66.1	68.8	69.5	66.9	75.0	68.2
Other, non-commercial organization	33.9	53.1	44.1	45.7	35.7	44.1
Own source	23.2	14.1	28.8	21.2	25.0	21.8
Private company	14.3	18.8	22.0	17.9	21.4	18.4
Source not indicated	5.4	7.8	5.1	7.3	–	6.1

Table 6.5 Access to data (multiple coding possible)

%	2013 (n = 56)	2014 (n = 64)	2015 (n = 59)	Not awarded (2013-2015) (n = 151)	Awarded (2013-2015) (n = 28)	Total (n = 179)
Access to data not indicated	35.7	43.8	52.5	46.4	32.1	44.1
Publicly available data	39.3	43.8	40.7	41.7	39.3	41.3
Requested data	21.4	15.6	28.8	19.2	35.7	21.8
Own data collection	8.9[1]	1.6[1]	16.9[1]	7.3	17.9	8.9
Scraped data	5.4	7.8	5.1	6.6	3.6	6.1
Leaked data	1.8	4.7	3.4	2.6	7.1	3.4

[1] $\chi^2 = 8.929$; df = 2; $p < 0.05$; Fisher's exact tests for pairwise comparisons with adjusted α-levels (Bonferroni–Holm correction) revealed only one significant difference between years 2014 and 2015 ($p < 0.01$).

between data journalism and investigative reporting (Parasie 2015). Nonetheless, the portion of leaked information as well as the shares of requested and self-generated data are larger than those found by Knight (2015) in her sample of data journalism in UK national newspapers or by Tandoc and Oh (2015) in their study of the *Guardian's Datablog*. Furthermore, in our sample, stories with requested or leaked information were significantly more likely to have a critical edge or a call for public intervention.[4] It is surprising that, despite data journalism's oft-cited association with openness and transparency, in over two-fifths of the pieces, journalists did not indicate at all how they accessed the data they used. In 2015 this was true for even more than half of the analysed pieces.

The data analysed in the stories refers to a range of geographical scales. Most notably, we found that while the share of projects drawing on international data has grown significantly over the years (10.7 per cent in 2013, 15.6 per cent in 2014, 32.2 per cent in 2015)[5], that of pieces based on regional data varies considerably (41.1 per cent, 9.4 per cent and 47.5 per cent respectively)[6].

In the vast majority of cases (88.3 per cent, see Table 6.6), the data is analysed with a focus on comparing values (e.g. to show differences between men and women or neighbourhoods) and half of the pieces show changes over time (e.g. *Climate Change: How Hot Will It Get in My Lifetime?*). Connections and flows are illustrated in about a third of all projects. Much less frequent are pieces that use data to show hierarchies, as in *Women as Academic Authors* which ranks the most important female scientists.

As expected, there are stories that combine at least two of these different ways to analyse and present data. In fact, although not statistically significant, there was an increase in the average number of analytical techniques in a story (2013: $M = 1.75$; 2014: $M = 1.80$; 2015: $M = 2.03$), which indicates that data journalists increasingly combine different approaches and perform more complex analyses.

The visualization elements of DJA-nominated projects

If we think of data journalism as a distinct style of reporting, it is crucial to learn about the particular ways it tells stories. Here, one of the most distinctive elements of data-driven pieces is the use of visualization techniques. Table 6.7 shows that there is a more or less stable set of visualization elements which mainly includes images and simple static charts (62.6 per cent each) as well as maps (48 per cent) and tables (33.5 per cent). Animated visualization was much rarer (16.8 per cent). The proportion

---

[4] Requested data: 87.2%, $n = 39$ vs. 38.6%, $n = 140$, Fisher's exact test: $p < 0.001$; leaked data: 100.0%, $n = 6$ vs. 47.4%, $n = 173$, Fisher's exact test: $p < 0.05$.

[5] $\chi2 = 9.412$, $df = 2$, $p < 0.01$; Fisher's exact tests for pairwise comparisons with adjusted α-levels (Bonferroni–Holm correction) revealed only one significant difference between years 2013 and 2015 ($p < 0.01$).

[6] $\chi2 = 23.712$, $df = 2$, $p < 0.001$; Fisher's exact tests for pairwise comparisons with adjusted α-levels (Bonferroni–Holm correction) revealed two significant differences between years 2013 and 2014 ($p < 0.001$) as well as between 2014 and 2015 ($p < 0.001$).

Table 6.6 Focus of data analysis (multiple coding possible)

%	2013 (n = 56)	2014 (n = 64)	2015 (n = 59)	Not awarded (2013-2015) (n = 151)	Awarded (2013-2015) (n = 28)	Total (n = 179)
Compare values	82.1	87.5	94.9	86.8	96.4	88.3
Show changes over time	46.4	46.9	57.6	49.7	53.6	50.3
Show connections and flows	32.1	35.9	28.8	31.8	35.7	32.4
Show hierarchy	14.3	9.4	22.0	17.2	3.6	15.1

Table 6.7 Visualization (multiple coding possible)[I]

%	2013 (n = 56)	2014 (n = 64)	2015 (n = 59)	Not awarded (2013-2015) (n = 151)	Awarded (2013-2015) (n = 28)	Total (n = 179)
Image	46.4[II]	71.9[II]	67.8[II]	58.9[III]	82.1[III]	62.6
Simple static chart	55.4[IV]	53.1[IV]	79.7[IV]	63.6	57.1	62.6
Map	51.8	46.9	45.8	49.7	39.3	48.0
Table	25.0[V]	28.1[V]	47.5[V]	31.8	42.9	33.5
Combined static diagram	19.6	17.2	22.0	18.5	25.0	19.6
Animated visualization	10.7	20.3	18.6	14.6	28.6	16.8
Other visualization	-	-	8.5	3.3	-	2.8
No visualization	-	-	1.7	0.7	-	0.6

[I] The numbers do not reflect whether elements of the same kind were included more than once: Several pictures, for instance, were counted as one visualization of that kind.
[II] χ^2 = 9.284; df = 2; p < 0.01; Fisher's exact tests for pairwise comparisons with adjusted α-levels (Bonferroni–Holm correction) revealed two significant differences between years 2013 and 2014 (p < 0.01) as well as 2013 and 2015 (p < 0.05).
[III] Fisher's exact test: p < 0.05.
[IV] χ^2 = 11.040; df = 2; p < 0.01; Fisher's exact tests for pairwise comparisons with adjusted α-levels (Bonferroni–Holm correction) revealed two significant differences between years 2013 and 2015 (p < 0.01) as well as 2014 and 2015 (p < 0.01).
[V] χ^2 = 7.803; df = 2; p < 0.05; Fisher's exact tests for pairwise comparisons with adjusted α-levels (Bonferroni–Holm correction) revealed no significant differences.

vof images, charts and tables also grew significantly from 2013 to 2015, echoing the findings of Appelgren and Nygren (2014) as well as Knight (2015, 65) that charts and maps are 'the most common form of data information presented'.

On average, nominated pieces contained more than two different kinds of visualizations ($M = 2.46$). This number grew significantly over the years (2013: $M = 2.09$; 2014: $M = 2.38$; 2015: $M = 2.90$)[7]. This indicates that Knight's conclusion from a sample of UK newspapers – that data journalism 'is practiced as much for its visual appeal as for its investigative qualities' (2015, 55) – might apply to this high-profile group of DJA nominees. Typical combinations of visualizing elements include simple static charts with images (39.7 per cent of all cases) or with maps (31.3 per cent) as well as maps coupled with images (28.5 per cent).

The interactive features of DJA-nominated projects

Elements that allow users to interact with the data presented[8] are often discussed as another 'key characteristic' of data journalism (e.g. Coddington 2015; Gray, Bounegru and Chambers 2012, 15). In our sample 15.1 per cent of cases offer no data-related interactive functions at all (Table 6.8). However, the average piece contains 1.67 different interactivity features and only one award-winning project provides no interactive feature at all. This leads us to speculate that interactivity is considered a quality criterion, although the overall results are more in line with Tabary, Provost and Trottier's (2016, 67) finding that 'data journalists focus on finding good quality data but engage very little with ... interaction or reader participation' and often only 'integrate minimum formal interactivity'.

The interactive features most often integrated into DDJ articles are zoom functions for maps, details on demand (e.g. the number of victims in each case of reported school shootings) and filtering functions which allow the user to filter the provided data with respect to different variables (e.g. to select voting results from only one state or one year). Personalization tools – where the user enters personal data like their ZIP code or age to tailor the piece with customized data – are less common (17.3 per cent of cases). Only six projects in the three-year sample include an opportunity for a gamified interaction (e.g. *Heart Saver*, a game in which the user must send ambulances as fast as possible to fictional characters having a heart attack). Looking at developments over the years, we find that the share of interactive features dropped in 2014 but rose again in 2015, which suggests that data journalism might be becoming more interactive again.

---

[7] $\chi 2 = 16.207$; $df = 2$; $p < 0.001$; Kruskal–Wallis test because of heteroscedasticity (Levene test). Games–Howell test revealed significant differences between: 2013 and 2015 ($p < 0.001$), 2014 and 2015 ($p < 0.05$).

[8] Features for follow-up communication that are often called interactive features as well, for example, comment sections, fall into a different category ('opportunities for communication'; see Table 6.2) which is not discussed in this paper.

Table 6.8 Interactive functions (multiple coding possible)

%	2013 (n = 56)	2014 (n = 64)	2015 (n = 59)	Not awarded (2013-2015) (n = 151)	Awarded (2013-2015) (n = 28)	Total (n = 179)
No interactive functions	12.5	23.4	8.5	17.2[I]	3.6[I,II]	15.1
Zoom/details on demand	57.1[III]	54.7[III]	78.0[III]	62.9	64.3	63.1
Filtering	53.6	50.0	66.1	56.3	57.1	56.4
Search	30.4	23.4	27.1	28.5	17.9	26.8
Personalization	23.2	14.1	15.3	15.2	28.6	17.3
Playful interaction	3.6	1.6	5.1	3.3	3.6	3.4

[I] Fisher's exact test: $p < 0.05$ (*one-sided*).
[II] One project: 'Reshaping New York'.
[III] $\chi^2 = 8.401$; $df = 2$; $p < 0.05$; Fisher's exact tests for pairwise comparisons with adjusted α-levels (Bonferroni–Holm correction) revealed only one significant difference between years 2014 and 2015 ($p < 0.01$).

Discussion: Retracing the development of DJA-nominated stories

This chapter has traced the development of the emerging reporting style of data-driven journalism through a content analysis of the pieces nominated for a *Data Journalism Awards* in the years 2013–15. We examined the actors producing DDJ, the topics they cover and, most importantly, the means they employ to do so (i.e. the structural elements and forms of presentation). The results show that data journalism is still evolving with ample space for flexibility: different types of data, analyses and visualization strategies are combined – or omitted – when they suit the story and its topic. This echoes Coddington's observation that data journalists subordinate the use of data 'to the professional journalistic value of narrative and the "story"' (2015, 339). However, the set of potential elements to be combined appears to be stable and finite as over the three years we did not have to add new categories or variables to our initial codebook developed in 2013 to make sense of novel components. Instead, the new reporting style is (still) firmly characterized by those features that cursory observations, literature reviews and actor studies have already hinted at.

Despite DDJ's general flexibility, we did find some 'typical combinations' reoccurring over the years. For instance, political stories are based significantly more often on polls and surveys than pieces on other subjects while business and economy topics are correlated with financial information, societal issues are covered using socio-demographics and geo-data, and health and science reports draw on measured values.

In terms of key players, data journalism, at least in our sample of high-profile projects, continues to be dominated by legacy print media and their online departments. The only other major players are investigative journalism organizations like *ProPublica* or the *ICIJ*. Stories produced by such investigative organizations and by private or public broadcasters appear relatively more often among the award winners than among only nominated projects. In contrast, projects by print magazines and news agencies, so far, have not been awarded at all.

The growing variety of the nominees' home countries over the three years suggests that data journalism is spreading around the globe. However, projects from the United States and, to some extent, from the UK consistently make up the largest proportion of the nominees. Moreover, we found that stories increasingly build on data gathered on an international scale and that they are more often published in two or more languages (one of them is usually English). On this basis, data journalism seems to have the potential to foster the internationalization of journalistic coverage and its distribution.

The average number of contributors to a data-driven piece has risen consistently while the share of projects involving partners from outside the newsroom has fallen. This suggests to us that the production of data journalism, at least at the level of high-quality pieces, is progressively personnel-intensive while the skills for it are increasingly being acquired within media organizations. This finding might go some way in explaining why only a few organizations have managed to be among the nominees every year.

In terms of news topics, data-driven reporting is characterized by an unchanged focus on political, societal and economic issues. Topics that have been relatively neglected so far – education, culture and sports – represent opportunities for expansion and innovation in the future.

Visualization, the storytelling element assumed to be most important in data-driven coverage, has maintained the same level of importance over the years while the average number of visual elements in a piece is still growing. However, there is an emphasis on rather simple types like maps and tables as well as on images with visual appeal but little relation to the actual data. This opens up further avenues for innovation and distinction in the field. The findings are similar for interactive features: over the three years, they remain restricted to zooming into maps, showing details on demand or filtering data by predetermined categories, while more sophisticated or gamified applications are rare.

Half of the projects in our sample assumed a watchdog role, containing elements of criticism or even calls for public intervention. This stance of holding power to account through monitoring the decisions and activities of politicians, corporations and other socially important actors appears to be strengthened, as indicated by the growing shares over the three-year study period of stories using more than one type of data (e.g. financial and socio-demographic data) or combining data from different sources (e.g. official institutions and NGOs). There was also some evidence that data journalists are looking more to other data sources besides official, openly accessible ones. However, the watchdog function could be fostered more by contrasting data from different social domains (e.g. pitching numbers indicating worsening school achievements against historical governmental spending on education) or by analysing data from their differing perspectives (e.g. looking at rising energy costs from both a business and an environmental perspective).

Our findings illustrate how much data-driven journalism with a certain critical or watchdog attitude is appreciated by the DJA committee. Yet, these pieces are, more often than not, based on publicly available data that does not even need to be investigated or 'uncovered' as such. Investigative approaches could be furthered by requesting data from institutions (e.g. through Freedom of Information requests) more often or collecting data oneself. This is especially essential because there is a branch of data public relations, or 'data-spin', that tries to influence coverage. As Nguyen and Lugo-Ocando (2016, 4–5) observe: 'Today, … all major social, economic and political institutions have integrated numbers as a central part of their public communication … strategies. [And] some sources … make efforts to hide data that might work against their interest and reputation.' One can only imagine the potential of an investigative and critical data journalism to expose exploitation, corruption and the failures of power as some projects by the *Guardian,* the *New York Times,* and *ProPublica* among others have already demonstrated. Above that, public communicators often underpin their claims with data that are 'inappropriately produced or improperly interpreted for all sorts of benign or malicious, objective or subjective reasons' (ibid., 4). That is why critical data-driven reporting and journalists capable of statistical reasoning are needed to check claims and the statistics they are based on so that audiences 'can function effectively, either as citizens or consumers, in their increasingly data-driven daily environment' (ibid., 5).

Having said all this, we must, again, point to the double bias of our sample we have already discussed in the methodologies section: first, the analysed pieces are based on self-selection, and, second, they are likely not to represent 'everyday' data journalism. Moreover, while some of our findings could be interpreted as suggesting that data journalism is becoming more complex, we should bear in mind that the opposite might also be true: the 'everyday' data-driven piece is increasingly easy to produce as more tools become available for journalists to get started (e.g. Datawrapper at https://datawrapper.de/). Nonetheless, we can assume that the analysed cases, as DJA nominations, fulfil a certain quality threshold and are considered best practice examples in the field. As such, they are likely to influence the shape of data journalism in the years to come.

References

Anderson, C. W. (2013), 'Towards a sociology of computational and algorithmic journalism.' *New Media & Society*, 15(7), 1005–21.

Appelgren, E. and G. Nygren (2014), 'Data journalism in Sweden. Introducing new methods and genres of journalism into "old" organizations.' *Digital Journalism*, 2(3), 394–405.

Baquet, Dean (2016), A note to staff from Dean Baquet. *The New York Times*. Retrieved 6 March 2017 from: http://www.nytco.com/a-note-to-the-staff-from-dean-baquet/

Borges-Rey, E. (2016), 'Unravelling data journalism. A study of data journalism practice in British newsrooms.' *Journalism Practice*, 10(7), 833–43.

Boyles, J. L. and E. Meyer (2016), 'Letting the data speak. Role perceptions of data journalists in fostering democratic conversation.' *Digital Journalism*, 4(7), 944–54.

Coddington, M. (2015), 'Clarifying journalism's quantitative turn. A typology for evaluating data journalism, computational journalism, and computer-assisted reporting.' *Digital Journalism*, 3(3), 331–48.

De Maeyer, J., M. Libert, D. Domingo, F. Heinderyckx and F. Le Cam (2015), 'Waiting for data journalism. A qualitative assessment of the anecdotal take-up of data journalism in French-speaking Belgium.' *Digital Journalism*, 3(3), 432–46.

Fink, K. and CW Anderson (2015), 'Data journalism in the United States. Beyond the 'usual suspects.' *Journalism Studies*, 6(4), 467–81.

Gray, J., L. Bounegru and Chambers L. (eds) (2012), *The Data Journalism Handbook. How Journalists can use Data to Improve the News (Early Release)*. Sebastopol: O'Reilly.

Holtermann, H. (2011), 'Datenjournalismus: eine neue Form der journalistischen Wertschöpfung aus Daten [Data journalism: a new form of journalistically creating value from data].' Master Thesis, University of Hamburg.

Jenkins, J. and Y. Volz (2016), 'Players and contestation mechanisms in the journalism field. A historical analysis of journalism awards, 1960s to 2000s.' *Journalism Studies*, online first, 15 November 2016, DOI:10.1080/1461670X.2016.1249008.

Karlsen, J. and E. Stavelin (2014), 'Computational journalism in Norwegian newsrooms.' *Journalism Practice*, 8(1), 34–48.

Knight, M. (2015), 'Data journalism in the UK: A preliminary analysis of form and content.' *Journal of Media Practice*, 16(1), 55–72.

Krippendorff. K. (2013), *Content Analysis: An Introduction to Its Methodology.* Los Angeles: SAGE.

Lanosga, G. (2014), 'New views of investigative reporting in the twentieth century.' *American Journalism*, 31(4), 490–506.

Nguyen, A. and J. Lugo-Ocando (2016), 'The state of data and statistics in journalism and journalism education – issues and debates.' *Journalism*, 17(1), 3–17.

Parasie, S. (2015), 'Data-driven revelation? Epistemological tensions in investigative journalism in the age of "big data".' *Digital Journalism*, 3(3), 364–80.

Parasie, S. and E. Dagiral (2013), 'Data-driven journalism and the public good. 'Computer-assisted-reporters' and "programmer-journalists" in Chicago.' *New Media & Society*, 15(6), 853–71.

Royal, C. and D. Blasingame (2015), 'Data journalism: an explication.' *#ISOJ*, 5(1), 24–46.

Shoemaker, Pamela and Vos, Timothy. (2009), *Gatekeeping Theory*. New York, NY: Routledge.

Tabary, C., Provost, A. M. and A. Trottier (2016), 'Data journalism's actors, practices and skills: A case study from Quebec.' *Journalism: Theory, Practice, and Criticism*, 17(1), 66–84.

Tandoc, E. C. and S. K. Oh (2015), 'Small departures, big continuities? Norms, values, and routines in *The Guardian*'s big data journalism.' *Journalism Studies*, online first, 5 November 2015, DOI:10.1080/1461670X.2015.1104260.

Tow Centre for Digital Journalism. (2014), Tow Research Conference, 'Quantifying Journalism', Panel 1: Beyond Clickbait.[Video file]. Retrieve from: https://www.youtube.com/watch?v=3EqiEmSny-8

Wahl-Jorgensen, K. (2013), 'The strategic ritual of emotionality: A case study of Pulitzer Prize-winning articles.' *Journalism: Theory, Practice, and Criticism*, 14(1), 129–45.

Weinacht, S. and R. Spiller (2014), 'Datenjournalismus in Deutschland. Eine explorative Untersuchung zu Rollenbildern von Datenjournalisten [Data-journalism in Germany. An exploratory study on the role conceptions of data-journalists].' *Publizistik*, 59(4), 411–33.

7

Numbers *Behind* the News: Audience Metrics and the Changing Nature of Gatekeeping

An Nguyen, *Bournemouth University, UK*
Hong Tien Vu, *University of Kansas, USA*

Introduction

As the news media evolve along the datafication of society, they have seen not only a rigorous move towards data-driven journalism but also an unprecedented intrusion of audience-tracking data (called web metrics or audience analytics) into news production and distribution. As the move and, to some extent, the mood of news users in the digital sphere can be effectively and efficiently tracked and quantified in real time, journalists today are equipped with an ever-powerful set of tools to inform themselves of what audiences want and do with the news, thereby to make decisions to optimize the way they select, produce and deliver the news. This represents probably the most profound transformation of the gatekeeping process in recent times, bringing both hopes and fears for the future of news and journalism. This chapter critically reviews the penetration of web metrics in the past decade or so in order to (a) assess their real and potential impact on newsroom processes and relationships and (b) identify some good principles and practices that might help journalism and journalists to make the most from such data.

We will begin with a brief discussion of the relatively powerless position of the audience in traditional gatekeeping. Against this backdrop, we explain how and why web metrics have penetrated into the newsroom, as well as assess how they have impacted on both the gatekeeping process and gatekeepers' service orientation, work ethos and autonomy. As news moves from being exclusively 'what newspapermen make it' to also something that the crowd wants it to be, newsroom processes and relationships are being transformed in ways that invite more misgivings and reservations than hopes and innovations. This is because of an entire new set of metrics-induced challenges that journalism and journalists have to face today, including (a) the real risk of journalism being further tabloidized and 'dumbed down', (b) the growing unhealthy addiction to futile numbers such as clicks and views and (c) the pervasive frustrations, tensions and conflicts that result from a constant chase for such numbers. Observing developments in the past few years, however, we conclude with a more positive and optimistic note: the power of metrics can be well harnessed for a better and more sustainable future for journalism, if newsrooms devise and enforce the right audience data philosophies and policies. Such policies must go beyond crude head-counting measures to focus more on variables that help to get into deeper layers of news interest, consumption and engagement. In order for that to happen, a strong professional culture that fosters journalists' public service, rather than market service, ethos as well as the autonomy they need to exercise their own news judgement for the sake of that professional ethos. Some promising examples and practices in using metrics will be introduced along the discussion.

The powerless news audience in traditional gatekeeping

Since David Manning White (1950) first brought Kurt Lewin's gatekeeping metaphor to journalism scholarship, news institutions have gone through numerous social, economic, political and technological transformations. But the professional nature of the gatekeeping process – understood here to include not just news selection (as in White's original use of the term) but also the 'writing, editing, positioning, scheduling, repeating, and otherwise massaging information to become news' (Shoemaker, Vos and Reese 2008, 73) – has barely changed. Scholars (Shoemaker and Reese 1996; Shoemaker and Vos 2009) have delineated a five-level model of micro and macro influences on gatekeepers. The *first level* – the lowest – focuses on the individual factors of the communicator (e.g. personal background, experiences, attitudes and beliefs). The *second level* is media routines (e.g. audience orientation, newsroom norms and formats such as the inverted pyramid). At the *third level* are the organizational influences (e.g. internal structure, ownership, goal and policy). Extra-media forces, or social institutional factors extrinsic to the media (e.g. sources, advertisers, audience, government control, market competition and technology), constitute the *fourth level*. The *fifth – and highest – level* is the social system, particularly ideology and culture. Shoemaker and Reese (1996) posited that although each level in the model has its own range of influences, the lower is subsumed by the higher ones, which must take the lower ones into account.

As can be seen, gatekeeping is by and large an arbitrary process that bears the direct impact of all but one key stakeholder: the audience, which is subsumed under the broad concept of media routines. This might be surprising for many outsiders: how on earth can the people who claim to be a profession administering a service in the public's interest stay so aloof from their main client, the public? To be fair, when asked to think about what news is, many journalists do place the audience squarely in their definition – for example, 'news is anything that makes a reader say "Gee Whiz!"' (William Randolph Heart), 'news is anything that makes people talk' (Charles Dana), or news is 'that which has been previously unknown to the reader, that which surprises him, informs him, titillates him' (Australian journalist in Baker 1980, 138). In addition, some of the most important news values that journalists share – such as relevance, proximity and impact – are implicitly developed and identified from a user-centric perspective.

The problem, however, is that such audience orientation is in practice based primarily on gut instincts. News is identified, selected and produced from the journalist's top-down vantage point, on the assumption that what interests them will interest their audience, rather than any empirical evidence of audience wants, needs, behaviours and attitudes. Audience research, in most of journalism's history, does not 'come often enough to help' news people to bring an audience dimension into gatekeeping (Shoemaker and Reese 1996, 105). Whenever results of expensive audience research are available, they reach people at managerial levels rather than individual journalists, who simply do not care and 'tend to be highly sceptical of claims made on the basis of market research' (Allan 2010, 123). As such, what people do with the news, why they do it and how they evaluate it – have over time become weightless in the journalist's news decision (Allan 2010; Gieber 1960; Green 1999; Schlesinger 1987). Even worse, many journalists do not seem to be bothered. 'I know we have twenty million viewers but I don't know who they are,' one American journalist told Gans (1980). 'I don't know what the audience wants, and I don't care.' Meanwhile, the minimal direct audience feedback – in the form of letters to editors and reporters – is often dismissed as 'insane and crazy' crap (Wahl-Jorgensen 2007) 'from cranks, the unstable, the hysterical and the sick' (Gans 1980).

Consequently, research has found a substantial gap between what journalists offer and what people need and/or want. More than half a century ago, Gieber (1960, 124) found that 'news selection has no direct relationship to the wants of readers'. More recently, Boczkowski and Peer (2011), comparing placement of stories on news sites with most-viewed articles on these web pages, found that journalists' and audience members' choices of news do not intersect. Wendelin, Engelmann and Neubarth (2017) compared German news output and its consumer preferences to discover the same thing: although journalists and audiences had much to share in terms of their perceived news values, there was quite a choice gap in news topics (more on this below). Chyi and Lee (2013) found only a third of the news that Americans receive from their media was perceived as noteworthy – that is, 'relevant or interesting' – to them. This could help to explain the long declining trend of news consumption because, as the authors further discovered, the noteworthiness of news was a strong and consistent predictor of how much people enjoy reading/watching the news, how much time they spend on it, how willing they are to pay for news online and in print. Chyi and Lee (2013, 11) calls for an audience-driven approach to the economic value of news:

> To be economically viable …, the value of news should be conceived from the audiences' perspective. Notwithstanding geographic boundaries – as long as journalism needs an audience, it will only be financially successful when the gap between what is considered newsworthy to news producers and what is considered noteworthy to audiences is narrowed.

In the past decade or so, such calls have indeed been listened and acted upon in a serious and rigorous way, as a response by the news industry and profession to declining readership, tough competition, fragmented audiences, plummeting advertising revenues, downsizing newsrooms and, above all, the power of audience-monitoring technologies (Lowrey and Woo 2010; McKenzie et al. 2011; Nguyen 2013; Shoemaker et al. 2010; Singer 2011).

The penetration of web metrics into the newsroom

In revising the traditional gatekeeping model, Shoemaker and Vos (2009) call scholars to look and go beyond the two traditional primary channels of gatekeeping – journalists (media channel) and sources (source channel) – to pay for more attention to the increasing influence of the 'audience channel'. Information from the audience channel, in this revised model, comes in the form of various non-purposive audience feedback that can be easily collected, processed, analysed and quantified. In the digital world, technological constraints – the major excuse for journalists to turn their blind eyes to client's wants and needs (Schlesinger 1987) – are no longer an obstacle.

On the one hand, the interactive and immediate digital environment enables, empowers and encourages journalists to engage in direct communication with news consumers. Boczkowski (2004, 183) found from an early ethnographic study of online journalism that digitizing the news is somewhat equal to moving from 'being mostly journalist-centered, communicated as a monologue ... to also being increasingly audience-centered, part of multiple conversations'. Under the increasing pressure of driving more traffic to their sites, professional gatekeepers now are more willing to give up their autonomy by passing off some of their tasks to audience members (Chung and Yoo 2008; Singer 2011). These include allowing audiences to personalize settings – such as creating profiles, tracking stories of their interests, or changing layouts of the websites – and enabling readers to leave feedback and engage in news production and distribution – through, for example, emailing journalists, commenting on stories generating news on their own news, sharing stories on social media and so on. In the digital world, as Singer (2011, 4) argued, for any news story to emerge from the vast pool of articles, 'the gatekeeping role must necessarily be shared far more broadly than in a traditional media environment'.

On the other hand, and of exclusive focus in the rest of this chapter, is the many opportunities for journalists to understand and quantify their audiences in the concrete form of web metrics. As each and every user's IP address and mouse click can be easily and constantly tracked, recorded, stored, aggregated and fed into newsrooms, the resulting data are quite natural and reliable. Many shortfalls of traditional audience measurement methods – such as the use of unrepresentative panels to extrapolate to general audiences, or the inability of television people-meters to distinguish between a turned-on and actually watched TV – seem to be eliminated in the digital environment. As such, the ability to track the move and the mood of audiences has been hailed as one of the greatest advantages of online journalism since its inception in the 1990s. But it was not until the past decade that metrics became influential at the forefront of newsroom operations, thanks substantially to the increasing versatility, efficiency and effectiveness of user-friendly audience-tracking technologies. Where it is third-party (e.g. Chartbeat, Omniture, Visual Revenue, Google Analytics) or in-house tracking software (e.g. Ophan at the *Guardian*, Lantern at the *Financial Times*, Stela at the *New York Times*), these tools carry with them two major powerful features.

First, they can collect and deliver real-time audience data with a high accuracy. Chartbeat, whose clients spread over thirty-five countries, including 80 per cent of the most trafficked online publishers in the United States (Petre 2015), markets itself as the tool for frontline newsroom teams to 'track the second-by-second, pixel-by-pixel attention of your audience, wherever they are' (as of May 2016).

Second, such tools can collect, analyse and provide a much more diverse range of data, usually hundreds of variables, about audience attributes, offering rich insights into what audiences do (and do not do) before, during and after visiting a specific news site (see Nguyen 2013 for a typology of metrics). For editors and reporters, such data provide a strong and somewhat palpable sense of – among other things – who are interested in what story/topic/section, how they attend to such content, when they do so, where they are from, and, to some extent, why they do so. Some software allows editors to do other things such as experimenting and testing different headlines for the same story. For those on the business side of the newsroom, metrics form the currency of online news, being sold to advertisers, either as individual indicators or as composite indices of overall performance (such as 'audience engagement' or 'audience growth'). Some tracking software, for better or worse, can use real-time data to pin down to the pennies the advertising income that a particular story generates, based on the number of clicks on advertisements on the page.

As such, metrics have penetrated into the newsroom and its gatekeeping process on a large scale and at an unprecedented rate. As *metrics have been institutionalized in most leading newsrooms, it* is no longer a novelty for editors to begin news meetings with a rundown of audience data. At some places, emails are sent every day to all staff, with dozens of performance numbers for each and every story on the day. The physical space of some newsrooms has also been reconfigured so that audience development teams work in tandem with the editorial body. The latest change to the editorial structure at the *Dallas Morning News*, for example, is the addition of three teams – audience, breaking news and photo/video – which are positioned alongside each other at the centre of the newsroom to reflect its shift to a strategic focus on online publishing (World Association of Newspapers and News Publishers 2016). Others erect fancy panels of data and graphics on the walls so that reporters can 'crunch the numbers' in real time to remain atop their individual and collective performance throughout the day. Hung over the reception desk on the editorial floor of the now-defunct Gawker Media, for example, were two big data panels: the Big Board featuring top posts (those with the most concurrent visitors) and the Leader Board top writers (those with the most unique visitors in the previous thirty days), with red/green arrows showing their relative position change over that period (Petre 2015).

In quantitative terms, Newman (2016) found from a survey that three quarters of CEOs, news editors and digital strategists across twenty-five countries saw better use of web metrics to understand and serve audiences a critically important part of their future. A survey we conducted with 318 US editors in 2013 found 84 per cent of online editors monitored traffic on a regular basis, with more than half (52 per cent) doing so on at least once a day.[1] Of the seven potential editorial adjustments listed in the questionnaire, 70 per cent were likely or very likely to increase the prominence of traffic-generating stories on the homepage of the site or the front page of the newspaper. This was followed by finding additional multimedia elements for the most-viewed/most-read (61 per cent), finding possible follow-ups for the most-viewed/most-read (60 per cent), updating most-viewed/most-read articles (56 per cent), running stories similar to the most-viewed/most-read, finding possible editorials for the most-viewed/most-read and, at the bottom, reducing the prominence of low-hit articles.

Most studies that examine the influence of metrics on actual content have also found evidence of that influence. A time-lagged analysis of three US news sites by Lee, Lewis and Powers (2014) found that news story placement was continually adjusted in parallel with audience clicks, especially as the news day goes on, although the inverse effect (news positioning on clicking) is minimal. Bright and Nichols (2014) analysed a six-week sample of 40,000 stories from five leading UK news sites to find that most-read ones were 25 per cent less likely to be removed from the front page in the short term. This 'turn toward populism', which they found in both entertainment and politics news, was more remarkable among the quality than the popular news outlets in the sample. Karlsson and Clerwall (2013) found from a Swedish sample of broadsheet, tabloid and public-service news output that metrics play a significant part in editorial judgement, especially among commercial media, although journalists try hard to strike

Table 7.1 Prevalence of different types of metrics-driven editorial adjustment (% of US editors who reportedly monitored traffic, n = 247)

'Likely' or 'very likely' to …	
… make articles that drive more traffic more prominent on the homepage or on the front page.	70
… look for possible additional elements (video, pictures, sounds etc.) for most-viewed/most-read articles.	61
… look for possible follow-up articles for the most-viewed/most-read ones.	60
… try to update most-viewed/most-read articles more often to attract audiences.	56
… run articles of the same kind as the most-viewed/most-read ones.	45
… try to look for possible editorials for most-viewed/most-read articles.	33
… make articles with low hit less prominent.	26

[1]   See Vu (2014) for the detailed methodological design of this survey.

a balance between professional autonomy and audience demands. Similarly, Welbers et al. (2015) studied five national Dutch newspapers during January–July 2013 to find that, although their interviewed journalists 'denied or strongly nuanced' the impact of clicks on editorial decisions, such impact was clear and significant in both print and online versions. The only exception that we find is Zamith (2016), which content-analysed fourteen US news sites over a two-month period and found limited evidence that a news item's popularity leads it to more prominent areas on the homepage or increases its risks being removed from such areas.

Against the historical backdrop of journalists' 'deliberate, technologically enabled ignorance' of audience by journalists (Anderson 2011, 553), such recent moves from gut feelings to web metrics – or the 'rationalisation of audience understanding' as Napoli (2010) calls it – represents quite a dramatic, radical transformation in the way journalists perceive and relate themselves to audiences. This transformation might have been driven by many factors, but the most oft-quoted driver is economic necessity. Lowrey and Woo (2010) found that uncertainties in newsrooms caused by recent financial woes were a key reason for editors to use metrics more often to compare and contrast their editorial decisions with consumer preferences. Bright and Nicholls (2014) observes the same thing: 'At a time when the business model of media outlets as a whole is under great strain, and many formerly profitable outlets are facing difficulty or closure, the impetus to "follow" traffic may be considerable.'

Editors, however, probably due to their professional pride, tend to downplay the role of commercial factors. In the above survey with US editors, we asked an open-ended question about the specific reasons for them to monitor metrics. Their responses were coded and classified into five distinctive categories:

- 58 per cent for *audience scrutiny* (e.g. 'to see what people are looking at and what is keeping them on our site,' 'to gauge the stories that interest and involve readers,' 'to see what kinds of stories, videos, databases draw traffic and readers' preferences');
- 31 per cent for *content planning/adjustment* (e.g. 'to help in constantly adjusting mix and display on site,' 'to help determine coverage,' 'to allocate our resources,' 'to get an idea of how to play future stories,' 'to pursue follow-ups that have high readership,' 'to decide what kinds of stories to cover');
- 6 per cent to *follow corporate agenda* (e.g. 'corporate wants us to,' 'to please corporate');
- 3 per cent for *advertising and/or marketing purpose* (e.g. 'to compile data for advertising'); and
- 2 per cent for *audience resonation* (e.g. 'just try to get all the feedback I can to make sure we're doing what we should,' 'to make sure that what we assign resonates with readers').

As can be seen, economic incentives were barely acknowledged. However, further hierarchal regression analysis of the data show a notable influence of economics on

Table 7.2 Hierarchal regression for monitoring traffic and making editorial changes actions on editors' demographics, publication circulation and perceived economic benefits of attracting more audience

	Traffic monitoring		Editorial changes	
	Model 1	Model 2	Model 3	Model 4
Block 1				
Age	0.08	0.07	−0.04	−0.05
Journalism training	−0.14*	−0.15*	−0.03	−0.05
Newspaper circulation	0.00	0.01	0.06	0.04
Block 2				
Economic benefits		0.03		0.23***
R^2	0.03	0.04	0.01	0.06***

- Model 1 (first block): Editors' demographics and circulation as predictors of traffic monitoring.
- Model 2 (first block): Editors' demographics and circulation and perceived economic benefits as predictors of traffic monitoring.
- Model 3 (second block): Editors' demographics, circulation and perceived economic benefits as predictors of traffic monitoring.
- Model 4 (second block): Editors' demographics, circulation and perceived economic benefits as predictors of editorial changes
- $p < 0.05$, *** $p < 0.001$
- Perceived economic benefits were measured by agreement on a seven-point scale statement with 'Getting more readers is necessary because more readers mean high revenues' (one = strongly disagree and seven = strongly agree).
Source: Adapted from Vu (2014).

the use of audience metrics: while perceived economic benefits did not seem to drive the editors to monitor web metrics, they did have a statistically significant effect on the willingness to make metrics-based editorial adjustments (see Table 7.2). This is in line with other research: Tandoc (2014), for example, found from his structural equation model of survey data with 206 American editors two key factors driving how they use analytics: perceived competition and perception of audiences as capital. Such influence of economic factors, especially the 'market logic', on news decision-making has been a primary cause of concern for academic and industry observers. Below we will discuss this, alongside other risks posed by metrics, before turning our attention to how to minimize the potential negatives of the rising 'audience agenda' on gatekeeping.

The many risks of metrics-driven audience agenda

With the growing ubiquity of metrics and their ensuing new 'audience agenda' have come some misgivings about the future of gatekeeping. Although metrics' influences on journalism and journalists still evolve, three more or less immediate risks have been observed:

First, there is the high possibility that metrics be misunderstood and misused in the newsroom. This is because journalism not only has little knowledge about the socio-psychological dynamics of news audiences but also, as many of the chapters in this book show, has long been known for its systemic hostility to and incompetence in using data and statistics. The speedy penetration of metrics into the often slow-to-adapt newsroom represents a fundamental transformation out of necessity, but not always with good preparation. As Nguyen (2013) argues, whether they love or loathe statistics and whether they want to understand or ignore audiences, journalists will have to accept a constant exposure to audience metrics in their daily job and to develop a click-thinking routine and culture among them. But with their traditional knowledge and skill deficits about both audiences and statistics, could they ever put metrics into good use? Matthew Ingram (2013) puts it eloquently:

> Media companies aren't trying to bring back something they already had by using analytics – it's more like they were remote villagers hidden in the rain forest who had never seen a ruler or a scale for measuring weight before, and suddenly when the web came along they were handed these tools and didn't really know what to do with them. So naturally, they ran around measuring the length and height and weight of everything in sight, without really knowing why.

The risk becomes more critical when considered in light of the contrast between the diverse, complex nature of web metrics and the intense, time-scarce newsroom. As Debrouwere (2013), using Google Analytics as an example, observes: 'It's a smorgasbord of numbers. You've got your demographics, your page views, referrers, time on site, time on page and frankly not much time before your head explodes trying to figure out what everything means.' All this can lead journalists to falling into the trap of faithfully and uncritically following the sentiment of the crowd that metrics carry (Nguyen 2013).

Second, such unwise uses of metrics could shift journalism towards a direction in which news is data-driven rather than data-informed, causing it to be dumbed down instead of being educational and informational (Petre 2015; Tandoc and Ferrucci 2017). The presence of technologists and audience development teams along journalists in the newsroom has been construed as paving the way to knock off the wall between the commercial and editorial sides in news organizations (Sullivan 2016). If journalists let economic considerations wade too deeply into their autonomous judgement under the guidance of metrics, they might end up doing nothing but to attract the largest possible audience attention. But the kind of news that can maximize audiences is often the so-called 'news you can use' – news that caters to the lowest common denominator of all tastes, addressing the most basic, least sophisticated and least sensitive level of lifestyles and attitudes. In practice, it often means soft news with high entertainment and low information values, not the sort of hard news about public affairs that people need to function well in democratic societies (Bird 2010; Boczkowski 2010; McManus 1992). In the aforementioned analysis of German news sites, for instance, Wendelin, Engelmann and Neubarth

(2017) found that the focus on providing public affairs content – such as political organization, economy and culture – did not pay off in terms of audiences: users paid far more attention to less serious (and less supplied) stories, such as sports and everyday services.

In other words, if metrics were to reign the increasingly intense and deadline-driven newsroom, journalists would have to think about what people *want* to consume and can consume *at ease*, rather than what they *need* to consume and must consume *with effort* to become informed and self-governed citizens. That would translate into an intensification of an already perennial problem of journalism: the tabloidization of news, which encompasses a range of unhealthy practices such as, among other things, 'the sensationalisation of news, the abbreviation of news stories, the proliferation of celebrity gossip, and the more intensive visual material such as large photographs and illustrations' (Rowe 2010, 351). Ubiquitous metrics, as Tandoc and Thomas (2015, 249) observe, bolster 'a media ecosystem that panders to, rather than enlightens and challenges its audiences'.[2]

Third, a click-thinking culture can add to the already intense settings of news production an entirely new set of moral and psychological tensions and conflicts. Such tensions and conflicts, which are real and pervasive in many newsrooms, come primarily from the uneasy negotiation between journalism's control over the profession's boundaries and the power of 'external factors' such as audience interests and demands. This third set of risks can be divided into three key issues.

One, there are legitimate concerns that web metrics have been used in a growing number of newsrooms (e.g. America Online, Bloomberg, Forbes) as the basis to evaluate staff performance, calculate story royalties, determine bonuses and/or set development targets. Some – including incumbents such as the *Washington Post* – have reportedly downsized news teams producing low traffic to reallocate resources to more popular content areas. If this continues on a large scale across the news sphere, what would be the future of serious, often less popular, areas of news and current affairs? Of course, not all newsrooms have opted for this model: *Huffington Post*, a 'digital native' whose success has been built largely on intensive and extensive use of metrics, does not run a metrics-based payment and staff evaluation system because, in the words its former managing editor, Jimmy Soni, 'tracking someone to a number … seems to suck the soul out of that creative process'. But what, if metrics-based payment and evaluation systems, which are in essence a newsroom discipline mechanism, became more common in the future (and, we must stress, the likelihood of that is not low at all)? The idea of journalists striving and competing for audience attention just to gain monetary rewards, rather than to fulfil their public duties, is quite scary, with rather unpredictable chilling prospects.

---

[2] It should be noted that from an economic perspective, soft news has another appeal to the industry: it is often much less expensive to produce than hard news. This creates a 'perfect combination' for those on the business side: it maximizes the output (audience attention) while minimizing the production cost at the same time. For an ailing news industry, that combination is certainly a rather appealing force.

Two, occupational stresses have come to a new height and on a more permanent basis. 'At a paper, your only real stress point is in the evening when you're actually sitting there on deadline, trying to file,' said Jim VandeHei, the executive editor of Politico.com. 'Now at any point in the day starting at five in the morning, there can be that same level of intensity and pressure to get something out' (quoted in Peters 2010a). That is not to mention that, as a result of the above performance evaluation system, competitiveness has become the name of the game in some newsrooms: almost all the former Gawker staffers interviewed by Petre (2015), for instance, saw this as the single most important personal quality to survive and thrive in their company. It is not surprising, therefore, to see journalists at metrics-driven newsrooms have often quit jobs or even changed careers for being unable to stand the constant pressure of producing news to the chart (Peters 2010a).[3]

In fact, the 'rationalization of audience understanding' has morphed into a new emotionalization of the newsroom. Emotion is in itself an aim of some audience-tracking software. As one Chartbeat employee told Petre (2015): 'It's not the identity of the number (but) the feeling that the number produces ... that's important.' At the former Gawker Media, Petre observes that editorial work along constant ups and downs of Chartbeat figures can be an 'emotional roller coaster' and can be as 'addictive' as gaming or gambling. Some reporters and writers, in dealing with the unpredictability of traffic figures, attempt to produce and post stories as frequently as possible as a strategy to improve their chance to appear on one of the boards (just like lottery playing). Anyone trying to escape the tyranny of those big panels of data, in the words of a Gawker writer, is like 'a cocaine addict on vacation in Colombia'. Tandoc (2014, 9) quoted the managing editor with a similar observation: 'It's like crack. You can sit here and watch it popping all night.' Reflecting on this, Tandoc made a worthy observation: 'The metaphor is funny, but it also has deeper implications. The reason illegal drugs are outlawed is because a drugged person might pose danger to herself and to others as she loses control and becomes unable to function normally.'

For some journalists, the most painful impact of metrics might lie in the 'conscience crisis' that they experience. Research by Anderson (2011) and Tandoc (2014) in the United States, Boczkowski (2010) in Latin America and MacGregor (2007) in the United Kingdom has produced substantial evidence that dilemmatic situations arising from the tension between serving people with the news they need and the news they want are now commonplace. One Philly.com reporter, citing a thoroughly researched story about a local army firm that 'just bombed ... and did terribly' on the site, lamented: 'You want to throw fear into the heart of journalism professionals? That's a way' (quoted in Anderson 2011, 559). In the three newsrooms studied by Tandoc

---

[3] Petre (2015), however, observed that some Gawker writers and editors decided to leave only to return later – a pattern dubbed the 'Gawker boomerang' by Capital New York – because they are too used to the Gawker way of work that they could no longer fit into those newsrooms that are less obsessed by traffic figures.

(2014), striking a balance between the perceived professional duty and pride of a journalist and the constant urge to generate traffic with the so-called click-baits is a tricky thing to do. One editor told Tandoc that it has become a 'luxury' for him/her to think along the normative dichotomy between producing quality journalism and drawing the largest traffic 'because if the company's not making money, then I might get laid off ... (and) that's just the way it is' (12). 'Sometimes you have to hold your nose,' said another (Tandoc 2014, 12).

Beyond head counting: Harnessing the power of audience metrics

Where does all this lead us? This chapter is by no means a call for a dismissal of metrics in news work. Neither is it to encourage the continuation of journalists' traditional ignorance of audience in the digital era. Rather, it is a call for journalists to take to the challenges of metrics and integrate them into their editorial processes before it is too late. Almost indispensably, metrics will continue to be woven into news organizations as technological, commercial and editorial solutions, and a slow move by journalists could give those on the commercial and technological sides a dangerous advantage over the editorial. The good news is that, after a period of being dismissive or at least ambivalent about metrics, recent years have seen journalists more curious and interested in learning about them (Cherubini and Neilsen 2016). In that learning process, we suggest, the most important thing for journalists is to reflect on the past decade or so to reconfigure their relationship with audiences and develop an 'audience data philosophy' that guides their overall approach to metrics. Such a philosophy, in our views, should entail the following.

First, metrics are above all a positive addition to journalism: a direct, real-time access to such data, by nature, adds an unprecedented, healthy element that can work to the advantage of journalism, both as a profession and as a business. As discussed above, these natural data provide a considerable amount of accurate and reliable information for journalists and news executives to understand certain aspects of the audience and use that understanding to serve them in a more considered, and perhaps more scientific, manner. The enhanced presence of audiences in the newsroom is a move towards a more caring, more inclusive and more relevant journalism. Nikki Usher (2010) – a former journalist and now an academic – argues that audience tracking 'turns journalism from elitism of writing for itself and back to writing what people are actually looking for'. News, at the end of the day, is only relevant if it serves the purpose of informing and educating the public. It would fail or would not be even considered 'news' if, as Ross (2016, 1) bluntly puts, 'nobody gives a shit' about it.

Second, it is the professional ideal of journalism as a public service – as opposed to a market service – that plays the key role in preventing the negatives and promoting the positives of metrics (Nguyen 2013). If journalists prioritize the market service, they would risk using metrics uncritically and unwisely for the mere sake of economics, reducing the audience to no more than a homogenous set of mere consumers that can be turned into a soulless commodity to sell to advertisers. But if they think firmly of themselves as administers of a specialized and complicated public service to humanities, they would find ways both to contain the weaknesses of metrics and to exploit their strengths to the fullest extent. They would understand that their 'gut feelings' in deciding what's news and what ought to be news continue to be, as they have always been, essential in maintaining journalism as an indispensable component of democratic life. They would know their immediate job is not to dismiss metrics altogether but, in the word of the deputy editor of a US news site, is 'to sit down on the table and have an honest conversation about what the goals are' and to strike a balance between 'having the money and being a respected journalist organisation' (as quoted in Tandoc 2014, 13).

Third, the use of web metrics must be bolstered a professional ethos that breeds, fosters and protects journalists' autonomy in exercising their specialist knowledge, skills, values and standards. This autonomy would keep journalists in healthy distances from the crowd sentiment that metrics might instruct, even 'force', them to follow. In the survey of US editors, we found clear evidence that more professionally educated journalists – that is, those with a journalism degree – were less likely to let economic concerns to get in their use of metrics (Vu 2014). In the broader landscape, for journalism to make the most from web metrics, we need a professional culture in which journalists are educated and encouraged to take confidence and pride in, among other things, their own news judgement and are, if necessary, able to stand up for it against market or management forces. Indeed, tracking tool developers understand this very well. To appeal to the news industry, for example, Chartbeat 'expends considerable energy and effort' on designing a dashboard that not only communicates rigorous data but also 'must demonstrate deference for traditional journalistic values and judgment, … must be compelling, … must soften the blow of bad news, and finally … must facilitate optimism and the celebration of good news' (Petre 2015). Indeed, in their daily direct interactions with clients, Chartbeat staff make it a working principle to defer to journalistic authority.

What, then, could be the first steps for journalism to put metrics into good use? Here we agree with Cherubini and Nielsen (2016) who call on journalists to work hard to find ways to turn mere audience metrics into 'editorial analytics' – that is, to go beyond crude, generic and additive real-time data such as clicks, page views and visits to tailor the diverse range of metrics to strategic editorial priorities, goals and imperatives. It is worth remembering that despite their high reliability, crude measures such as clicks or page views might not always be valid measures of audience interests. Like the traditional TV viewer who tunes in the television news bulletin to pass the time rather than watching it, many people click on stories not necessarily to actively seek the relevant information but simply because, for example, they feel saddened or annoyed by a headline, or the links are available at a convenient/free time. Similarly, 'no click' does not always imply a lack of interest – maybe the user does not want to disrupt the reading of an ongoing story, or he/she has just too many choices to click on at the same time. Kormelink and Meijer (2017) found from observing digital news consumption behaviours of fifty-six users that there are at least thirty distinct cognitive/mental, affective/emotional and pragmatic/practical reasons for clicking or not clicking on news stories. As they conclude, while clicks are not meaningless, 'they just capture a limited range of users' interests or preferences' and therefore are 'a flawed instrument' to even seek a rough estimate of people's news interests (2017, 14).

Hence, although page views and clicks are important, they are only superficial data that need to be considered in relation to dozens, or hundreds, of available audience variables. Audience-monitoring strategies must evolve into a more complex process of tracking and making the most from different types of available audience data – such as how much time people spend on the page, how they interact with content, what devices they use, whether they only stumble upon the page or are loyal subscribers, where they come from, where they go after the visit, and so on. Combined in a wise and rigorous manner, under the right organizational audience data philosophy, analytics can inspire the invention and use of many cutting-edge practices and techniques in news production and distribution. In the short term, they can be used to optimize the effectiveness and impact of daily editorial work – though activities like headline testing and writing, story positioning, day-parting (publishing certain stories in accordance to the typical audiences across different times of the day) and platform-parting (choosing which platforms to publish certain stories or which styles should be used to adapt stories from one platform to another).[4] Over time, such composite use of audience metrics can help build long-term strategies such as tailoring content to specific audience groups, or offering the right content and platforms to reach and appeal to those who have not visited or are not loyal readers of news sites (Cherubini and Neilsen 2016; Zamith 2015; Trilling, Tolochko and Burscher 2016).

In fact, the past few years have seen such strategies being effectively adopted and institutionalized by an increasing number of major news organizations around the world (*e.g. the Wall Street Journal, the Huffington Post, National Public Radio, CNN, Guardian, Financial Times, Quartz*). Even a 'stubborn laggard' like the *New York Times*, which had been known for being 'publicly dismissive, even scornful, of the idea of using metrics to inform editorial processes' (Petre 2015, 9) until a year or two ago, has begun to see the positive impact, professional *and* economical, of metrics since it started to open up Stela, its in-house audience-tracking tool, across the entire newsroom in the autumn of 2015 (Byers 2016; Wang 2016). Within a short time under its new executive editor, Dean Baquet, the legacy newspaper went from restricting its reporters from accessing metrics to offering a 'desk-by-desk digital training regimen' and turning such data 'an integral part of our daily conversations' (as seen in one of Baquet's memos to staff in 2016). Although it is not easy to pinpoint how the use of 'editorial analytics' contributes to its recent success, the fact is that the *The Times* has seen its online paid readership surge at an impressive rate in the past year or so (100,000 new digital subscriptions in the fall quarter of 2016 alone).

---

[4] See Hanusch (2016) for a more detailed discussion of day-parting and platform-parting.

One common thing that we observe at the above news organizations is that *they often start the* institutionalization of metrics with the establishment of analytics teams to go beyond head counting and work in tandem with the editorial body *to debunk the myth of audience data* and *to identify opportunities.* At the *Financial Times*, which launched its own analytics tool, Lantern, in early 2016, its head of audience engagement, Renee Kaplan, said that his team's objective is to capture a more holistic picture of the readership and how audience members are interacting with FT's content beyond page views and clicks (Lichterman 2016). At the *New York Times,* its director of news analytics, James Robinson, told a Tow Centre conference that his job was not to use 'metrics for data' but to bring insights about readers into the newsroom so that journalists can better understand audiences in order to align editorial decision-making towards them (Tow Centre for Digital Journalism, 2014). That includes, according to him, helping journalists identify potential readers, their interests and reading habits so that they can work out how a story should be covered in order to be relevant to its target audience.[5] At the *Wall Street Journal*, its executive emerging media editor, Carla Zanoni, reports that her team is involved in every step from the conception and inception to promotion to post-publication revitalization of a story. Its ultimate role is to make sure that 'everyone in the newsroom understands the data and learns how to build a narrative from the numbers' (Cherubini and Neilsen 2016, 29).

If these examples suggest anything, it is this: a serious investment, coupled with a systematic structural change and the right philosophy about metrics, can help journalists to effectively 'datafy' news work and harness the power of 'big (audience) data' to prevent dwindling audiences and build new loyalty, without losing their traditional precious grip on gatekeeping. The future might be already now.

References

Allan, Stuart (2010), *News Culture.* 3rd edn. Maidenhead: Open University Press.

Anderson, C. W. (2011), 'Between creative and quantified audiences: Web metrics and changing patterns of newswork in local US newsrooms.' *Journalism: Theory, Practice and Criticism*, 12(5), 550–66.

Bird, Elizabeth (2010), 'News practices in everyday life: Beyond audience response.' In Stuart Allan (ed.), *The Routledge Companion to News and Journalism*, 417–27. London: Routledge.

Boczkowski, Pablo (2010), *News at Work: Imitation in an Age of Information Abundance.* Chicago: University of Chicago Press.

Boczkowski, Pablo (2004), *Digitizing the News: Innovation in Online Newspapers.* Cambridge, MA: The MIT Press.

Boczkowski, Pablo J. and Limor Peer (2011), 'The choice gap: The divergent online news preferences of journalists and consumers.' *Journal of Communication*, 61 (5), 857–76.

---

5 Watch this at https://www.youtube.com/watch?v=3EqiEmSny-8.

Bright, J. and Nicholls, T. (2014), 'The life and death of political news: Measuring the impact of the audience agenda using online data.' *Social Science Computer Review*, 32(2), 170–81.

Byers, D. (2016), *New York Times adds 41,000 subscriptions after Trump's election. CNNMoney*. Retrieved 3 December 2016 from https://tinyurl.com/hmzz5ep.

Calderone, M. (2016), 'New York Times eyes ambitious overhaul in quest for "journalistic dominance".' *The Huffington Post*. Retrieved 3 December 2016 from https://tinyurl.com/zyzyp6g.

Cherubini, F. and Neilsen, R. (2016), *Editorial Analytics: How News Media Are Developing and Using Audience Data and Metrics*. Oxford: Reuters Institute for the Study of Journalism. Retrieved 30 April 2016 from http://tinyurl.com/jqrvsu2.

Chung, Deborah S. and Chan Yun Yoo (2008), 'Audience motivations for using interactive features: Distinguishing use of different types of interactivity on an online newspaper.' *Mass Communication and Society*, 11(4), 375–97.

Chyi, H. I. and Lee, A. M. (2013), 'Online news consumption: A structural model linking preference, use, and paying intent.' *Digital Journalism*, 1(2), 194–211.

Debrouwere, S. (2013), 'Cargo cult analytics.' Available at https://tinyurl.com/ztdwwo3.

Feola, Christopher J. 'By the Numbers,' *The Quill*, January/February 1995, 16.

Gans, Herbert (1980), *Deciding What's News*. New York: Vintage Books.

Gieber, Walter. 'How the "gatekeepers" view local civil liberties news' (1960), *Journalism and Mass Communication Quarterly*, 37(2), 199–205.

Green, Kerry (1999), 'How newsroom failures limit readership gain.' *Australian Studies in Journalism*, 8, 18–36.

Hanusch, F. (2016), 'Web analytics and the functional differentiation of journalism cultures: Individual, organizational and platform-specific influences on newswork.' *Information, Communication and Society*. OnlineFirst article.

Howard, A. (2014), *The Art and Science of Data-Driven Journalism*. Tow Centre for Digital Journalism. Retrieved 12 February 2015 from http://tinyurl.com/jmxskoz.

Houston, Brant. 'When Numbers Talk, Journalists Help People Listen,' *Nieman Reports*, spring 1999, 51.

Ingram, M. (2013), 'Some advice for media: Just because you can measure something doesn't make it important.' *Gigaom*, August 29. Available at https://tinyurl.com/hbewwkp.

Jones, Stacy. 'Numbers expose truth: Demographer urges newspapers to take advantage of powerful analytical tool.' *Editor & Publisher*, 3 May 1997, 41.

Karlssons, M. and Clerwall, C. (2013), 'Negotiating professional judgement and clicks: Comparing tabloid, broadsheet and public service traditions in Sweden.' *Nordicom Review*, 34(2), 65–76.

Kogut, T. and Ritov, I. (2005) 'The "identified victim" effect: An identified group, or just a single individual?' *Journal of Behavioral Decision Making*, 18, 157–67.

Kormelink, T. G. and Maijer, I. C. (2017), 'What clicks really mean: Exploring digital user news practices.' *Journalism*. OnlineFirst article.

Lee, A. M., Lewis, S. and Powers, M. (2014), 'Audience clicks and news placement: A study of time-lagged influence in online journalism.' *Communication Research*, 41(4), 505–30.

Lichterman, Joseph (2016), *The FT is launching a new analytics tool to make metrics more understandable for its newsroom. Nieman Lab*. Retrieved 4 December 2016 from https://tinyurl.com/jdc3kmu.

Lowrey, Wilson and Woo, Chan Wan. (2010), 'The news organization in uncertain times: Business or institution?' *Journalism and Mass Communication Quarterly* 87(1), 41–61.

MacGregor, Phil (2007), 'Tracking the online audience.' *Journalism Studies*, 8(2), 280–98.

Maier, Scott R. (2003), 'Newsroom numeracy: A case study of mathematical competence and confidence.' *Journalism and Mass Communication Quarterly*, 80(4), 921–36.

Maier, Scott R. (2005), 'Accuracy matters: A cross-market assessment of newspaper error and credibility.' *Journalism and Mass Communication Quarterly*, 82(3), 533–51.

McKenzie, Carly T., Wilson Lowrey, Hal Hays, Jee Young Chung and Chang Wan Woo. 'Listening to news audiences: The impact of community structure and economic factors.' *Mass Communication and Society*, 14(3), 375–95.

McManus, John (1992), 'Serving the public and serving the market: A conflict of interest?' *Journal of Mass Media Ethics*, 7(4), 196–208.

Meyer, Philip. *The Vanishing Newspaper: Saving Journalism in the Information Age* (Columbia: University of Missouri Press, 2004).

Napoli, Phillip (2010), *Audience Evaluation: New Technologies and the Transformation of Media Audiences*. New York: Columbia University Press.

Newman, N. (2016), *Journalism, Media and Technology Predictions 2016*. Oxford: Reuters Institute for the Study of Journalism Retrieved 30 April 2016 from http://tinyurl.com/h86ak9t.

Nguyen, A (2013), 'Online news audiences: The challenges of web metrics.' In Karen Fowler-Watt and Stuart Allan (eds), *Journalism: New Challenges*. Bournemouth University Centre for Journalism and Communication Research. Retrieved 23 July 2013 from https://tinyurl.com/jbuw9ms.

Peters, Jeremy (2010), 'In a world of online news, burnout starts younger.' *New York Times*, 18 July. Retrieved 23 July 2012 from http://tinyurl.com/cv2sl4n.
Petre, C. (2015), *The Traffic Factories: Metrics at Chartbeat, Gawker Media, and The New York Times*. New York City: The Tow Center for Digital Journalism. Retrieved 23 December 2015 from http://tinyurl.com/h7pkblu.

Radding, Alan. 'Strength in Numbers,' *Writer's Digest*, August 1994, 34–35.

Ross, A. (2016), 'If nobody gives a shit, is it really news?' Changing standards of news production in a learning newsroom.' *Digital Journalism*, Online First version, 1-18.

Rowe, David (2010), 'Tabloidisation of news.' In Stuart Allan (ed.), *The Routledge Companion to News and Journalism*, 350–61. London: Routledge.

Schlesinger, Philip (1987), *Putting Reality Together: BBC News*. London: Methuen.

Shoemaker, Pamela J. and Reese, Stephen D. (1996), *Mediating the Message*. New York: Longman.

Shoemaker, Pamela J., Philip R. Johnson, Hyunjin Seo and Xiuli Wang (2011), Readers as gatekeepers of online news: Brazil, China, and the United States. *Brazilian Journalism Research*, 6(1), 55–77.

Shoemaker, Pamela J., Vos, Tim P. and Reese, Stephen D. (2008), Journalists as gatekeepers. In Wahl-Jorgensen K. and Hanitzsch T. (eds), *Handbook of Journalism Studies*. New York: Routledge, 73–87.

Singer, Jane B. (2011), Crowd control: Collaborative gatekeeping in a shared media space. *The Association for Education in Journalism and mass communications annual conference*, St. Louis. 10–13 August.

Sullivan, M. (2016), 'One Year Later, 11 Questions for Dean Baquet.' *Public Editor's Journal*. Retrieved 4 December 2016 from https://tinyurl.com/jfoa9td.

Tandoc, E. C. (2014). 'Journalism is twerking? How web analytics is changing the process of gatekeeping.' *New Media and Society*, 16(4): 1–17.

Tandoc, E. C. and Ferrucci, P. R. (2017), Giving in or giving up: What makes journalists use audience feedback in their news work? *Computers in Human Behavior*, 68, 149–56.Tow Centre for Digital Journalism. (2014).

Tandoc, E. C. and Thomas, R. J. (2015), 'The ethics of web analytics.' *Digital Journalism*, 3(2), 243–58.

Trilling, D., Tolochko, P. and Burscher, B. (2016), 'From newsworthiness to shareworthiness: How to predict news sharing based on article characteristics.' *Journalism and Mass Communication Quarterly*. OnlineFirst version. Retrieved 30 August 2016 from https://tinyurl.com/h8uw6o2.

Usher, Nikki (2010), 'Why SEO and audience tracking won't kill journalism as we know it.' *Nieman Journalism Lab Blog*, 14 September. Retrieved 30 August 2012 from http://tinyurl.com/26fbbo9.

Vu, Hong T. (2014), 'The online audience as gatekeeper: The influence of reader metrics on news editorial selection.' *Journalism*, 15(8), 1094–1110.

Wahl-Jorgensen, K. (2007), *Journalists and the Public: Newsroom Culture, Letters to the Editor, and Democracy*. Creskill, NJ: Hampton Press.

Wang, S. (2016), 'The New York Times is trying to narrow the distance between reporters and analytics data.' *Nieman Lab*. Retrieved 3 December 2016 from https://tinyurl.com/h637o6k.

Welbers, Kasper, Van Atteveldt, Wouter, Kleinnijenhuis, Jan, Ruigrok, Nel and Joep Schaper (2015), News selection criteria in the digital age: Professional norms versus online audience metrics.' *Journalism*. OnlineFirst article. doi:10.1177/1464884915595474.

Wendelin, M., Engelmann, I. and Neubarth, J. (2017), 'User rankings and journalistic news selection: comparing news values and topics.' *Journalism Studies*, 18(2), 135–53.

White, David Manning (1950), 'The "gate keeper": A case study in the selection of news.' *Journalism Quarterly*, 27, 383–90.

World Association of Newspapers and News Publishers (2016), *Dallas Morning News Editor Mike Wilson on their evolving newsroom - World News Publishing Focus by WAN-IFRA*. (2016). *Blog.wan-ifra.org*. Retrieved 4 December 2016 from http://blog.wan-ifra.org/2016/10/31/dallas-morning-news-editor-mike-wilson-on-their-evolving-newsroom.

Yarnall, L., Johnson, J. T., Rinne, L. and Ranney, M. A. (2008), 'How post-secondary journalism educators teach advanced CAR data analysis skills in the digital age.' *Journalism and Mass Communication Educator*, 63(2), 146–64.

Zamith, R. (2015), *Editorial Judgment in an Age of Data: How Audience Analytics and Metrics are Influencing the Placement of News Products* (PhD). University of Minnesota.

Zamith, R. (2016), 'On metrics-driven homepages: Assessing the relationship between popularity and prominence.' *Journalism Studies*. OnlineFirst article, doi:10.1080/1461670X.2016.1262215.

Section Two

Data and Statistics in News Consumption

8

The Power of Numbers, Reconsidered[1]

Scott R. Maier, *University of Oregon, USA*

Introduction

It has long been held in journalism practice and research that numbers are as essential as words to explain what is happening in the world. The literature, including my own published findings, cites the importance of numerical data in giving a news story authority, context and perspective (e.g. Maier 2003; Paulos 1996). In a call for newsroom numeracy, I wrote: 'Virtually every aspect of modern life – from the quality of the air we breathe to the safety of the cars we drive – is measured and influenced by statistics. The decisions on how to interpret and react to the data are too important to be left solely to the scientists and the politicians. If the media do not have the ability to sort through the conflicting, confusing morass of numbers, who will?' (Maier 2002, 597). Accurate statistical reporting thereby contributes to civic discourse and engagement; when presented statistical evidence of compelling need, readers are moved to action.

But what if this common wisdom is wrong?

Experimental psychology research has shown that statistical information can actually diminish empathy and discourage humanitarian response. In one such study, for example, charitable contributions to rescue a starving seven-year-old girl in Africa fell significantly when statistics disclosed that millions of other children faced a similar plight (Small, Loewenstein and Slovic 2007). Behavioural research documenting 'psychophysical numbing', in which lives are valued less as their numbers increase, holds important implications for journalism. While statistics can be used to convey the enormity of a crisis, they also may overwhelm and undermine civic response (e.g. Fetherstonhaugh et al. 1997; Friedrich et al. 1999).

To test how these findings in social psychology apply to journalism, I conducted a year-long examination of reader response to the use of numbers by *New York Times* columnist Nicholas Kristof (Maier 2015). In addition, co-researchers and I administered an experimental study assessing how numbers-based stories compared to other story forms in their influence on reader emotions (Maier, Slovic and Mayorga

---

[1] The chapter includes previously published research by the author in a reappraisal of the power – and limitations – of quantitative content in the news.

2016). The findings, presented here from previously published research, offer a cautionary lesson: selective use of statistics may heighten awareness and provide important context, but the news media cannot rely on the numbers alone to drive an audience's response to calamity and mass injustice.

The power of numbers

The journalistic value of numbers has been widely recognized. Numbers have been shown to confirm, refute and qualify claims (e.g. Feloa 1995), to reveal underlying social and economic trends (Jones 1997), and to give credibility to news reporting (Houston 1999; Radding 1994; Maier 2003). Numbers have become an essential component of modern reporting. In a study of 500 local news articles, I found that nearly half of the stories involved some sort of mathematical calculation or numerical point of comparison (Maier 2002). Moreover, numbers-based stories got the best play, dominating the front page of news, metro and business sections. As the late Victor Cohen noted in his authoritative guide *News & Numbers*, 'We journalists like to think we deal mainly in facts and ideas, but much of what we report is based on numbers. Politics comes down to votes. Budgets and dollars dominate government. The economy, business, employment, sports – all demand numbers … Like it or not, we must wade in' (Cohn 1989, 3–4).

Yet, by many accounts, journalists use numbers poorly. More than seventy years of accuracy research has documented mathematical errors in newspapers and television news (e.g. Charnley 1936; Berry 1967; Brown 1965; Galdieri 1999; Maier 2003; Porlezza, Maier and Russ-Mohl 2012). A mathematics audit of a regional US newspaper identified a unique type of math error about every other day; most errors were self-evident and involved elementary math (Maier 2002). In a survey of 4,800 news sources in fourteen newspapers, the largest cross-market assessment of newspaper accuracy, researchers found 'numbers wrong' to be the third most common error type (Maier 2005). Errors of all types eroded credibility and willingness of sources to cooperate with the press (Meyer 2004; Maier 2005; Porlezza, Maier and Russ-Mohl 2012). Recognizing the need to prepare aspiring journalists to use mathematics effectively, the Accrediting Council for Education in Journalism and Mass Communication classifies 'the ability to apply basic numerical and statistic concepts' as a core competency of undergraduate journalism education (ACEJMC 2000).

In *A Mathematician Reads the Newspaper*, John Allen Paulos (1996) contends the most grievous outcome of newsroom innumeracy is neither blunder nor ignorance but lack of appreciation of how mathematics can enrich journalism. Paulos (1996, 3–4) holds that statistics provide a 'way of thinking' that sharpens and broadens the reader's view of the world:

> Numbers stories' complement, deepen, and regularly undermine 'people stories.'
> Probability considerations can enhance articles on crime, health risks, or racial

and ethnic bias. ... And mathematically pertinent notions from philosophy and psychology provide perspective on a variety of public issues. All these ideas give us a revealing, albeit oblique, slant on the traditional Who, What, When, Why and How of the journalist's craft.

Psychic numbing and compassion fatigue

As I explore in my recent article 'Compassion fatigue and the elusive quest for journalistic impact' (Maier 2015), how a story is told can be as important as the information it conveys. Story framing – defined by Entman (2010, 391) as 'selecting a few aspects of a perceived reality and connecting them together in a narrative that promotes a particular interpretation' – influences understanding of the information and the response, if any, that follows. By depicting the characteristics and dimensions of a public issue, story frames are found to define problems, diagnose causes, make moral judgements and suggest remedies (Entman 1993; Scheufele 2009). The lens used to report humanitarian crises – often distant and hidden from public view – is considered especially important because public knowledge is largely mediated by what is presented by newspapers, television and the internet. However, while knowledge is essential, awareness is not necessarily sufficient to engender response (Pruce 2012). 'What is sufficient?' asks Tristan Borer, political scientist and human rights scholar. 'The answer is not simple, despite the widely held view that "if only people knew, they would do something." There is nothing automatic about the link between information and action' (Borer 2012, 36).

When disaster strikes, atrocities occur, or war escalates, journalists, policymakers and the public all want to know: How many people died? How many are injured? How many are missing? Numbers indicate the magnitude of the calamity and, one could rationally presume, guide society's response according to need. But cognitive research reveals a paradoxically inverse relationship between the enormity of crisis and civic reaction – people more readily respond to the aid of a person or two in need than to mass suffering. 'As numbers get larger and larger, we become insensitive; numbers fail to trigger the emotion or feeling necessary to motivate action,' writes psychology scholar Paul Slovic (2009, 30) in a sobering commentary titled 'The More Who Die, the Less We Care.'

One explanation for insensitivity to mass suffering is 'psychic numbing', a term coined to characterize the shutdown of feelings experienced by rescue workers dealing with the countless victims of the Hiroshima atomic bombing. 'Psychic numbing', defined by BehaveNet as the 'reduced emotional responsiveness associated with exposure to traumatic events', has since been used to explain why caring people go to great lengths to help an individual in dire need but are seemingly indifferent to mass suffering. Both scale and distance have been found to influence human capacity to experience affect – 'the most basic form of feeling ... that something is good or bad' – and to sway people's judgements, actions and decisions (Slovic 2007). As Slovic (2007, 84) explains, there is considerable evidence that the value placed on saving human lives 'may follow the same sort of "psychophysical function" that characterizes our diminished sensitivity to a wide range of perceptual and cognitive entities – brightness, loudness, heaviness, and money – as their underlying magnitude increase.'

Experimental studies document the phenomenon in which people relate to the suffering of one as a tragedy but tune out the loss of thousands as a statistic (e.g. Slovic 2007; Slovic 2009; Slovic et al. 2013). Identified as the 'singularity effect', a strong human preference exists for helping a single identified victim over a group of victims. For example, Israeli researchers solicited donations for a child requiring life-saving medical treatment and then repeated the process asking for donations to aid eight children needing the treatment. Even though the collective need obviously is greatest for the group of dying children, donations to the single child were nearly twice that given to the group of eight (Kogut and Ritov 2005a). In a follow-up study, researchers found that donations to a starving seven-year-old African child went down dramatically when donors were told that the child was one of millions desperately needing food (Small, Loewenstein and Slovic 2007). At what point does the psychic numbing begin? Research suggests that the blurring occurs with as few as *two* individuals. In a study by Slovic and Västfjäll (2010), people donated less to a pair of needy children in Africa than to either a girl or boy posed singly.

This numbing effect also has been described as 'compassion fatigue', which first was used to portray burnout of service workers but quickly widened to include a jaded public 'weary of unrelenting media coverage of human tragedy and ubiquitous fund-raising appeals' (Kinnick Krugman and Cameron 1996, 687). Just as journalists know to give a 'human face' to their stories, charities such as Save the Children have long modelled campaigns on a single needy child. 'When it comes to eliciting compassion, psychological experiments demonstrate that the identified individual, with a face and a name, has no peer', researchers assert with the proviso, '*providing the face is not juxtaposed with the statistics of the larger need* [italics added]' (Slovic and Västfjäll 2016, 36). These findings are congruent with prior social psychology research showing that depiction of human triumphs over adversity raises viewers' sense of efficacy – that they too can and should make a difference (Bandura 1997).

But personalizing a story does not necessarily evoke an empathetic response, especially if the individual's plight is unsympathetic or seemingly hopeless. What matters, Oliver and her colleagues (2012) report, is how the story is told: 'Narrative format does not invariably create either favorable or unfavorable reactions in consumers. Rather, relative to non-narrative format, it amplifies various responses in ways that are consistent with the story line' (217).

Combining personification with statistics: Nicholas Kristof's approach to psychic numbing

A two-time Pulitzer-winning reporter for *The New York Times*, Nicholas Kristof turned to the work of psychology scholars for guidance on overcoming compassion fatigue. 'To me, the lessons of this research are twofold,' Kristof explained. 'First, tell an engaging individual story to suck people in. Second, show that it's not hopeless, but that progress is possible' (as cited in Chong 2012). But Kristof stops short when it comes to the behavioural research finding that statistical information stifles emotional response. 'If you follow this research, you would leave out the context,' Kristof said in an interview on National Public Radio. 'All you would do would be telling individual stories, and that would be one step too far for me. I do want to connect with people and inform them about these larger problems. So my compromise is that I do try to find a story that will resonate with people' (Gladstone 2009).

In doing so, Kristof provided the opportunity for me to conduct a case study testing whether these reporting strategies drawn from psychology research make a difference in real-world journalism (Maier 2015). In a content analysis of Kristof's columns in a one-year period, I measured the extent that Kristof's reporting focuses on the individual and on victims who overcome adversity. The study also assessed how story personification was counterbalanced with statistical information and mobilizing information. In particular, I sought to investigate whether numbers rouse or desensitize readers to some of the world's most troubling conflicts. Using a variety of media analytics, the study assessed readers' online reactions, gauging response on Twitter, Facebook, Google and in the blogosphere.

The study confirmed that Kristof 'practises what he preaches', that is, he provided a strong focus on individual stories counterbalanced with quantitative information. Nearly half of the columns provided three or more paragraphs with quantitative information. Only 13.5 per cent of his columns lacked statistical data. Drawing from experimental research indicating that quantitative information diminishes civic response, the study predicted that reader response negatively corresponds with Kristof's focus on statistics. Contrary to the hypothesis, however, Google+ users were somewhat *more likely* to share Kristof's columns with other viewers when an abundance of quantitative data was provided. No other statistically significant correlations were found between Kristof's use of numbers and propensity of readers to comment, 'like', email, or otherwise share his columns. However, the correlations were uniformly positive (Maier 2015).

The analysis described above examines whether numbers and statistics of any kind had a numbing effect. But how do readers respond when given statistical information that explicitly expresses the magnitude of the issue or event described? Nearly 60 per cent of Kristof's columns provided specific quantitative information regarding lives at stake (e.g. number of victims killed, deported, hospitalized), longevity (e.g. years of conflict) and cost (e.g. financial losses, governmental expenditure). When comparing the stories that provided and did not provide statistics indicating magnitude of the issue or event, the apparent effect on reader response comes into somewhat sharper focus. Independent-samples t tests indicated that when given statistics on the magnitude of a problem, readers more frequently shared the column on Facebook, referred to the column on Twitter or other social media, viewed the story online, clicked on 'like' on Google+, shared on Google and commented on Google. In contrast, few of the digital metrics studied offered a significant relationship with personification of story, triumph over adversity, or mobilizing information.

Proximity and story topics were found to be by far the strongest predictors of reader response. It is no surprise that Kristof's column on Lady Gaga's campaign against bullying elicits greater response than a column about a young woman who had escaped from a Cambodian brothel. But the results underscore why the media gravitate to superstars and sensational topics while remote, difficult distant issues such as human trafficking get relatively little notice. Seeking to overcome these challenges, Kristof purposely turns to individual narratives when the storyline is least likely to have popular appeal. Kristof explains in an interview with the Poynter Institute:

> The issues I care most passionately about, from Sudan to sex trafficking, aren't ever going to do well on the most emailed list, because they're off the agenda. That's precisely the reason I'm writing about them, trying to get people to care more about neglected issues. (as cited in Beujon 2012)

This also may be one reason that the relationship between personification and reader response is statistically inconclusive and often in the negative direction predicted. Another feasible explanation is that the *New York Times* draws a distinctly engaged readership that doesn't need the personified priming that a narrative provides; instead, these highly primed readers are prone to share and comment to the news when the enormity of the humanitarian crisis is backed by numerical evidence.

Numerically versus non-numerically based news: An experimental study on reader responses

Exploring this line of inquiry further, I joined Slovic and doctoral student Marcus Mayorga in an experimental study examining how statistics and other story elements influence reader response to news accounts of rampant violence in Africa. In an online survey with embedded experimental conditions, a panel of 900 US adults were randomly assigned a news story emphasizing one or more storytelling elements: personification, statistical documentation, mobilizing information and photographs portraying people in need. A stripped down 'just the facts' news version served as a control story. After reading the assigned story, participants answered a questionnaire designed to assess their feelings of empathy. A Kristof column reporting on renewed violence in Darfur served as one of two news story prototypes used by this study.

The 'numbers story' focused on quantifiable information providing figures and statistical references indicating the enormity of the strife. The statistical story highlighted the hundreds of thousands of civilians displaced by renewed fighting, leading to overflowing refugee camps and near-starvation conditions. To amplify the statistical format, the numbers story included a 'By the numbers' sidebar presenting estimates of deaths (more than 300,000), refugee camp population (more than 1 million), number of Sudanese dependent on humanitarian assistance (more than 3.2 million) and other numerical information.

The advantage of the experimental study is that we could draw on a broad-based readership and control the content presented by each participant. To avoid respondent bias, the test stories were formatted as a generic newspaper article authored under the pseudonym 'Mark Nester', revealing neither the true identity of the writer nor the *New York Times* as the story source. To minimize artificiality, the original wording of each article was largely preserved, though the story's focus was manipulated through editing and rearranging content and, on occasion, by providing additional information. Hence, each treatment offers a similar storyline but is distinct in emphasis and scope. Six variations of each story served as the independent variable.

Participants also were entered into a drawing providing a US$50 award to one randomly selected winner. To assess whether the story forms influence charitable giving, respondents were given the opportunity to donate part or all of the prospective lottery prize to Doctors Without Borders, a relief agency that provides medical assistance to troubled areas such as South Sudan and Darfur. In US$5 increments, respondents designated how much, if any, of the US$50 award they would be willing to donate.

The study, 'Reader reaction to news of mass suffering', showed that the news can strongly influence reader's mood (Maier, Slovic and Mayorga 2016). Asked at the onset of the online questionnaire, 'Overall, how do you feel?' respondents were overwhelmingly positive; only 8.3 per cent reported feeling negative. But when the question was repeated after respondents completed reading their assigned news story on the violent conflict in Africa, nearly 70 per cent reported feeling negative. Only 9 per cent reported no change. On a 7-point Likert scale ranging from '*very negative*' to '*very positive*', the mean well-being score dropped from an upbeat 5.45 rating to a '*feeling negative*' 2.96 rating. A paired-samples *t*-test indicated the change was statistically significant.

Not surprisingly, the straight fact-based news story – the predominant form of newspaper and wire-service reporting – evoked the weakest emotional response by almost every measure. This underscores that simply 'reporting the news', providing objective information about a current event or issue, is often not sufficient to arouse a strong response. As predicted, the study found that above and beyond all other story forms, personification elicited significantly stronger reader interest, concern and sympathy for the victims of the distant crisis in Africa. Indicative of conflicting feelings, readers of the personified news story reported significantly greater motivation and sense of urgency to act – as well as heightened hopelessness.

The results, unexpectedly, indicated numbers had only a small effect on reader response. Contrary to the psychology research, journalistic use of numbers didn't seem to either numb or overwhelm readers; on the other hand, quantification, even when presenting an accounting of mass suffering, did not profoundly enhance audience impact. Combining story forms did not offer a multiplying effect. For example, adding quantification and mobilizing information to the human narrative elicited slightly weaker reader response compared to the personified story focused on a sole refugee's plight. These findings should be viewed with caution; further study is needed to examine how different ways of conveying statistics and other contextual information might alleviate rather than accentuate psychic numbing.

A potential shortcoming of self-reported emotions is that respondents may not be aware of – or candid about – what they are feeling. Arguably a more indicative measure of engagement tracks whether respondents are willing to back their humanitarian response with money. In this study, respondents designated on average US$17.70 of their potential US$50 prize money to Doctors Without Borders. Respondents reading the statistical story pledged the highest amount (m = US$20.35) to the relief agency while, ironically, those reading the story with mobilizing information were among the least generous (m = US$15.75). Only those reading the combined elements story (without photo) pledged less (m = US$15.30), but an analysis of variance indicated that the differences in donations among treatments were not statistically significant.

The direct link between story form and donations was not apparent, but what about the influence of reader affect on charitable giving? A Pearson correlation coefficient was calculated examining the relationship between self-reported emotional feelings and pledged donations to Doctors Without Borders. A weak but statistically significant correlation was found between donation amounts and overall feelings reported after reading the article (the worse they felt, the more they gave). The correlations were also weak but significant for the other affective variables. The strongest association with giving was with the extent to which the story held interest, extent story made the reader sad, and extent story elicited sympathy. The weakest associations with donations were whether the story left readers feeling hopeless or, conversely, inspired.

Conclusion: The quantitative paradox

It is telling that even a single news story can influence reader mood. The study's panel of US adults, almost universally upbeat at the onset of the study, widely reported 'feeling negative' after reading about the plight of refugees in Sudan and Darfur. Across all story forms, the negative change in 'how do you feel' was statistically and substantively significant – on average, a 2.5 drop on a 7-point scale. Clearly, the news stories had emotional impact. But if the primary effect is to dispirit readers, it's little wonder that reporting horrific news does not necessarily lead to public action. Social psychology research documents that negative feelings do not induce action unless there is an effective remedy available (e.g. Västfjäll et al. 2015). In this light, it is also instructive that readers given a plan of action reported significantly less despair than those left without apparent solutions. These findings are congruent with prior research showing that the failure by media to give mobilizing information leads to a sense of inefficacy, the perceived inability to effectively respond even when the need is great (e.g. Bandura 1997; Whitehorn 1989).

This study also examines the collective influence of story treatments. What happens when personification and mobilizing and statistical information are provided in a single news story? The answer: not much. The synergy of combining story forms did not evoke reader response significantly different from when the story treatments were provided separately. Only when a photograph was added to the mix did the story draw significantly stronger response on any of the emotional measures. This suggests that combining the personal story with factual context – classic attributes of news feature writing – may be considered good journalism, but the emotional impact of this approach is not apparent.

The study documents a weak but significant association between reader affect and charitable giving – those reporting the strongest emotional response to the news stories also were the most generous in designating donations to Doctors Without Borders. While the direct association between story form and charitable giving was not evident, mediation analysis indicates that story personification influences emotional response and, in turn, emotional response influences charitable giving. In other words, emotions served as mediators of the framing effect. The human exemplar fosters a more intense emotional response, which subsequently strengthens the story form's impact on charitable giving. However, indirect effects for other story forms were not significant. This suggests that key to action is the emotional response that the news story triggers, not the form that the story takes.

The primary purpose of journalism, media analysts Bill Kovach and Tom Rosenstiel proclaim in *The Elements of Journalism*, is to provide information, context and perspective in a way that readers can 'make order of it, make it useful, to take action on it' (2014, 26). Numbers clearly play a role in that sorting process. Numbers are needed to convey cost, time, gains and losses in both remunerative and human terms. Without quantification, how can news audiences evaluate and appropriately respond to humanitarian crisis – so often remote and unfathomable?

However, cognitive research shows that numbers also pose a conundrum for journalists: how to fulfil the 'paradoxical need for numbers' without succumbing to 'the numbing, desensitizing effects of quantitative discourse' (Slovic and Slovic 2015, 1). As previously discussed, the evidence is abundant that the personal story carries greater emotional power than the stark use of statistics. In their book *Numbers and Nerves*, Slovic and Slovic observe: 'We require data in order to describe such phenomenon as contamination, genocide, species extinction and climate change. But the data alone, while bolstering the authority of journalists and scientists, tend to wash past audiences with minimal impact' (2015, 1).

Fortunately, there appears to be middle ground. In the context of news stories, numerical data did not have the profound desensitizing effect so vividly demonstrated in the research lab. On the other hand, numbers-based news stories failed to arouse readers even when the figures documented monumental death and disruption. This suggests that prudent use of numbers may serve as a mortar of authoritative news reporting but statistics alone are unlikely to move the public to action. Kristof probably had it right: seek the personal story but don't forsake the numbers.

References

Accrediting Council for Education in Journalism and Mass *Communication* (2000), *'Principles of Accreditation'* (Lawrence, KS: ACEJC). Available online: https://www2. ku.edu/~acejmc/PROGRAM/PRINCIPLES.SHTML (accessed 19 December 2016).

Bandura, A. (1997), *Self-efficacy: The Exercise of Control*. New York: W. H. Freeman.

Beujon, A. (2012), 'Kristof: "The U.S. is losing interest" in foreign reporting.' *Poynter Online*, 25 September. Available online: http://www.poynter.org/latest-news/ mediawire/189532/kristof-the-u-s-is-losing-interest-in-foreign-reporting/ (accessed 11 April 2014).

Berry, F. (1967), 'A study of accuracy in local news stories of three dailies.' *Journalism Quarterly*, 44(autumn), 482–90.

Borer, T. A. (2012), 'Introduction: Willful ignorance – news production, audience reception and responses to suffering.' In T. A. Borer (ed.), *Media, Mobilization, and Human Rights: Mediating Suffering*, 1–41. New York, NY: Zed Books.

Brown, C. (1965), 'Majority of readers give papers an A for accuracy.' *Editor and Publisher*, (13 February), 13, 63.

Charnley, M. V. (1936), 'Preliminary notes on a study of newspaper accuracy.' *Journalism Quarterly*, 13(December), 394–401.

Cohn, V. (1989), *News and Numbers: A Guide to Reporting Statistical Claims and Controversies in Health and Other Fields*. Ames: Iowa State University Press.

Chong, R. (2012), 'How Nicholas Kristof uses his pulpit to engage people with empathy.' *Co.Exist*, 26 November. Available online: http://www.fastcoexist.com/1680966/ how-nicholas-kristof-uses-his-pulpit-to-engage-people-with-empathy (accessed 15 October 2013).

Entman, R. M. (1993), 'Framing: Toward clarification of a fractured paradigm.' *Journal of Communication*, 43(4), 51–8.

Entman, R. M. (2010), 'Media framing biases and political power: Explaining slant in news of Campaign 2008.' *Journalism*, 11, 389–408.

Epstein, S. (2003). 'Cognitive-experiential self-theory of personality.' In T. Millon and M. J. Lerner (eds), *Handbook of psychology: Personality and social psychology* (Vol. 5, pp. 159–84). New York, NY: Wiley.

Fetherstonhaugh, D., Slovic, P., Johnson, S. M. and Friedrich J. (1997), 'Insensitivity to the value of human life: A study of psychophysical numbing.' *Journal of Risk and Uncertainty*, 14, 283–300.

Friedrich, J., Barnes, P., Chapin, K., Dawson, I., Garst, V. and Kerr, D. (1999), 'Psychophysical numbing: When lives are valued less as the lives at risk increase.' *Journal of Consumer Psychology*, 8, 277–99.

Galdieri, C. (1999), *Bad Habits: Six Ways Stations Undercut Their Stories*. New York: Project for Excellence in Journalism.

Gladstone, B. (2009), *On the Media* (broadcast) 11 December. Transcript available online: http://www.onthemedia.org/2009/dec/11/fol- low-for-now/transcript/ (accessed 17 October 2013).

Kinnick, K. N., Krugman, D. M. and Cameron, G. T. (1996), 'Compassion fatigue: Communication and burnout toward social problems.' *Journalism and Mass Communication Quarterly*, 73, 687–706.

Kovach, B. and Rosenstiel, T. (2007), *Elements of Journalism*. New York: Random House.

Maier, S. R. (2002), 'Numbers in the news: A mathematics audit of a daily newspaper.' *Journalism Studies*, 3(4), 507–19.

Maier, S. R. (2015), 'Compassion fatigue and the elusive quest for journalistic impact: A content and reader-metrics analysis assessing audience response.' *Journalism and Mass Communication Quarterly*, 92(3), 700–22.

Maier, S. R., Slovic, P. and Mayorga, M. (2016), 'Reader reaction to news of mass suffering: Assessing the influence of story form and emotional response.' *Journalism: Theory, Practice and Criticism*, (August): 1–18 (online edition, print edition forthcoming).

Oliver, M. B., Dillard, J. P., Bae, K. and Tamul, D. J. (2012), 'The effect of narrative on empathy for stigmatized groups.' *Journalism and Mass Communication Quarterly*, 89, 205–24.

Paulos, J. A. (1996), *A Mathematician Reads the Newspaper*. New York: Doubleday.

Porlezza, C., Maier, S. R. and Russ-Mohl, S. (2012), 'News accuracy in Switzerland and Italy: A trans-atlantic comparison with the U.S.' *Journalism Practice*, 6(4), 530–46.

Pruce, J. R. (2012), 'The spectacle of suffering and humanitarian intervention in Somalia.' In T. A. Borer (ed.), *Media, Mobilization, and Human Rights: Mediating Suffering*, 216–39. New York, NY: Zed Books.

Reinard, J. C. (1988). The empirical study of the persuasive effects of evidence: The status after fifty years of research. *Human Communication Research*, 15, 3–59.

Reyna, V. F., Woodruff, W. J. and Brainerd, C. J. (1987). Attitude change in adults and adolescents: Moderation versus polarization: Statistics versus case histories. Unpublished manuscript, University of Texas, Dallas.

Scheufele, D. (2009), 'Framing as theory of media effects.' *Journal of Communication*, 43(4), 103–22.

Slovic, P. (2009), 'The more who die, the less we care.' In E. Michel-Kerjan and P. Slovic (eds), *The Irrational Economist: Decision making in a Dangerous World*, 30–40. New York, NY: Public Affairs Press.

Slovic, P. and Västfjäll, D. (2010), 'Affect, moral intuition and risk.' *Psychological Inquiry*, 21, 387–98

Slovic, P. and Fästfjäll, D. (2015), 'The More Who Die, the Less We Care: Psychic Numbing and Genocide.' In Slovic, S. and Slovic, P. (eds), *Numbers and Nerves: Information, Emotion, and Meaning in A World of Data*, 27–41. Corvallis: Oregon State University Press.

Slovic, S. and Slovic, P. (eds) (2015), *Numbers and Nerves: Information, Emotion, and Meaning in A World of Data*. Corvallis: Oregon State University Press.

Small, D. A., Loewenstein, G. and Slovic, P. (2007), 'Sympathy and callousness: The impact of deliberative thought on the donations to identifiable and statistical victims.' *Organizational Behavior and Human Decision Processes*, 102, 143–53.

Västfjäll D, Slovic, P. and Mayorga, M. (2015), 'Pseudoinefficacy and the arithmetic of compassion.' In Slovic S. and Slovic P. (eds), *Numbers and Nerves: Information, Emotion, and Meaning in a World of Data*, 42–52. Corvallis: Oregon State University Press.

Whitehorn K. (1989), 'The use of the media in the promotion of mental health.' *International Journal of Mental Health*, 18, 40–6.

Zillmann, D., Perkins, J. W. and Sundar, S. S. (1992). Impression-formation effects of printed news varying in descriptive precision and exemplifications. *Zeitschrift für Medienpsychologie*, 4, 168–85.

9

Big Data, Little Insight: Anecdotal and Quantitative Descriptions of Threatening Trends and their Effects on News Consumers

Charles R. Berger, *University of California, Davis, USA*

Introduction

In addition to providing the usual dramatic story elements associated with annual summer wildfires in California, a large wildfire that occurred in the vicinity of a community 90 miles northeast of San Francisco during the summer of 2016 involved an additional dramatic feature: area police arrested a person suspected of being an arsonist who set the fire. During the previous summer, the same area had experienced several wildfires, raising suspicions that an arsonist might have set them. As part of its coverage of the 2016 wildfire, a Sacramento, California newspaper published an article featuring an extended interview with a law enforcement official specializing in solving arson cases (Buck 2016). The piece focused on the motivations that may lead individuals to commit arson. Included in the story was the graph shown in Figure 9.1, taken from a 2014 report available on the California Department of Forestry and Fire Prevention's website (California Department of Forestry and Fire Prevention 2016).

At first blush, the numbers included in this bar graph seem quite impressive. In some of the years, the graph includes, literally hundreds of California wildfires were set by arsonists. Surely, recognizing the scope of the arson problem, we should make it a top priority to understand the motivations of arsonists and to prevent arson-related wildfires.

However, before reaching such a dramatic conclusion, one fraught with potential 'public policy implications', consider the two graphs in Figure 9.2 and Figure 9.3, which also appeared in the same 2014 California wildfires report but were neither included nor mentioned in the article:

The data depicted in Figures 9.2 and 9.3 place arson-caused wildfires in a statistical context that tends to minimize arson as a primary cause of such fires. The data shown in Figure 9.1 suggest that wildfires occur in California every year with considerable frequency. But if Figures 9.1 and 9.2 are combined to compute the yearly percentage or rate of arson-caused wildfires (by dividing the number of arson-caused wildfires for each year by the total number of wildfires for that year), the results vary between

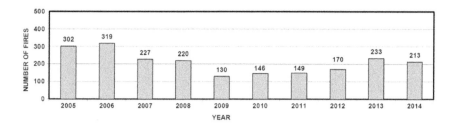

Figure 9.1 Arson Fire Occurrence 2005–14.
Source: '2014 Wildfire Activity Statistics', California Department of Forestry and Fire Protection.

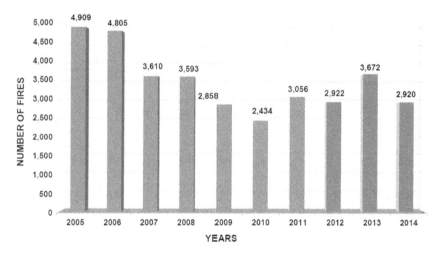

Figure 9.2 Number of Fires 2005–14.
Source: '2014 Wildfire Activity Statistics', California Department of Forestry and Fire Protection.

4 per cent and 7 per cent for the years included in the graph. This range of rates comports with the 7 per cent arson percentage shown in Figure 9.3, which shows data for 2014 only. As Figure 9.3's data also suggest, rather than being a potentially burgeoning phenomenon, arson-caused wildfires are relatively rare, although it is possible that additional arson cases are included in Figure 9.3's relatively large (23 per cent) 'undetermined' category. Given the statistical distribution of the various causes in Figure 9.3, the prevention of wildfires would seem to be best served by inducing citizens to be more careful in wildfire-prone areas rather than focusing intensely on ferreting out unsavoury arsonists lurking among the trees and bushes (although this latter and somewhat more sinister alternative may have significantly greater attention-getting appeal for some news consumers). To be sure, arson-caused wildfires happen in California, and some can be very serious, but they appear to be

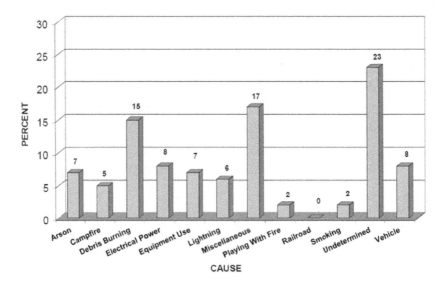

Figure 9.3 Per cent of Fires by Cause 2014.
Source: '2014 Wildfire Activity Statistics', California Department of Forestry and Fire Protection.

much rarer than the potential impression fostered by the data in Figure 9.1, the only one graph that appeared in the news story.

This brief case serves to illustrate how, in the age of 'big data', journalists may present quantitative data in ways that serve to promote distorted perceptions of the causes underling various hazards and threats; moreover, such statistical depictions may encourage news consumers to form distorted estimates of their risk of exposure to such hazards and threats. Although these misuses of quantitative data may be inadvertent, such presentations may affect news consumers' perceptions of the phenomena depicted by the data. This chapter will focus on the problems associated with numerical depictions of risk, but since individual cases are frequently used to exemplify hazards and threats, the impact of such anecdotes on risk perceptions will first be considered. In some instances, stories that focus primarily on individual cases or exemplars are augmented by statistical data; thus, the quantitative data depicted in such hybrid stories may suffer from problems similar to those illustrated by the arson story.

Anecdotal depictions of hazards and threats

Communication researchers have devoted considerable attention to the inferential problems associated with news reports featuring individual cases. In addition to noting that the media traffic heavily in anecdotes, some observers contend that news characterizations of social phenomena generally include collections of individual cases

(Zillmann and Brosius 2000). However, rather than being representative samples of cases that adhere to the canons of statistical sampling theory, these collections consist of arbitrarily selected special cases that focus mostly on the extraordinary. Such non-representative collections result in 'the inaccurate perception, if not plain misperception, of the projected phenomena' (Zillmann and Brosius 2000, viii). The reasons for the news media's proclivity to feature non-representative individual cases and collections cannot be addressed in detail here. However, beyond the often-heard cliché that news is not found in the routine of everyday life but in unique events, a potential explanation for this tendency may be that statistically rare cases serve to attract the attention of potential audience members, not unlike the attention-getting calculus underlying 'freak shows' popular in England and the United States during the nineteenth century. The resulting increases in circulation figures and television viewer ratings may be positive both for journalists' careers and for media organizations' financial bottom lines.

That anecdotal, case-based news accounts are potentially misleading because they do not represent the prevalence of the phenomena they depict is but one aspect of the problem. Because of limitations in their ability to process information, humans may be prone to base inferences about a general state of affairs in the physical or social world on non-representative individual cases or collections of them. News consumers may have neither the time nor the available mental resources to process news reports in the detail necessary to pose critical questions regarding their representativeness. Rather, hurried and sometimes harried citizens may rely on mental shortcuts or heuristics to generalize about the size and scope of a problem. The availability heuristic may play an important role in this regard (Kahneman and Tversky 1973). As considerable experimental evidence has suggested, items that are more likely to be recalled from memory exert an inordinate impact on frequency and probability judgements. Thus, news consumers exposed to stories that include vivid, individual cases are more likely to judge the associated phenomenon to be more prevalent than readers of accounts that do not include such exemplars. The central problem is that stories featuring only individual cases ignore the depicted phenomenon's base-rate –that is, the actual frequency of its occurrence. Thus, a dramatic story about a single, large, arson-caused wildfire in California might encourage news consumers to infer, erroneously, that many or perhaps most large wildfires are caused by arsonists, an inference clearly at variance with comparative base-rate data in Figure 9.3.

We have conducted a number of studies to determine the conditions under which news consumers become concerned with the probative value of news events to which they have been exposed. Probative value is the degree to which news consumers judge a story to be a reliable indicator of the prevalence of the phenomenon depicted in the report. In the typical experiment, after reading a story about a specific threat incident (for example, a theft), participants judge the degree to which the story is a reliable indicator of the threat's prevalence,that is, to assess the degree to which the events actually take place within the locale in question. We have found that merely engaging in the process of making probative value judgements tends to reduce the amount of fear news consumers report in response to reports about individual crime incidents (Berger and Lee 2011). In particular, students who read stories about individual incidents of computer theft and other campus crimes and then made judgements about the stories' probative value subsequently reported less fear of crime victimization than those who did not judge the stories' probative value before reporting their levels of apprehension and victimization fear. One explanation for this reduction in reported fear is that making a probative value assessment after reading the story serves to remind consumers that the story they have just read is only one case and that, without any statistical evidence, it may be an isolated and unrepresentative incident (an outlier in statistical terms).

Thus, merely raising questions about a story's probative value may be one way to overcome the availability heuristic's potential effects on prevalence judgements. Another way to counter the effects of availability might be for journalists to provide base-rate data that serve to place the individual case or collection of cases in a broader statistical context. Thus, a story about a specific crime incident could be accompanied by statistical data about the incidence of the crime in the story's locale. A low rate of the reported crime might serve to assuage the fear of victimization induced by exposure to the story depicting the individual incident. One variation of this approach would be to provide statistical data about other, more prevalent hazards and threats as a way of placing the individual case in a statistically comparative threat context.

Research conducted during the anthrax scare that occurred in the fall following the World Trade Center terrorist attack (9/11) examined this latter possibility (Berger, Johnson and Lee 2003). In the wake of the 9/11 attack, during the fall of 2001, the media reported a few cases of letters containing the anthrax toxin that were intercepted at postal facilities and a few individuals who were exposed to anthrax. Among the few exposed to anthrax, a particularly mysterious case involved a hospital worker in New York City. Exhaustive investigation of the victim's work place and home revealed no traces of anthrax and this prompted some to speculate that the worker may have been exposed to the toxic agent while walking in public, a possibility that might stoke considerable public fear. An edited version of the hospital worker story was used in a series of experiments designed to determine the degree to which base-rate data about another hazard might serve both to raise the salience of the anthrax story's probative value and, at the same time, reduce the amount of apprehension reported in response to the anthrax story (Berger, Johnson and Lee 2003).

In the first experiment, some individuals read only the anthrax story and then responded to it. Others, before reading the same anthrax story, read one about the number of traffic deaths in the United States in the preceding year (approximately 40,000). In addition to indicating the degree to which they felt fearful, all participants completed a measure of the degree to which they were skilled at and favoured rational and intuitive thinking styles (Epstein et al. 1996). This experiment found that individuals favouring a rational thinking style who first read the traffic fatalities story reported less apprehension in response to the story featuring the individual anthrax case than did low rational individuals who read both stories. Providing base-rate information about the more frequently occurring traffic death hazard reduced the amount of apprehension induced by the anthrax story but only among those with highly rational thinking styles. The same experiment was repeated on different research participants several months later, when the anthrax threat had subsided considerably. In contrast to the first experiment, no difference in reported fear emerged between the high- and low-rationality groups with respect to their exposure to the traffic deaths base-rate story. However, in general, those with highly rational thinking styles reported less victimization fear than did the lows.

In sum, news stories that include only single cases or skewed collections of individual cases not only provide news consumers with a non-representative picture of the phenomena they depict, but also play into the miserly information-processing proclivities of news consumers. The confluence of two consequent tendencies – the use of cognitive short cuts or heuristics to inflate the probability of events depicted in the news stories and the distraction from the critical question of the depicted phenomena's base-rate – tends to fuel excessive fear when story events depict hazards and threats. Our research demonstrates that by making news stories' probative value salient and by providing a comparative base-rate context, journalists can help news consumers overcome some of the potentially distorted inferences that may follow from exposure to stories featuring individual cases. Of course, one solution to the larger problem might be to depict threatening phenomena exclusively with statistical data and to avoid reporting individual cases that might encourage distorted, availability-based inferences. Although this may seem to be a straightforward solution, especially in the current era of 'big data' and data-driven journalism, its implementation in news reporting has also been problematic, as the following discussion will reveal.

Quantitative depictions of threatening trends

Frequency versus rate data

Beyond news reports that adduce individual cases to depict hazards and threats are stories that attempt to portray the behaviour of threatening phenomena over time. These stories use statistical data, usually in graphical forms, to illustrate trends through time. One problem that arises in such stories is the presentation of raw frequency data to illustrate the purported trend. The frequency comparisons can involve two or more points in time and may include percentage increase figures. Thus, if 1,000 instances of the threat occurred during year 1 and 2,000 instances during year 2, the report may note that there has been a 100 per cent increase in the frequency of the phenomenon.

A story dealing with a state referendum concerning banning plastic grocery bags and the effects of plastic bag bans in California cities illustrates the problem of reporting per cent change statistics (Luna 2016). This story included the graphs in Figure 9.4. As the top graphs indicate, after the plastic grocery bag ban was enacted in the City of San Jose, there were substantial increases in the percentages of people using reusable bags and those using no bags. Although not surprising, these data are properly represented. However, the lower graph depicting reductions in plastic bag littering in different locations is potentially misleading because readers have no idea how many bags were found in these locations before and after the ban was enacted. That is, readers have no idea on what frequencies the percentage reduction values are based. For example, an 89 per cent reduction from 100 plastic bags found in storm drains before the ban would be considerably less impressive than an 89 per cent reduction from 100,000 bags of the same sort. In the absence of the before-ban frequency data, the per cent reduction figures are, at best, ambiguous on this point. Unfortunately, nowhere in the article's text were readers provided with the before- or after-ban frequencies. The consequence is that readers would not be able to form an estimate of the problem's magnitude before the ban was enacted and how effective the ban was in reducing the problem.

Another potentially serious problem with stories that include frequency data depictions of threatening trends over time is the possibility that the base-rate of the depicted threat has changed over the same time period. This difficulty is illustrated by the following data about traffic fatalities (list of motor vehicle deaths in the United States by year 2016). In the graph in Figure 9.5, I plotted the number of traffic fatalities

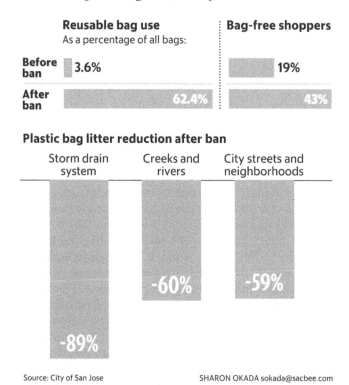

Figure 9.4 A news graph in the Sacramento Bee showing the effect of plastic bag ban in San Jose, California. © The Sacramento Bee, 2016. Reprinted with permission.

in the United States between 1921 and 2014, inclusive. As can be seen, there was a rather steady increase in the number of fatalities from 13,253 in 1921 to a peak in 1972 of 54,589 fatalities. There was also a secondary peak that exceeded 50,000 fatalities during the years 1978, 1979 and 1980. Since that time, the number of fatalities has declined.

One problem with the Figure 9.5 graph is that the frequency data included in it fail to take into account two very important parameters that have changed dramatically during the period 1921–2014. First, the population of the United States was 108,538,000 in 1921, but this number tripled to 316,129,000 by 2014. Second, in addition to this large population increase, people travelled many more miles in 2014 than they did in 1921 and thus increased the probability that they might become accident victims. In 1921, the first year that such estimates were made available, people travelled 55.03 billion vehicle miles. By 2014, this figure had increased to 3,026 billion vehicle miles. Given the large changes in these two base-rates, one straightforward explanation for the rising frequency of traffic fatalities in the years following 1921 might simply be the fact that more people travelling more miles in the United States increased the likelihood that accidents would occur. Another way to look at the traffic fatalities trend is to compute the rate of fatalities in terms of millions of miles of vehicular travel for each year. The results for this rate measure appear in the graph in Figure 9.6.

The data in Figure 9.6 paint a very different picture of the traffic fatalities trend in the United States from 1921 to 2014 than does the Figure 9.5 graph. As Figure 9.6 shows, there were 24.09 traffic fatalities per million vehicular miles travelled in 1921,

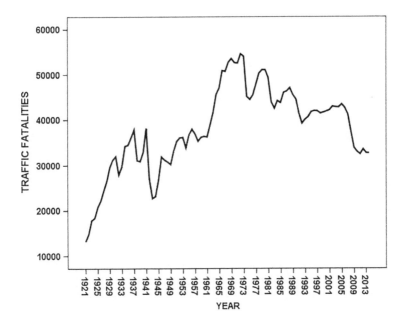

Figure 9.5 Number of Traffic Fatalities in the United States: 1921-2014.

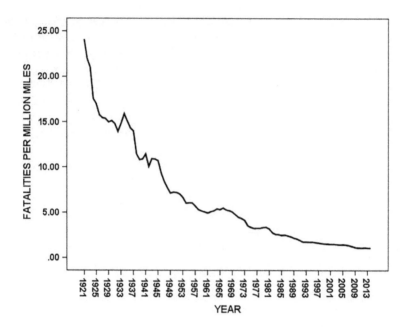

Figure 9.6 Traffic Fatalities Per Million Miles Travelled in the United States: 1921-2014.

but by 2014 this figure had diminished to 1.08 fatalities per million miles travelled. Furthermore, the decline during the period was quite steady with minor upticks in 1930 and 1941. Figure 9.5 frequency data lead viewers to the conclusion that during the period 1921–72, there was a significant worsening trend of the traffic fatalities. But Figure 9.6 rate data tell the opposite: taking into account the number of miles travelled in vehicles, there was a marked and steady decline in the fatality rate during the 1921–72 period. Moreover, even though Figure 9.5's frequency data show a decline in the number of traffic fatalities after 1972, Figure 9.6's rate data indicate that the decline in the rate of traffic fatalities began much earlier and continued to decrease over the entire 93-year period.

In absolute terms, the Figure 9.6 rate graph suggests that vehicular accident fatalities are a relatively rare phenomenon, although the garden-variety local television news broadcast in the United States somehow finds such cases to air on a daily basis. Setting this proclivity aside, the current rate of just over one fatality for every one million miles travelled in vehicles seems to be a relatively low one and represents a substantial improvement over the 1921 rate, a fact that very likely has a relatively low probability of being reported in the news. Parenthetically, in the face of data depicting an improving state of affairs such as these, 'advocates' and 'activists' who seek to reduce the levels of such hazards and threats will discount the obvious improvement by suggesting 'even one victim is one victim too many', implying that no risk is the only acceptable risk level (Berger 2003).

On the basis of this traffic fatalities example, one might be tempted to conclude that raw frequency data should never be used to characterize the behaviour of phenomena over time, especially those that involves hazards or threats. Although this is probably sound advice, it is possible for frequency data to be representative of trends if relevant base-rates in the population have not changed over time. Thus, in the present example, if US population had not grown since 1921 and citizens did not travel in vehicles any more miles in 2014 than they did in 1921, the frequency data depiction would be informative. In addition, an unchanging base-rate over time would enable reporters to use per cent change calculations involving frequency data to make meaningful per-cent-change estimates. However, when characterizing the behaviour of such phenomena as accidents, crime and disease over time, it is almost always the case that the base-rates of these phenomena are changing because the size of the population on which the statistics are based is changing.

In the traffic fatalities example, base-rates increased significantly from 1921 to 2014; however, it is possible for base-rates to diminish over time. The traffic fatalities data illustrate this possibility. Inspection of the Figure 9.5 frequency graph for the years 1941–5 shows a distinct drop in the number of traffic fatalities, from 38,142 fatalities in 1941 to 27,007 in 1942, and further to 22,727 in 1943. By 1946, fatalities rose to 31,874. Similarly, the number of vehicular miles travelled also declined significantly from 333.61 billion in 1941 to 268.22 billion in 1942, to 208.19 billion in 1943, but rose to the pre-Second World War levels (340.88 billion miles) by 1946. This systematic variation in fatalities and vehicular miles travelled is explained by the fact that after the United States entered the Second World War in December 1941, gasoline rationing was imposed nationally and thus curtailed vehicular travel. When the frequency and rate graphs are compared for the Second World War period, the former shows a steep decline in fatalities while, by contrast, the latter a less dramatic decrease. In general, rate data tend to mute such changes while frequency data tend to amplify them. Journalists may choose frequency data because they tend to exaggerate changes and serve to increase story drama. However, unless it is clear that base-rates remain stable over time, it is safer to live by the maxim that rate data rather than frequencies should be used to depict and characterize trends.

The scope of the problem

If we were to be satisfied by the examples adduced thus far in this chapter, we would be making the mistake of generalizing the reporting of quantitative data on the basis of a few, potentially non-representative cases.

Fortunately, in the case of the just-discussed frequency versus rate data problem, at least one systematic study based on randomly selected and thus potentially representative newspaper articles and television news broadcasts has been reported to address this question (Berger 2001). During randomly selected weeks within the months of March, June, September and December of 1998, coders identified a total of 2,075 accident, crime and health stories that appeared in the *Los Angeles Times*, New York Times, Sacramento Bee and USA Today or were aired on the nightly national and local news broadcasts of ABC, CBS, NBC and CNN's World Today. Of the 2,075 stories, 84 reported quantitative data for at least one trend, that is, data for at least two points in time. The majority of the trend stories (77 per cent) came from newspapers. Sixty-six per cent of the 84 trend stories concerned health-related topics, 24 per cent focused on crime and 10 per cent dealt with accidents. Because some individual stories reported multiple statistical trends –for example quantitative trends for different types of crime, the 84 stories yielded a total of 214 trends for analysis. The 214 trends were classified with respect to whether they portrayed an improving or worsening condition and whether the data were presented in frequency or rate form.

For both television and newspaper stories, those stories that reported quantitative trends were over twice as long, in terms of time or column inches, as those stories that did not do so. Among those stories that employed frequency data to illustrate the trend, 65.2 per cent depicted a worsening trend; in contrast, rate data were employed to depict worsening trends in only 25.7 per cent of the cases. Thus, this study revealed a distinct proclivity to depict worsening trends with frequency data and to illustrate improving trends with rate data. Although additional analyses showed no difference between frequency and rate data trends in the rate at which graphs were used to illustrate the trend, among those trends that depicted a worsening condition, 42 per cent included a specific exemplar, but among those illustrating an improving condition, 18 per cent included a specific exemplar. Television and print news stories did not differ with respect to their proclivities to report improving versus worsening trends; however, while newspaper stories were found to devote about equal amounts of space to the two types of trends, television stories that reported worsening trends were substantially longer than stories that depicted improving trends. While newspapers devoted approximately equal amounts of space to stories when reporting frequency versus rate data, television news stories reporting trends based on frequency data were substantially longer than television news stories reporting rate data.

Although 69 per cent of the 214 trends included in the study were illustrated with rate data, there were large differences among the three story types with respect to this parameter. Specifically, 73 per cent of the health trends and 63 per cent of the crime trends were couched in rate data; however, 28 per cent of the accident trends were illustrated with rate data. The fact that only 10 per cent of the stories reporting quantitative trends dealt with accidents may, at least in part, explain the observed overall distribution of frequency versus rate trends. Moreover, these data suggest that within certain topic domains, such as health and crime, rate data may be more available than they are within other domains. Consequently, when reporters seek numerical data to include in their stories, certain topic domains may afford more opportunities to obtain rate data than others, for example, the FBI Uniform Crime Statistics are reported in rate form (crimes/100,000 people). Nonetheless, even when only frequency data are available to depict trends of interest, obtaining relevant base-rate data, such as population numbers over time, is usually not a difficult task and the calculations involved in computing rates given frequency and base-rate data require only primary school arithmetic skills. Since this study was conducted, rate data presentations of trends may have become somewhat more common in news reports; however, the contemporary exemplars presented earlier in this chapter suggest that the practice of presenting inappropriate frequency data remains a problem in news reporting.

Correcting for base-rate neglect

One could argue that depicting the behaviour of threatening phenomena over time with frequency data is not as serious an epistemological sin as it may appear, if it is assumed that those who consume news reports of quantitative trends based on frequency data recognize that the frequency data fail to take into account changes in base-rates over time. The gist of this line of reasoning is that awareness of the lack of a base-rate allows news consumers to 'correct' the trend as represented by the frequency data. The question is whether it is reasonable to assume that such a 'correction' process actually takes place when individuals are presented with frequency data depictions of trends. Fortunately, a series of experimental students has addressed this issue (Berger 1998, 2000, 2002).

The experimental paradigm employed in these studies was one in which all research participants were exposed to news reports about crime trends that were illustrated by frequency data for a period of several years, culminating in the year before the study was conducted. The data depicted increasing frequencies of specific crimes over the years. After reading these stories, research participants completed a number of measures including one that assessed the degree to which the story they read made them feel apprehensive and another that asked them to estimate the risk of being a potential victim of the crime about which they had just read. These two measures were moderately and positively correlated. Before reading the crime stories, some research participants read a news report concerning increases in the population of the locale in which the crime story was based. The population increase stories presented yearly population figures for the same years encompassed by the crime stories. Other research participants did not read the population increase story but responded only to the crime story. Thus, those who first read the population story were provided with a strong base-rate cue. The general expectation was that exposure to the base-rate data would attenuate the victimization risk judgements of those in this group. Before reading the crime increase story, in one of the experiments some participants read a story about the number of strangers people typically come into face-to-face contact during an average day. The article reported that the average number is hundred strangers. Although this story did not report population increase data, it might serve to make the base-rate issue salient.

Typically, crime statistics are based on crimes/100,000 people; however, this way of computing the crime rate ignores the interaction base-rate or the number of people individuals typically encounter face-to-face who could do them harm. If this interaction base-rate were used to compute crime statistics, rather than crimes/100,000 people, some crime rate estimates would be dramatically reduced. In conducting their social commerce during a year's period, many citizens experience face-to-face contact with countless others. The number of such face-to-face social contacts per year far exceeds the number of people in the population. Employing the interaction base-rate would provide a more accurate assessment of victimization risk because it takes into account the degree to which people are exposed to others who could perpetrate certain crimes against them. Although obtaining estimates of the number of face-to-face encounters citizens have with each other during particular time periods presents a monumental research challenge and one that remains to be addressed; in fact, viewing crime incidence statistics from this purview probably would provide a more accurate assay of the magnitude of crimes typically perpetrated during face-to-face interactions among citizens.

Although the predicted effect was observed in several experiments, in each instance it was qualified by the sex of the individual who read the story. Specifically, men first exposed to the population increase story evinced significantly lower victimization risk judgements than men who did not first read to the population increase story; however, among women, exposure to the population increase story had no effect on their apprehension and victimization risk judgements (Berger 1998, 2000, 2002). This well-replicated interaction effect was also obtained when individuals first read the oblique story about the number of strangers people typically encounter in their daily lives. Men who first read the oblique story reported lower levels of apprehension and victimization risk in response to a crime story featuring frequency data than men who did not first read the oblique story. Among women, there were no differences between those who read the oblique story and those who did not (Berger 2000, Experiment 2). Although there are a number of alternative explanations for this sex difference, evidence suggests that although both men and women may heed relevant base-rate data, the cognitive responses they have to them differ (Berger 2002). For example, among women, those first exposed to the population increase story evinced significantly less shock and surprise in response to the subsequent crime story than did women not first exposed to the population increase story. Among men, this difference was not significant. Most tellingly, although both men and women increased their production of the critical cognitive response 'burglary increases are caused by population increases' in response to being exposed to the population increase base-rate story, men's rate of production of these responses was substantially greater than that of women. Specifically, men exposed to the population increase story generated 5.19 times more of these responses than did men who did not read the population story, but among women this figure was 1.98. Thus, women show evidence of heeding relevant base-rate information but the information appears to affect women's cognitive responses, and thus their reported levels of apprehension, less remarkably than it does men's (Berger 2002).

The results of these studies should give considerable pause to those who argue that news consumers can second-guess frequency depictions of threatening trends in such a way that they are able to factor in relevant base-rates that may act both to dampen the perceived increase in the depicted threatening trend and their levels of apprehension and victimization risk in response to the threatening trend data. Even when population increase data are provided to news consumers before they encounter a threatening trend depicted with frequency data, they do not necessarily uniformly experience reductions in their apprehension in response to the threatening frequency data depictions; moreover, individuals vary considerably in their numerical skills or numeracy. Berger (2002) found that although men scored significantly higher than women on a measure of numeracy (Schwartz et al. 1997), numerical skills did not account for the observed interaction between sex and base-rate information on apprehension; however, in general, those who scored high in numeracy reported less apprehension than did those who scored low. Thus, in contrast to individuals with low levels of numeracy, those with high numeracy levels are more prone to respond to frequency depictions of threatening trends in ways that reduce their apprehension.

Although the numeracy finding may provide some fuel to the argument that at least some news consumers can discount potentially misleading frequency depictions of threatening trends such that they experience less apprehension, another issue arises with respect to overall numeracy levels in the United States' population. There has been considerable research interest in the degree to which numerical skills influence patients' understanding of quantitative medical information and their decision-making in health-related contexts (Reyna et al. 2009). Patients, some of them with life-threatening illnesses, may be confronted with probabilistically couched medical treatment choices presented in ways they may not fully understand. Research has sought to identify ways such information can be presented so that patients can make informed medical decisions. In addition, there have been large-scale national and international surveys of representative samples of adults that have assessed levels of numeracy in the general adult population. For example, the 1992 National Adult Literacy Survey (NALS) (Kirsch et al. 2002) assessed the prose, document and quantitative literacy of more than 26,000 respondents in the United States. Quantitative literacy was indexed using a five-levels system of arithmetic and mathematical problems in increasing order of difficulty from a rank of one (easiest) to five (most difficult). Results revealed that 47 per cent of the sample was unable to solve quantitative problems above the first two difficulty levels (levels one and two) and 21 per cent of the sample was able to solve problems at the two most difficult levels (levels four and five). In general, those with higher education levels demonstrated higher numeracy levels. Similar studies have been conducted in different countries. For example, the Program for International Assessment of Adult Competencies (PIAAC) done under the auspices of the Organisation for Economic Development and Co-Operation (OECD) conducted a study of 16–65 year olds in thirty-four countries to assess literacy and numeracy skills (OECD 2016). This study found that the United States ranked 27th in numeracy. In general, these studies attest to the relative lack of numeracy skills in the United States' adult population, both on an absolute and comparative basis. Consequently, the notion that United States' news consumers are able to correct for the statistical shortcomings of news reports, at least on a widespread basis, seems to be an implausible one.

Individual and social consequences of statistical distortions

The preceding discussion has provided ample evidence that the news media sometimes present quantitative data in ways that are likely to mislead unwary news consumers. As noted previously, this chapter cannot present detailed analyses of the causes of these distorted presentations; however, the adult literacy studies suggest that one reason such errors may be made, at least in the United States, is the low levels of numeracy of the population from which journalists come. Moreover, those attracted the journalism profession may be more inclined to hone their literary skills rather than their mathematical and statistical skills, thus further exacerbating their numeracy skill deficits. In any case, although the reasons for these statistical missteps in news reporting remain to be elucidated in greater detail, some have suggested that news reporting of hazards and threats has played a major role in the inculcation of unwarranted public fear and anxiety. Although misleading presentations of threatening trends using frequency data are not solely responsible for stoking excessive public apprehension and fear, they make a perceptible contribution to this undesirable state of affairs. The idea that Americans have become excessively apprehensive is not a new one. Nearly four decades ago, Wildavsky (1979, 32) averred, 'How extraordinary! The richest, longest-lived, best-protected, most resourceful civilization with the highest degree of insight into its own technology is on its way to becoming the most frightened.' Two decades later, others echoed similar sentiments about Americans and Western Europeans (Cohl 1997; Furedi 1997; Glassner 1999) and similar observations have been made in the new millennium (Altheide 2002; de Becker 2002; Siegel 2005). These theorists and researchers argue that the induction of inordinate fear levels in the public fosters a number of undesirable consequences, some of which are enumerated below.

A potential effect of exaggerating the risks associated with hazards and threats in the news is increased anxiety in the population at large, thus creating a potential public health problem. Mental health specialists have specifically cited the news media's constant flow of threat-related stories as a potential source of excessive public fear and anxiety (Siegel 2006). Although there are several different types of anxiety disorders, generalized anxiety disorder (GAD) is particularly germane as a potential news exposure effect. GAD is characterized by excessive worry about a variety of everyday problems for a period of at least six months. People suffering from GAD may worry excessively about and anticipate problems with their finances, health, employment, and relationships, topics that the news media frequently cover. GAD sufferers usually have difficulty assuaging their concerns, even though they realize that their anxiety is more intense than situations warrant. Among American adults, 3.1 per cent report having GAD within the past 12 months and 5.7 per cent indicate they have suffered from GAD sometime during their life (National Institute of Mental Health 2016). Unfortunately, longitudinal data are not available to determine whether these rates have changed over time, but, given the content of news coverage, especially that airing on local television news, it seems reasonable to suppose that high rates of exposure to such news might well make a contribution to the development or intensification of GAD, especially among those who are susceptible to it.

Exposure to unrepresentative and skewed news reports featuring threatening phenomena may foster other undesirable effects. Some have suggested that individuals who come to believe that they are living in a very dangerous world are more likely to lower their achievement expectations (Furedi 1997). Being a mere 'survivor' in an ostensibly dangerous world itself becomes a major life achievement for such individuals. In addition, individuals may become more risk-averse, as they tend to perceive relatively low risk hazards and threats to be more probable than they actually are. The combination of attenuated achievement expectations and increased risk aversion may, in turn, act to stifle innovation. Individuals who are content to count mere survival as a significant life achievement and who are unwilling to take the risks inherent in the adoption of new ideas are not likely to be innovators.

The misperception of risk also may lead to the development and enactment of public policies that address relatively low risk hazards and threats and ignore those hazards and threats that are much more likely to occur. Such miscalculations could lead to the squandering of scarce resources. In addition, a fearful and cowering public may be more likely to cede excessive power to government officials and thus invite the emergence of authoritarian political regimes in which government officials assume the role of paternalistic 'protectors' of the public (Altheide 2002). Of course, these individual and social effects are not solely the product of exposure to inappropriate presentations of quantitative data used to illustrate threatening trends in the news. Other features of news reporting may prompt such effects. For example, Altheide (2002) reported significant increases in the use of the term 'fear' in news reports during the period leading up to the new millennium. Of course, the news media are not solely responsible for producing these effects, but they are a significant social amplifier that highlights numerous hazards and threats (Kasperson et al. 2003).

Conclusion

Since the inception of communication science during the years preceding the Second World War, an era dominated by the study of mass communication effects (Delia 1987), communication theorists have proposed a number of functions that communication serves for society. Among the functions posited by early theorists is surveillance of the environment, a function carried out in the service of identifying threats and opportunities within communities (Lasswell 1948). This function in particular served as a keystone construct for early models of communication and journalistic praxis. In these models, reporters were portrayed as agents who search the environment for potential threats and opportunities, report the details of these threats and opportunities to editors who, in turn, act as gatekeepers who mediate the flow of information to the public (Westley and MacLean 1958). These models postulated that most members of the public do not have the time or the resources necessary to identify such threats on their own; consequently, reporters and editors serve as 'watch dogs' on their behalf.

The advent of the internet and citizen journalism has challenged the accuracy of these early models. Although the surveillance function continues to be a significant feature of contemporary conceptualizations of journalists' role definitions, there is controversy over the degree to which watchdog journalism is effective in rooting out government corruption and other threats to functioning democracies (Coronel 2010). Setting these issues aside, traditional print and electronic news organizations still function as gatekeepers and portrayers of reality for those who consume their news products. As noted above, the news media continue to be accorded a central role in risk communication models that depict the processes subserving the social amplification of risk (Kasperson et al. 2003).

It is within this general context that the present chapter has examined specific problems associated with the depictions of potential threats and hazards in news reporting. It is one thing to identify potential threats and hazards. It may turn out that initial depictions of a phenomenon as 'threatening' may prove to be ill-founded, for example, the catastrophic Y2K computer failures that were widely predicted to occur at the millennium's turn. However, even granting the reality of purported threats, their depiction in news reports may take several forms, some of which may serve to exaggerate their likelihood and severity. Of course, certain ways of depicting risk may encourage underestimates of likelihood and severity. For example, some may discount the threat of a highly fatal and highly contagious disease because of its current statistical rarity in the population; however, because the disease is highly contagious, the number of cases may increase both dramatically and quickly. While this chapter has focused on data-based exaggerations of risk, such underestimates are possible.

An important construct in the domain of electronic communication systems is signal/noise ratio. This construct is used to evaluate the degree to which receiving equipment in communication systems is able to detect or pull out signals from the noise that accompanies them. Noise may come from the receiving equipment's own circuits or from external sources such as the atmosphere or emissions from nearby electrical equipment. Too much noise will overwhelm the incoming signal and communication will not occur because noise renders the signal imperceptible to the listener. Journalists who have become enamoured with the 'big data' construct would do well to remember that news reports based on 'big data' may contain considerable noise from at least three sources. First, most statistics contain inaccuracies due to recording and other errors. These inaccuracies and errors may not be explicitly reported in the statistical sources that journalists may consult. A second source of noise is that associated with the nature of the measures being used. For example, estimates of the percentage of the United States' workforce that is currently unemployed do not include individuals who are currently unemployed and have stopped looking for work. Thus, the monthly unemployment rate figures routinely reported by the news media underestimate the size of the unemployed population in the United States. A third noise source is that associated with the mode of presentation of statistics in news reports. Presenting data, be they 'big' or 'little', in misleading formats only serves to undermine and perhaps swamp any signal value they may have. It is not simply a question of such data presentations adding little information to a story but, perversely, a question of their potential contribution to an increasingly noisy world.

References

Altheide D. L. (2002), *Creating Fear: News and the Construction of Crisis*. New York: Walter de Gruyter.

Berger, C. R. (1998). 'Processing Quantitative data about risk and threat in news reports.' *Journal of Communication*, 48, 87–106.

Berger, C. R. (2000), 'Quantitative depictions of threatening trends in the news: The scary world of frequency data.' *Human Communication Research*, 26, 27–52.

Berger, C. R. (2001), 'Making it worse than it is: Quantitative depictions of threatening trends in the news.' *Journal of Communication*, 51, 655–77.

Berger, C. R. (2002), 'Base-rate bingo: Ephemeral effects of population data on cognitive responses, apprehension and perceived risk.' *Communication Research*, 29, 99–124.

Berger, C. R. (2003), 'Effects of discounting cues and gender on apprehension: Quantitative versus verbal depictions of threatening trends.' *Communication Research*, 30, 251–71.

Berger C. R. and Lee K. J. (2011), 'Second thoughts, second feelings: Attenuating the impact of threatening narratives through rational reappraisal.' *Communication Research*, 38, 3–26.

Berger, C. R., Johnson, J. T. and Lee, E. J. (2003), 'Antidotes for anthrax anecdotes: The role of rationality and base-rate data in assuaging apprehension.' *Communication Research*, 30, 198–223.

Buck, C. (2016), 'Expert offers insight on arsonists' motivations.' *Sacramento Bee*, 18 August, 1B.

California Department of Forestry and Fire Prevention (2016, 2014) wildfire activity statistics. Available at: http://www.fire.ca.gov/downloads/redbooks/2014Redbook/2014_Redbook_Graphs-Charts.pdf.

Cohl, H. A. (1997). *Are we Scaring Ourselves to Death? How Pessimism, Paranoia, and a Misguided Media are Leading us Toward Disaster*. New York: St. Martin's.

Coronel, S. S. (2010), 'Corruption and the watchdog role of the news media.' In P. Norris (ed.), *Public Sentinel: News Media and Governance Reform*, 111–36.Washington, DC: World Bank.

De Becker, G. (2002), *Fear less: Real Truth About Risk, Safety, and Security in a Time of Terrorism*. Boston: Little, Brown & Co.

Delia, J. G. (1987) 'Communication research: A history.' In C. R. Berger and Chaffee, SH (eds), *Handbook of Communication Science*, 20–98. Newbury Park, CA: SAGE Publications.

Epstein, S., Pacini, R., Denes-Raj, V. and Heier H. (1996), 'Individual differences in intuitive-experiential and analytic-rational thinking styles.' *Journal of Personality and Social Psychology*, 71, 390–405.

Furedi, F. (1997), *The Culture of Fear: Risk Taking and the Morality of Low Expectation*. London: Cassell.

Glassner, B. (1999), *Culture of Fear: Why Americans are Afraid of the Wrong Things*. New York: Basic Books.

Kasperson, J. X., Kasperson, R. E., Pidgeon, N. and Slovic, P. (2003), 'The social amplification of risk: Assessing fifteen years of research and theory.' In N. Pidgeon, R. E. Kasperson and P. Slovic (eds), *The Social Amplification of Risk*, 13–46. Cambridge, UK: Cambridge University Press.

Kirsch, I. S., Jungeblut, A., Jenkins, L. and Kolstad, A. (2002), *Adult literacy in America: A first look at the findings of the National Adult Literacy Survey* (Third Edition). Washington, DC: U. S. Department of Education, National Center for Education Statistics.

Lasswell, H. D. (1948), 'The structure and function of communication in society.' In L. Bryson (ed.), *The Communication of Ideas*, 37–51. New York: Harper.

List of motor vehicle deaths by year (2016), Wikipedia, wiki, 2016. Available from: https:// en.wikipedia.org/wiki/List_of_motor_vehicle_deaths_in_U.S._by_year.

Luna, T. (2016), 'Nuisance or necessity? Voters to decide bags' fate.' *Sacramento Bee*, 8 October, p 1B.

National Institute of Mental Health (2016), Generalized anxiety disorder among adults, Available at: https://www.nimh.nih.gov/health/statistics/prevalence/generalized-anxiety-disorder-among-adults.shtml.

Organisation for Economic Development and Co-Operation (2016), *Skills Matter: Further Results from the Survey of Adult Skills*. Paris: OECD Publishing.

Reyna, V. F., Nelson, W. L., Han, P. K. and Dieckmann, N. F. (2009), 'How numeracy influences risk comprehension and medical decision making.' *Psychological Bulletin*, 135, 943–73.

Schwartz, L. M., Woloshin, S., Black, W. C. and Welch, H. G. (1997), 'The role of numeracy in understanding the benefit of screening mammography.' *Annals of Internal Medicine*, 127, 966–72.

Siegel, M. (2005), *False Alarm: The Truth About the Epidemic of Fear*. Hoboken, NJ: John Wiley.

Wildavsky, A. (1979), 'No risk is the highest risk of all.' *American Scientist*, 67, 32–7.

Westley, B. H. and MacLean, M. S., Jr. (1958), 'A conceptual model for communications research.' *Journalism Quarterly*, 34, 31–8.

Zillmann, D. and Brosius, H. B. (2000), *Exemplification in Communication: The Influence of Case Reports on the Perception of Issues*. Mahwah, NJ: Lawrence Erlbaum Associates.

10

Effects of Statistical Information in News Reports on Individuals' Recall and Understanding of Events and Issues: Implications for Journalistic Practices

Rhonda Gibson, *University of North Carolina at Chapel Hill, USA*
Coy Callison, *Texas Tech University, USA*

Introduction

Journalists routinely use numeric information in news stories, whether in telling how many workers lost their jobs when a local textile factory closed, showing that violent crime rates have dropped over the past year, or warning individuals about a local disease outbreak. Numbers are used to indicate the scope of an event or issue, make comparisons to similar situations, and to help audiences assess opportunities or risk. In addition to numeric information, journalists also commonly use exemplars – illustrative individual cases (Brosius and Bathelt 1994) – in their news reports to put a 'human face' on the issue under consideration. Exemplars provide details about someone involved with the focus of the news report and often contain vivid description and first-person testimony in the form of quotations (Gibson and Zillmann 1993).

Ideally, a journalist is able to produce simultaneously accurate and compelling news reports by combining scope-defining numeric information with scene-setting exemplars. This assumes, of course, that journalists are proficient in the deployment of numeric information and that audiences are capable of understanding it. Likewise, it assumes that journalists select representative exemplars that are in line with numeric information related to the issue in question, as opposed to exemplars that might be more interesting to audiences because of their extreme nature. And, lastly, it assumes that audiences are cognizant of the extent to which news report exemplars align with the numeric scope of the issue and thus are able to resist being swayed by extreme exemplars.

Not surprisingly, research suggests that neither journalists nor their audiences are consistently strong in their abilities to use or understand numeric information (Curtin and Maier 2001; Steen 2004). Likewise, evidence suggests that news consumers often form perceptions of issues addressed in news reports based on information contained in the reports' exemplars, even when those reports contain statistical information that may not align with the gist of the exemplars (Gibson and Zillmann 1994; Zillmann and Brosius 2000). This likelihood of being swayed by news exemplars increases for those who are low in numeric literacy (Callison, Gibson and Zillmann 2009; Gibson, Callison and Zillmann 2011; Zillmann, Callison and Gibson 2009).

The current chapter will take a look at the use of statistical information in news reports and its effects on individuals' understanding of that information and its implications for their lives. The first part of the chapter will briefly introduce readers to exemplification theory (Zillmann 2002), which makes predictions about the role of base-rate data and exemplars in news reports on audience perception of the issues addressed in those reports. It will then review research that measures the effects of various numeric presentation formats, drawing from studies that examined the effectiveness of using percentages, fractions, probabilities and primary vs. secondary ratios to present data in the news. The second part of the chapter will address more recent research that examines how an individual's level of quantitative literacy affects his or her willingness and ability to process numeric information in news reports. This section will also explore how individuals differing in numeric ability form quantitative impressions and affective reactions after exposure to statistical information and/or exemplifying case studies in news reports about risk-related topics, in addition to the subsequent effects on those individuals' levels of empathy with victims and personal risk assessment as well as their assessment of risk to others. The chapter will close with discussion of the implications of this research for journalists and others who create messages designed to inform the public.

Exemplification theory of media effects

Exemplification theory addresses how messages – usually but not always in the form of news reports – affect recipients' subsequent social judgements by influencing the accessibility of information that is deployed in the decision-making process (Busselle and Shrum 2003; Zillmann and Brosius 2000). As noted above, news reports about issues and social phenomena routinely contain two types of information: (1) statistical base-rate data that quantify the scope of the issue by detailing the number or proportion of people or things involved, and (2) exemplars that describe the experience of an individual somehow involved in the issue. Regardless of the specific topic, be it the risks of the Zika virus for child-bearing-age women or the ongoing violence in Syria, it is common for a news report to open with description of a relatively typical example, provide relevant statistical information about the situation, and then continue with details about additional exemplars (Gibson and Zillmann 1994; Zillmann and Brosius 2000). The numeric component of the report is a more objective – albeit pallid – means of describing a topic, providing information about its frequency, magnitude, and potential risk or opportunity. Exemplars, on the other hand, are more subjective and can be evaluated only in terms of their representativeness of the population in question rather than by statistical precision. Yet there is evidence that exemplifying cases may be selected because they are vivid and sensational (Bogart 1980; Haskins 1981) and often contain emotional first-person language that allows audience members to envision the individual's plight and to vicariously experience his or her reaction (Zillmann 2002; Greene, Campo and Banerjee 2010). As Gibson and Zillmann (1994) note, this all-too-common lack of exemplar representativeness can be exacerbated by their aggregation in news stories: 'Atypical exemplars may be amassed, and exemplars attesting to the existence of alternative and opposing cases might not be included at all' (604). Think of a news story covering a number of restaurant patrons who became ill because of an E. coli outbreak. The exemplars in a news story would almost certainly focus on the plight of the people who became ill, and few if any exemplars would be included that detailed people who dined at the restaurant but had no negative reaction.

Given that news consumers rarely pay close attention to news reports and the data contained within (Gunter 1987), the lumping together of atypical exemplars in news stories may prove detrimental to the ability of news consumers to draw accurate conclusions about the issues and social phenomena they read about or watch in news reports. Research from the fields of information processing and social cognition shows that people process information using heuristics – essentially mental shortcuts – especially under conditions of low involvement (Kahneman, Slovic and Tversky 1982; Sherman and Corty 1984). The *availability* heuristic applies most directly to the effects of aggregated atypical exemplars.

The availability heuristic (Higgins 1996; Tversky and Kahneman 1973) relies on examples that immediately come to mind when a person is evaluating a situation or making a decision about the frequency or likelihood of an event. This method produces relatively reliable judgements *as long as* a sufficient number of accurately distributed cases have been witnessed within a short period of time. Judgement reliability is diminished, however, when the assumed distribution of cases is altered by the prevalence of atypical and vivid exemplars that lead the individual making a judgement to overestimate the relative frequency of an occurrence. In this situation, information that is the most accessible tends to be disproportionately utilized in the formation of judgements, even when more relevant information is present in memory (Busselle and Shrum 2003; Taylor and Fiske 1978). One reason provided for the enhanced availability and accessibility of exemplars is their vividness, which is 'information that's emotionally interesting, concrete and imagery provoking' (Nisbett and Ross 1980, 45). These characteristics draw attention to the message, helping to maintain audience interest and increase the likelihood of activating similar constructs in memory (Baesler and Burgoon 1994; Nisbett and Ross 1980).

Conceptually, this suggests that the selective use of exemplars in news reports may lead to misperceptions of the issues reported on. And, indeed, findings from a number of studies designed to test the assumptions of and refine exemplification theory (cf. Zillmann and Brosius 2000) suggest that exemplars – especially in aggregation (Brosius and Bathelt 1994; Zillmann, Perkins and Sundar 1992; Zillmann et al. 1996) – exert more power over individuals' perceptions of issues addressed in news reports than do statistical base-rate data. Results from lab and field experiments that have varied the number, distribution and format of exemplars have consistently indicated that, even when presented with numeric information about the actual scope of an issue or event, message recipients pay disproportional attention to and base perceptions of risk and personal consequences on the more concrete, often vividly described individual cases, especially those that engage the recipients' emotions. This preference has held true regardless of how exemplars are presented: in direct or paraphrased quotation (Gibson, Hester and Stewart 2001; Gibson and Zillmann 1998), in photographs accompanying print news reports (Gibson and Zillmann 2000; Zillmann, Gibson and Sargent 1999), or in televised news reports (Aust and Zillmann 1996).

It is important to note, however, that not all investigations of the effects of numeric information versus exemplars have reported consistent results (Hornikx 2007). In the persuasion literature, researchers have examined the effects of statistical vs. narrative evidence in messages on readers' cognitive and affective reactions (Allen and Preiss 1997; Baesler and Burgoon 1994; Kopfman et al. 1998; Nisbett and Ross 1980). Research findings are mixed, with some studies showing the persuasive ability of numeric information (cf. Allen and Preiss 1997) and others suggesting that narrative messages are more persuasive (cf. Reinard 1988). Others have found that there is no difference in the persuasive effect of evidence type (Iyengar and Kinder 1987; Reyna, Woodruff and Brainerd 1987).

Presentation formats

Journalists and other message creators utilize numeric information to describe the scope of an issue or the likelihood of an event because numbers are precise, often convey an aura of scientific credibility, and can be converted from one format to another, such as a percentage to a probability (Lipkus 2007; Zillmann and Brosius 2000). Although reliance on the use of statistics to convey information, especially related to risk, has been criticized because of a perceived 'lack of sensitivity for adequately tapping into and expressing gut-level reactions and intuitions and problems people have understanding and applying numerical concepts' (Lipkus 2007, 699), journalists are nonetheless encouraged to populate their news stories with verified statistics (Cohen 2001).

Journalists can choose from a variety of methods to incorporate numeric data into news stories. Data can be presented as whole-number frequencies – for example, *The train wreck in Reading, Ohio, on Good Friday killed 36 commuters*. In this case, there is no means of comparison provided to use in evaluating the information. Ratio formats, however, involve comparison of two numbers, such as with fractions (less than 1/3 of the student body attended the police protest), traditional ratios (the high school's student–teacher ratio is 25:1), or verbalized ratios (4 out of every 11 millennials has substantial education-related debt). Ratio formats can also be presented in percentages (8 per cent of those infected by the virus had not been vaccinated) and probabilities (there is a 0.86 likelihood that the river will flood tomorrow), in which one of the two comparison numbers is implied (there is a 0.14 likelihood that the river will NOT flood tomorrow). Numeric information can also be hinted at, obviously less precisely, with terms such as *occasionally*, *rarely*, and *almost certainly*, which are subject to variability in reader interpretation (Gibson and Zillmann 1994; Lipkus 2007).

To determine how reporters most commonly present numeric information in news stories, Callison, Gibson and Zillmann (2009) conducted a content analysis of articles in *Time*, *Newsweek*, and *U.S. News & World Report*. They examined bylined stories, analysis pieces, and opinion columns from four issues of each of the three news magazines from the first two months of 2008, coding a total of 3,264 numeric presentations. Frequencies were by far the most common format, with 81.6 per cent of numeric presentations delivered through frequencies. Percentages accounted for 15.4 per cent of numeric presentations, with pure ratios, verbalized ratios, and probabilities making up the remaining 3 per cent.

In a subsequent experiment, Callison, Gibson and Zillmann (2009) tested the effects of numeric information presentation format in news stories on reader recall of information. The researchers hypothesized that frequencies would lead to the greatest recall because they are the most common – and thus likely the most familiar – method used in news stories. Frequencies are also less complex than other formats such as ratios because they do not refer to general prevalence in a larger population but rather address the observed frequency in a smaller reference class in question (Lipkus 2007). The researchers' prediction was not supported, however. Material presented as *percentages* actually yielded the highest recall scores, although they were not statistically different from frequencies or probabilities. Only verbalized ratios (e.g. 3 out of 15 children live in poverty) lead to statistically inferior recall of information. These results differed somewhat from a public health study of messages about the benefits of mammograms in which Schwartz et al. (1997) found that numeric information was more likely to be comprehended when presented as frequencies instead of probabilities and that presenting information visually as histograms or graphs also improved message comprehension.

Other research also has examined the benefits of using graphical displays to express numeric information. Lipkus (2007), in a summary of the research about numeric and visual presentation of health-related risk information, found that the advantages of such displays include their ability to summarize a great deal of data and reveal patterns that would otherwise go undetected using other methods. Graphical displays are also able to attract and hold people's attention because they display data in concrete, visual terms. Furthermore, graphs may be especially useful to help visualize part-to-whole relationships (Lipkus 2007), as, for example, in conditional probability reasoning (e.g. a pie chart showing a slice out of the whole, pictographs showing the number in a population who are affected, or Venn diagrams indicating the degree of overlap among clustered events). Lipkus (2007) also noted that certain types of graphical displays are well suited for specific tasks. For example, bar charts work well when making comparisons (e.g. comparing the magnitude of risk by age or ethnicity); line graphs (e.g. infant mortality curves) are good for showing trends over time; and pie charts allow readers to judge proportions. Bar graphs have been found to be an especially popular format for statistical information (Sanfey and Hastie 1998), one that aids in decision-making tasks and understanding proportions (Lipkus and Hollands 1999).

However, as Lipkus (2007) noted, there are disadvantages of visually displayed information, including the fact that it may often be poorly designed, and readers may lack the ability to interpret graphical elements (MacGregor and Slovic 1986). A study by Sanfey and Hastie (1998) found that individuals who received information in textual form were more accurate in subsequent judgements than were individuals who received information in graphic form, a finding the researchers attribute in part to the fact that text-based material may induce more of an explanation-based judgement strategy that leads individuals to more carefully scrutinize the material. Another possible drawback of visual presentation formats is a concern that graphs can mislead audiences by drawing attention to certain elements and away from others. The part of the graph that is attended to may unduly influence interpretation and subsequent decisions/actions (Lipkus 2007).

In addition to measuring which presentation format leads to higher levels of comprehension, research has also examined the effects on issue perception. Greater risk is perceived when situations are described in terms of verbalized ratios (20 of 100 patients die within three weeks) than when they are presented as percentages (20 per cent) (Lipkus 2007). Slovic et al. (2002) explain that this is likely because the frequency format is believed to bring to mind frightening images of the specific risk, whereas the percentage format leads to more abstract thoughts about small chances.

The role of arithmetic ability

For decades now, national education experts and policymakers have bemoaned a perceived deficiency in the ability of US citizens to use and understand numeric information, and both national and international studies have indicated their concern is warranted. More than 5,000 US residents were surveyed in 2011–12 as part of an international Survey of Adult Skills administered in twenty-three industrialized countries by the Organization for Economic Cooperation and Development (2012). Results ranked the United States 21st out of 23 in terms of numeracy skills. A nationwide study provided similarly dismal results: only 23 per cent of a national sample of 12th graders scored at the 'proficient' level on a standardized mathematics exam (National Center for Educational Statistics 2005). Even highly educated adults struggle with numeracy tasks that require basic conversions between different forms of ratios such as percentages, probabilities and fractions (Lipkus, Samsa and Rimer 2001).

Assessments of arithmetic ability – also referred to as arithmetic aptitude, numeracy and quantitative literacy – have shown that individual differences in such capacities exist when assessed through both objective and subjective measures (Lipkus, Samsa and Rimer 2001; Fagerlin et al. 2007). There are differences in opinion about whether numeracy should be considered a component of general intelligence or as a unique characteristic (Hoard, Geary and Hamson 1999; Zillmann, Callison and Gibson 2009), but regardless of how it is classified, researchers have long been concerned with citizens' apparent difficulties in using and understanding numbers, as well as the underlying causes of these difficulties (Kahneman 2001; Reyna and Brainerd 2007).

Scholars have used a variety of conceptual and operational definitions to measure numeric ability. Much of the work has emerged from the field of health decision-making, where scholars often take an expansive view of numeracy. For example, Peters (2012) suggests that numeracy is the 'ability to understand probabilistic and mathematical concepts' (31), arguing that it goes beyond simple comprehension of numbers and influences how individuals process and deploy both numeric and non-numeric information in judgements and decision-making. Lipkus, Samsa and Rimer (2001) and Peters et al. (2007) have developed two of the most popular numeracy scales, focusing primarily on risk assessment and heavily based on transpositions between probabilities and percentages. Zillmann, Callison and Gibson (2009) suggested that some existing scales may not be sufficiently demanding and as such may not adequately identify respondents of particularly low competence. To address this perceived shortcoming, Zillmann, Callison and Gibson (2009) developed their own measure of arithmetic ability designed to span more of a range of arithmetic difficulty from elementary through demanding.

Not surprisingly, research overwhelmingly shows that individuals with high numeracy understand numbers and mathematical principles better and are more likely to apply those mechanics in decisions (cf. Peters 2012). Low levels of numeric ability are also associated with lower comprehension and use of health-related information (Peters et al. 2007). Research from Reyna et al. (2009) and Peters et al. (2007) furthermore suggests that those higher in numeracy rely more on number-related intuitions and are less susceptible to effects of how numeric information is framed and formatted. Reyna et al. (2009) reviewed the literature concerning the effects of numeracy, and suggested that the aggregated evidence shows, not surprisingly, that those high in numeric ability reliably and correctly interpret statistical information, whereas those low in numeric ability tend to be less reliable and correct in their interpretations because they are more susceptible to irrelevant stimuli in their decision-making. In a separate review of the related literature, Peters (2012) reported that those high in numeracy aren't as likely to be influenced by non-numeric information in a message and show greater sensitivity to numbers and a stronger and more precise number-related affect.

More recent work has addressed the role of an individual's numeric ability on the processing of explicit numeric information in news reports, in addition to the recognition of quantitative information implicit in aggregations of news exemplars (Gibson, Callison and Zillmann 2011). Findings confirm previous research showing that individuals high in numeracy process, interpret, and retrieve numeric information more proficiently than do individuals low in numeracy (Blake and Daschmann 2008; Zillmann, Callison and Gibson 2009). Zillmann, Callison and Gibson (2009) examined the effects of individual differences in arithmetic aptitude on attention to and recall of statistical information in print news reports, with a focus on estimation of risk and benefit likelihoods. Individuals high in numeracy recalled ratios and frequencies more correctly, both in precise and approximate terms, than did individuals low in numeracy. In a separate study, Callison, Gibson and Zillmann (2009) found that individuals low in arithmetic aptitude experienced considerable difficulties with information presented as percentages as compared to those of intermediate and high arithmetic aptitude. Knobloch-Westerwick et al. (2015) manipulated online science-related messages to feature either exemplars or numeric information. The researchers predicted that readers would be attracted more to messages featuring exemplars versus those that contained numeric information; however, there was no evidence of such an effect. However, it is important to note that selective exposure to exemplar versus numeric messages depended on two individual differences: high trait empathy and high numeracy. Readers high in trait empathy (reactions of one individual to the observed experiences of another) spent more time reading exemplar-based messages, whereas those with higher numeric ability spent more time reading statistics-based messages.

In an effort to untangle some of the mixed results regarding the power of exemplars vs. statistical information in news reports on reader issue perception, Gibson, Callison and Zillmann (2011) designed an experiment to examine the role of one particular individual difference variable – quantitative literacy (i.e. numeracy) – in the effects of exposure to statistical information and/or sets of exemplars on estimations of risk, incidence assessments and affective disposition. They proposed that persons of low arithmetic ability would default to intuitive shortcuts for processing numeric information by focusing instead on the implications of individual cases, such as those provided through news report exemplars. Thus, when exposed to news reports, those low in numeracy would intuitively rely on exemplar aggregations in subsequent decision-making situations and pay less attention to any base-rate data contained in the report. In their study, a news report about American volunteer workers falling ill on a mission trip to Central America was manipulated to present information about illness rates either through (1) precise numeric information, (2) the distribution of exemplars, or (3) a combination of statistical information and exemplars that either supported or contradicted the statistics. All of the exemplars featured acute human suffering, and dependent measures included respondent estimations of the number of afflicted people, the magnitude of the victims' suffering, the risk of affliction for visitors in general, and respondents personal support for risk-reducing measures. Respondents' arithmetic aptitude was determined using the measure developed by Zillmann, Callison and Gibson (2009). Numeracy was not found to affect incidence estimates; however, it did influence affect-mediated assessments of empathy with victims, safety risks and protective concerns: exposure to news reports with exemplars lead to higher assessments by individuals lower in numeracy than by individuals higher in numeracy. Also, all variations of the news report that contained exemplars, with and without statistical information, fostered stronger affective assessments than did conditions with news reports containing only base-rate data and no exemplars. The authors suggest that variation in numeracy is associated with information-processing differences that leads to affect-mediated risk-related assessments. As they conclude: 'It might be argued, in fact, that exemplification is inextricably linked to affect instigation and that base rates can impact affect only to the extent that they are associated with well-articulated exemplars' (115).

Subsequent to the previous work that varied the level of support exemplars provided to the base-rate facts, researchers attempted to discern the extent to which numeric proficiency moderates attention to exemplars and base-rate information specifically in relation to estimates of risk. In particular, this line of research attempted to outline how individuals with the highest level of arithmetic aptitude would vary from their lower proficiency peers in terms of risk assessment when considering dangerous scenarios. In a 2013 study, Callison, Gibson and Zillmann asked participants to read a news story about gang violence directed at tourists visiting a Mexican resort. While manipulations of the severity of violence perpetrated on victims was varied across versions of the experimental stimulus material, the unique findings of this study centred on how numeracy influenced perceptions of risk of gang attack generally, regardless of level of violence outlined in the story. In fact, the study's data showed that those high arithmetic ability participants not only underestimated their own risk of attack compared to participants with lower numeracy scores but that the most apt also overestimated the risk to others. The unique contribution of this work was that it begins to shed light on the mechanisms that may underline how arithmetic proficiency alters how a reader incorporates into his/her worldview information gleaned from a new story. Specifically, the researchers when discussing their findings proposed that higher arithmetic aptitude, and the expected greater rationality and intellect that may accompany that quality, induces a self-perception that the more apt are better able to 'conceive superior strategies for diminishing and altogether eluding threatening conditions' (102–3) while predicting that their less intelligent peers would more easily fall prey to victimization.

Implications for journalists and other public communicators

The research reviewed thus far can be useful for news personnel and others who design informational and persuasive messages. First of all, journalists need to recognize and address the likelihood that they may actually be among those who struggle with numeracy. For years, surveys and content analyses have shown that journalists often exhibit discomfort with computational tasks and that news reports frequently contain number-related factual errors (Curtin and Maier 2001; Harrison 2016; Maier 2002). Fortunately, there are resources for journalists who need to improve their arithmetic skills; indeed, manuals have been developed to address the specific type of skills that reporters need in order to understand numeric information and present it clearly in news reports (Cohen 2001; Wickham 2003).

Journalists and other message creators should also pay close attention to research findings that address the effects of statistical information presentation format on individuals' ability to comprehend information. As noted above, Callison, Gibson and Zillmann (2009) found that individuals low in arithmetic aptitude experienced more difficulties with information presented as percentages than did those of intermediate and high arithmetic aptitude. The researchers themselves provided advice for reporters who are considering how to present numeric information: 'For audiences that may be relatively low in mathematical proficiency, news reports should avoid percentages and instead focus more on frequencies. For more mathematically proficient audiences, however, information may be presented in either frequency, percentage or probability format' (53). Their findings also showed that, regardless of level of arithmetic ability, readers had difficulty processing information presented as verbalized ratios – for example, 'four out of seven doctors recommended against a daily aspirin regimen' – suggesting that this format be avoided.

Similarly, Lipkus (2007) provided practical strategies for message creators who want to increase the likelihood that readers better understand reports related to risk: (1) present risk-related information both numerically and visually; (2) be consistent in use of numeric formats; for example, do not compare percentages with frequencies; (3) round numbers; and (4) provide context for numeric information (e.g. what represents high or low risk). In a discussion about the policy implications of numeracy research, Peters (2012) also called for more contextualization of numeric information, suggesting that information providers can improve less numerate individuals' comprehension by labelling numbers' evaluative meaning, for example, by using terms such as 'poor' or 'excellent' and by supplementing numeric information with some type of graphical display that allows for the visual comparison of quantities. Likewise, Peters et al. (2007) suggest that 'less is often more' (744) and that people more easily comprehend numeric information and make better decisions with that information when 'the presentation format makes the most important information easier to evaluate and when less cognitive effort is required' (745). Lastly, Hibbard et al. (2007) recommend that message creators reduce the amount of required inferences and calculations. So, rather than stating that a current fee is $150 and will increase by 10 per cent, it would be better to explain that the increased rate will be $165 (Hibbard et al. 2007).

Research into the effects of exemplar aggregation can also provide guidance for journalists, specifically that they should be aware of the extent to which the exemplars they select to illustrate the scope of an issue or phenomenon are representative of the quantitative nature of that issue or phenomenon (Brosius and Bathelt 1994; Gibson and Zillmann 1994; Zillmann et al. 1994). For example, if base-rate data indicate that only a small percentage of the general population is likely to suffer severe side effects from a new medication, a news report on the topic should not feature a host of exemplars who have suffered severe side effects, as this pattern of representation will lead readers to overestimate their personal risk and the risk of others. Extreme exemplars may be more appealing to journalists for their attention-getting and vivid nature, but their use is unwise unless it is representative of the phenomenon in question. Although they are often instructed that the use of interesting exemplars and anecdotes is expected, journalists have shown some concern about the influence of such techniques on their audiences. Hinnant, Len-Rios and Young (2013), in a series of in-depth interviews with and surveys of health journalists, found that reporters often select exemplars to draw attention and help readers connect to news stories but also that they recognize the ethical concerns that might emerge from such practices. The surveyed health journalists themselves ranked the use of exemplars below the use of numeric information and expert testimony when it comes to helping audiences understand complex health-related topics.

There are also takeaways from the research findings that indicate that individuals differing in arithmetic competence employ different message processing styles. As Gibson, Callison and Zillmann (2011) note, 'persons of lower arithmetic ability, almost as if they were cognizant of not being well served by concentration on numbers per se, apparently turn to exhibitions of pertinent concrete cases, process these cases attentively, and as a result tend to be more strongly engaged affectively' (114). This suggests that targeting the less numerate with messages that deploy representative aggregations of exemplars might lead to superior numeric assessments than would messages relying more heavily on explicit quantifications. Gibson, Callison and Zillmann (2011) also found that persons of low arithmetic competence were consistently more compassionate in assessments than were their higher-skilled peers. The researchers connected this finding to Epstein's (2003) proposal of two distinct information-processing styles: intuitive/experiential vs. analytical/abstract. 'Such stratification should prove useful in designing media strategies for apprising the public of issues of concern, especially in providing information that enables estimates of the likelihood and severity of danger to health and safety' (117). In creating informative and persuasive messages, appeals could be tailored to an individual's intuitive versus analytical tendency, thus increasing comprehension of information and broader understanding of the issue under consideration. Likewise, Knobloch-Westerwick et al. (2015) suggested that science writers tailor their messages to different population segments. Individuals high in numeracy may avoid reading online content that includes heavy exemplar use, whereas individuals high in trait empathy would be drawn to science-based messages that feature exemplars. It should also be noted that in tailoring messages to an audience, writers should be mindful of not only how reader numeracy may influence selection of content but also that high arithmetic aptitude and the high levels of intelligence that likely accompanies it may lead to particularly percipient readers being less prone to seeing themselves vulnerable to any risks outlined in text (Callison, Gibson and Zillmann 2013).

Lastly, there is advice for where in a message to include statistical information. In a study of the relationship between math anxiety and individuals' ability to process health-related numeric information (specifically regarding genetically modified foods), Silk and Parrott (2014) found that numeric information in a message tended to trigger crippling math anxiety; thus, they recommended that numeric information be presented later in messages so that resulting anxiety about the message is not triggered immediately.

References

Allen, M. and Preiss, R. W. (1997), 'Comparing the persuasiveness of narrative and statistical evidence using meta-analysis'. *Communication Research Reports*, 14, 125–31.
Aust, C. F. and Zillmann, D. (1996), 'Effects of victim exemplification in television news on viewer perception of social issues'. *Journalism & Mass Communication Quarterly*, 73, 787–803.

Baesler, E. J. and Burgoon, J. K. (1994). 'The temporal effects of story and statistical evidence on belief change'. *Communication Research*, 21(5), 582–602.

Blake, C. and Daschmann, G. (2008, August), Experts and base-rates, laymen and single case information: Can statistical knowledge prevent exemplification effects? Paper presented at the meeting of the Association for Education in Journalism and Mass Communication, Chicago.

Bogart, L. (1980), 'Television news as entertainment'. In P. H. Tannenbaum (ed.), *The Entertainment Functions of Television*, 209–49. Hillsdale, NJ: Lawrence Erlbaum.

Brosius, H.-B. and Bathelt, A. (1994), 'The utility of exemplars in persuasive communications'. *Communication Research*, 21, 48–78.

Busselle, R. W. and Shrum, L. J. (2003), 'Media exposure and exemplar accessibility'. *Media Psychology*, 5(3), 255–82.

Callison, C., Gibson, R. and Zillmann, D. (2009), 'How to report quantitative information in news stories'. *Newspaper Research Journal*, 30(2), 43–55.

Callison, C., Gibson, R. and Zillmann, D. (2013), 'Effects of differences in numeric ability on the perception of adversity risk to others and self'. *Journal of Media Psychology*, 25(2), 95–104.

Cohen, S. (2001), *Numbers in the Newsroom: Using Math and Statistics in News*. Columbia, MO: Investigative Reporters and Editors.

Curtin, P. A. and Maier, S. R. (2001), 'Numbers in the newsroom: A qualitative examination of a quantitative challenge'. *Journalism & Mass Communication Quarterly*, 78(4), 720–38.

Fagerlin, A., Zikmund-Fisher, B. J., Ubel, P. A., Jankovic, A., Derry, H. A. and Smith, D. M. (2007), 'Measuring numeracy without a math test: Development of the Subjective Numeracy Scale'. *Medical Decision Making*, 27, 672–80.

Gibson, R., Callison, C. and Zillmann, D. (2011), 'Quantitative literacy and affective reactivity in processing statistical information and case histories in the new'. *Media Psychology*, 14, 96–120.

Gibson, R., Hester, J. B. and Stewart, S. (2001), 'Pull quotes shape reader perceptions of news stories'. *Newspaper Research Journal*, 22(2), 66–78.

Gibson, R. and Zillmann, D. (1993), 'The impact of quotation in news reports on issue perception'. *Journalism Quarterly*, 70(3), 793–800.

Gibson, R. and Zillmann, D. (1994), 'Exaggerated vs. representative exemplification in news reports: Perception of issues and personal consequences'. *Communication Research*, 21(5), 603–24.

Gibson, R. and Zillmann, D. (1998), 'Effects of citation in exemplifying testimony on issue perception'. *Journalism & Mass Communication Quarterly*, 75, 167–76.

Gibson, R. and Zillmann, D. (2000), 'Reading between the photographs: The influence of incidental pictorial information on issue perception'. *Journalism & Mass Communication Quarterly*, 77(2), 355–66.

Greene, K., Campo, S. and Banerjee, S. C. (2010), 'Comparing normative, anecdotal, and scientific risk evidence to discourage tanning bed use'. *Communication Quarterly*, 58, 111–32.

Gunter, B. (1987), *Poor Reception*. Hillsdale, NJ: Lawrence Erlbaum.

Harrison, S. (2016), 'Journalists, numeracy, and cultural capital'. *Numeracy*, 9(2), Article 3. DOI:http://dx.doi.org/10.5038/1936-4660.9.2.3.

Haskins, J. B. (1981), 'The trouble with bad news'. *Newspaper Research Journal*, 2(2), 3–16.

Hibbard, J. H., Peters, E., Dixon, A. and Tusler, M. (2007), 'Consumer competencies and the use of comparative quality information: It isn't just about literacy'. *Medical Care Research and Review*, 64(4), 379–94.

Higgins, E. T. (1996), 'Knowledge activation: Accessibility, applicability, and salience'. In E. T. Higgins and A. W. Kruglanski (eds), *Social Psychology: Handbook of Basic Principles*, 133–68. New York: Guilford.

Hinnant, A., Len-Rios, M. E. and Young, R. (2013), 'Journalistic use of exemplars to humanize health news'. *Journalism Studies*, 14(4), 539–54.

Hornikx, J. M. A. (2007), 'Is anecdotal evidence more persuasive than statistical evidence? A comment on classic cognitive psychological studies'. *Studies in Communication Sciences*, 7(2), 151–64.

Iyengar, S. and Kinder, D. R. (1987), *News that Matters*. Chicago: University of Chicago Press.

Kahneman, D. (2001), 'A perspective on judgment and choice: Mapping bounded rationality'. *American Psychologist*, 58(9), 697720.

Kahneman, D., Slovic, P. and Tversky, A. (1982), *Judgment under Uncertainty: Heuristics and Biases*. Cambridge: Cambridge University Press.

Knobloch-Westerwick, S., Johnson, B. K., Silver, N. A. and Westerwick, A. (2015), 'Science exemplars in the eye of the beholder: How exposure to online science information affects attitudes'. *Science Communication*, 37(5), 575–601.

Kopfman, J. E., Smith, S. W., Ah Yun, J. K. and Hodges, A. (1998), 'Affective and cognitive reactions to narrative versus statistical evidence organ donation messages'. *Journal of Applied Communication Research*, 26, 279–300.

Lipkus, I. M. (2007), 'Numeric, verbal, and visual formats conveying health risks: Suggested practices and future recommendations'. *Medical Decision Making*, 27, 696–713.

Lipkus, I. M. and Hollands, J. G. (1999), 'The visual communication of risk'. *Journal of the National Cancer Institute Monographs*, 25, 149–63.

Lipkus, I. M., Samsa, G. and Rimer, B. K. (2001), 'General performance on a numeracy scale among highly educated samples'. *Medical Decision Making*, 21, 37–44.

Maier, S. R. (2002), 'Numbers in the news: A mathematics audit of a daily newspaper'. *Journalism Studies*, 3(4), 507–19.

MacGregor, D. and Slovic, P. (1986), 'Graphical representation of judgmental information'. *Human-Computer Interaction*, 2, 179–200.

National Center for Educational Statistics (2005), *The Nation's Report Card: 12th-Grade Reading and Mathematics*. Washington, DC: US Department of Education.

Nisbett, R. and Ross, L. (1980), *Human Inference: Strategies and Shortcomings of Social Judgment*. Englewood Cliffs, NJ: Prentice-Hall.

Organization for Economic Cooperation and Development (2012), *Survey of Adult Skills*. http://www.oecd.org/skills/piaac/surveyofadultskills.htm).

Peters, E. (2012), Beyond comprehension: The role of numeracy in judgments and decisions. *Current Directions in Psychological Science*, 21, 31–5. doi:10.1177/0963721411429960.

Peters, E., Hibbard, J. H., Slovic, P. and Dieckmann, N. F. (2007), 'Numeracy skill and the communication, comprehension, and use of risk and benefit information'. *Health Affairs*, 26, 741–8.

Reyna, V. F. and Brainerd, C. J. (2007), 'The importance of mathematics in health and human judgment: Numeracy, risk communication, and medical decision making'. *Learning and Individual Differences*, 17, 147–59.

Reyna, V. F., Nelson, W. L., Han, P. K. and Dieckmann, N. F. (2009), 'How numeracy influences risk comprehension and medical decision making'. *Psychological Bulletin*, 135, 943–73.

Sanfey, A. and Hastie, R. (1998), 'Does evidence presentation format affect judgment?: An experimental evaluation of displays of data for judgments'. *Psychological Science*, 9, 99–103.

Schwartz, L. M., Woloshin, S., Black, W. C. and Welch, G. H. (1997), 'The role of numeracy in understanding the benefit of screening mammography'. *Annals of Internal Medicine*, 127, 966–72.

Sherman, S. J. and Corty, E. (1984), 'Cognitive heuristics'. In R. S. Wyer and T. K. Srull (eds), *Handbook of Social Cognition*, Vol. 1, 189–286.

Slovic, P., Finucane, M., Peters, E. and MacGregor, D. G. (2002), 'The affect heuristic'. In T. Gilovich, D. Griffin and D. Kahneman (eds), *Heuristics and Biases: The Psychology of Intuitive Judgment*, 397–420. New York: Cambridge University Press.

Steen, L. A. (2004), *Achieving Quantitative Literacy: An Urgent Challenge for Higher Education*. Washington, DC: The Mathematical Association of America.

Taylor, S. E. and Fiske, S. T. (1978), 'Salience, attention, and attribution: Top of the head phenomena'. In L. Berkowitz (ed.), *Advances in Experimental Social Psychology*, Vol. 11, 249–88. New York: Academic Press.

Tversky, A. and Kahneman, D. (1973), 'Availability: A heuristic for judging frequency and probability'. *Cognitive Psychology*, 5, 207–32.

Wickham, K. (2003), *Math Tools for Journalists*. Portland, OR: Marion Street Press.

Zillmann, D. (2002), 'Exemplification theory of media influence'. In J. Bryant and D. Zillmann (eds), *Media effects: Advances in Theory and Research*, 2nd edn, 213–45. Mahwah, NJ: Lawrence Erlbaum.

Zillmann, D. and Brosius, H.-B. (2000), *Exemplification in Communication: The Influence of Case Reports on the Perception of Issues*. Mahwah, NJ: Lawrence Erlbaum Associates.

Zillmann, D., Callison, C. and Gibson, R. (2009), 'Quantitative media literacy: Individual differences in dealing with numbers in the news'. *Media Psychology*, 12, 394–416.

Zillmann, D., Gibson, R. and Sargent, S. L. (1999), 'Effects of photographs in news-magazine reports on issue perception'. *Media Psychology*, 1, 207–28.

Zillmann, D., Gibson, R., Sundar, S. S. and Perkins, J. W. (1996), 'Effects of exemplification in news reports on the perception of social issues'. *Journalism & Mass Communication Quarterly*, 73(2), 427–44.

11

Numbers in the News: More Ethos than Logos?

Willem Koetsenruijter, *Leiden University, the Netherlands*

Introduction

Our daily dose of news is full of numbers: casualties in a war, percentages of houses sold under a certain price, the likelihood of making a full recovery from cancer, the number of decibels produced by traffic on a highway, the average temperature last winter, the annual profit rates of a big multinational or the proportion of cars driving respectively on diesel, gas or electricity. In a quick scan of all the 226 headlines of the *Daily Mail Online* of 13 June 2016, we find almost one hundred numbers, varying from ages to amounts of money, and from casualties to heights and lengths. And these are only the headlines. Some headlines may not have any numbers, but the articles under them will be likely to be permeated with numerical data. An article headlined 'Passenger Jet Service from U.S. to Cuba starts Wednesday', for example, almost certainly contains information about the number of flights and passengers, et cetera.

More systematic research has shown the pervasiveness of numbers in the news. Wester, Pleijter and Hijmans (2006, 79) found descriptive statistics – averages, percentages, modes and proportions – in 53 per cent of a sample of 623 news articles. Maier (2002, 510) found in a one-month sample of news in a local newspaper that two-thirds of the stories contained 'some sort of mathematical calculation or numerical point of comparison was required'. Zillman and Brosius (2000, 42) report in their study of exemplification that precise quantities, such as frequencies, ratios or rates of change, occurred in 44 per cent of their collected news stories. The frequency of numbers, of course, is not the same across all news, but is related to news genres, news subjects and the news medium. Hard news probably contains more numbers than softer genres such as human interest. The economic section on a news website definitely contains more numerical information than the lifestyle or culture section. News on election is loaded with statistics, while a report about a visit by the pope most likely is not. And even different media generate different amounts of statistical data. A radio show on election polls can 'display' fewer numbers than a television item on the same subject, or an online section, which can be packed with infographics.

That is hardly surprising from a journalistic perspective. Data and statistics are believed to give news an aura of factuality and credibility. Facts expressed in numbers are more precise than facts without numbers. A journalist who writes about 'a crowd' could have been at home during the demonstration. But a journalist who writes about a crowd of 2,500 people has been there, has been counting and sounds therefore more reliable. Thus, it seems logical that journalists use numbers so extensively and intensively to boost the credibility and impact of their stories.

But what about the audience? How do news consumers receive and process numbers in the media? Based on their frequency in the news, one might expect that numbers are processed well by news consumers, that they understand them and remember at least some of the numbers, and that numerical information sometimes changes their minds about events and issues. The reality, however, is not that simple. This chapter will first review the relevant research literature to demonstrate that the impact of numbers on the way news consumers process, understand, recall and perceive news facts and issues might not be as strong as expected. It will, however, then present a series of my own experiments that show numbers do seem to have an impact on the perceived credibility of news stories among users. It concludes with a further discussion of this potential number paradox to ask whether numbers are not much more than a rhetorical device.

How well do consumers process numbers in the news?

Despite the pervasive presence of statistics in the news, there is some general evidence to suggest that we cannot be too hopeful about what people *understand* from them. Zillman, Callison and Gibson (2009, 395) evaluate the numeracy of the average American citizen as poor. Citing the National Mathematics Panel (2008), the National Center for Education Statistics (2007) and Steen (2004), these authors observe that 'surprisingly large numbers of citizens fail to become proficient in the elementary manipulation of frequencies into ratios, especially into percentages and probabilities' and 'much of the workforce is considered deficient in these arithmetic skills', with only about one quarter of 12th grade students meeting the criterion of proficiency in standardized math exams. And, according to Lipkus, Samsa and Rimer (2001, 37), 'that even highly educated participants have difficulty with relatively simple numeracy questions'. In Europe, too, there are significant reports that express concerns about the level of mathematical skills in general. The European Commission (2013, 13) concluded that 'the current situation of more than 20 per cent of young Europeans not reaching a minimum level of basic skills in mathematics and science is alarming'.

Moreover, when it comes to how much people actually learn and remember news facts – including numerical and non-numerical facts – from the media, the issue is not as straightforward as common wisdom might suggest. In their overview of research on news processing and retention, Price and Feldman (2008) conclude that audience characteristics, the kind of media, and message features are important factors that influence the recall of news. In another overview of research, Gunter (2015) mentions a number of important factors affecting television news recall, including pre-existing topic knowledge, story characteristics, the kind of people involved, and whether a news item is visualized. Petty and Cacioppo (1983) and others have also shown the moderating role of personal involvement in remembering information. We know, for instance, that multitasking has a negative effect on what we remember (Angell et al. 2016), that age has an effect on what we remember from the news (Hill et al. 1989), and that connected news facts are better remembered than single, unrelated facts (Lang 2000). Mesbah (2005) found from his experiment that the nature of the medium has some influence: written news is better memorized than televised news, not because written news is easier to process, but because 'reading news via a computer screen or a newspaper is a cognitively more demanding task that results in better levels of memory recognition' (2005, 4).

Although there has been much research into what *influence* our memory of news, not many studies are concerned with how much we *remember* from the news. Some existing evidence, however, suggests a little worrying picture. Van Gorp (2006, 221) found that only 1 per cent of the readers remembered the correct heading of news articles after half an hour, and 7 per cent knew part of the heading. Hargreaves, Lewis and Speers (2003) reported that only 38 per cent of their questions about news facts were answered correctly, even though the issues questioned received regular attention in the media and had serious implications for public policy. Their study also shows that people confuse details. In line with this and other studies (e.g. Gunter 1987, 15), Gunter (2015) concludes that 'when viewers are tested for what they can remember from a television news programme they recently watched, most of its contents are lost to them. Furthermore, when specific news stories are recalled or recognized afterwards, viewers' understanding of specific details can be confused'. Memory for details performed poorest overall in surveys about remembering television news (Gunter 2015, 57).

In light of this literature, it might be safe to assume that numbers, as story details, would suffer, probably even more severely, from those problems. Little research has been done on the more specific question of whether and how we remember *numbers* from the news, but some evidence has emerged. Barrio, Goldstein and Hofman (2016) showed how adding a perspective to numbers in the news helped readers to recall them. In one experiment, they investigated the ability to remember or at least to approximate numerical quantities that one has read. Subjects read news headlines and – after two minutes of playing a Tetris computer game – were asked to reproduce the numbers. Only 40 per cent of the subjects could recall the number exactly. That percentage can be judged as quite low, bearing in mind that the recall was not after an hour, or a few days, but after only two minutes while subjects knew that they were involved in an experiment, which probably made them more attentive and alert to the news content than news consumers normally are.

It should be noted that the above argument might apply only to numbers as news story details – for example, as evidence to support the main news fact, to show that something is bad, huge, not well distributed, too small or too big. In some cases, a specific number *is* the news itself – as in '400 new cases in Catholic Church Abuse' or 'Sun Energy Vehicle speeds up to 250 km/h'. Church abuse is not the news, but the fact of another 400 is. Just driving fast in a solar energy vehicle is not news, but passing the limit of 250 kilometres an hour is. In these cases, the central role of the number as the main news detail might provoke some differences in consumers' recall. But when numbers are partial details in the story, our expectations about what news consumers remember from and/or are influenced by them should not be too optimistic.

Numbers and public perception of news issues

If consumers tend to forget most numbers in the news and do not understand them well at all, could it be that numbers can still persuade people into making informed decisions about news and current affairs?

The way people are persuaded by complex numerical information can be expounded by dual-coding theories, such as Petty and Cacioppo's Elaboration Likelihood Model (Petty and Cacioppo 1984; Petty, Cacioppo and Schumann 1983), Chaiken's heuristic systematic model (Chaiken and Trope 1999) and system-1 and system-2 thinking in the work of Tverski and Kahneman (as summarized in Kahneman 2011). These theories predict that people will in many cases not use statistics to form their opinions. They are, as Kahneman states, just not used to it. And in many everyday situations it is simply too much effort or not possible because we do not have all the arguments on which to base a decision. The core idea of these two-system theories is a dichotomy in two modes of deliberating: system 1 is a fast, emotional and instinctive system, while system 2 is slower, more logical, deliberative, and based on arguments and statistics. In daily routines, such as buying a cucumber, reading a newspaper or choosing a shirt in the morning, people tend to use system 1. They use rules of thumb to decide what they think about something, or whether they think something is convincing or not. Such rules can be that they believe what their peer group think, what a journalist says or what they have seen in one appealing example. It is not seldom that decisions are based on these routines.

A large body of news research has by and large confirmed this. Although authors such as Hornikx (2005, 205) argue that 'statistical and causal evidence is more persuasive than anecdotal evidence', O'Keefe (2002) concludes otherwise from his overview of empirical research on persuasion. Noting some inconsistency across research results, O'Keefe (2002, 229) nevertheless concludes that 'despite the seemingly greater informativeness of statistical information', 'it is clear that a single exemplar can be more persuasive than parallel statistical summary information'. Olsen (2015) investigated how people make sense of 'hard data' when evaluating organizations. Although her research is not about news, the results are potentially transferable. She offered one group of respondents results from satisfaction reports of hundred former patients and another experience stories of one former patient. She found three differences between these groups: the respondents said they had a strong preference for statistical data if asked to evaluate an organization, but episodic information appeared to have a stronger impact on the subjects and appeared to be more emotionally engaging than statistics: 'If asked to immediately recall recent performance information about public services, citizens report more nuanced and elaborate information about personalized stories and experiences than about statistics or numbers' (Olsen 2015, 1).

Zillman and Brosius (2012) found similar responses to numbers from their investigation into the persuasive effects of exemplars against what they call base-rate information in the news. That is information on how often something normally occurs (e.g. if one out of 10.000 people commits a crime, the base rate for crime is 1 out of 10.000 or 0.01 per cent). Not always is this kind of information of any influence on the issue perception: when it comes to spectacular or sensational cases, according to Zillman and Brosius, adding quantitative information does not show a clear effect on user perception. As the researchers observe: 'Exemplification featuring atypical, spectacular, and sensational cases tends to foster distorted issue perceptions, even when supplemented with corrective base-rate information' (Zillman and Brosius 2012, 93).

It is noteworthy, however, that some studies show that adding visual numeric information, like bar charts, graphs, infographics or formulae, has a positive effect on the reader's judgement about the credibility of an ad (e.g. Kilyeni 2013) or general information about medical issues (e.g. Tal and Wansink 2016). In line with findings about news recall, the impact of numbers also depends on audience characteristics, for example people's knowledge of and/or involvement with the subject and their ability to process and apply numerical information, especially in the case of more complex statistics (Mérola and Hitt 2016).

In short, although the above review covers a very small part of a huge amount of research into this subject, it is safe to conclude that there is no common ground for the general statement that numerical information in news influences or persuades people better than episodic information or examples.

Numbers as a rhetorical device

Following on from the above, a paradox seems clear: journalists use a lot of numbers but their audiences neither remember nor understand them, nor are noticeably influenced by them. A big question then arises: Why do journalists use these numbers? One reason, as Kevin McConway observes in Chapter 13, is a special status of numerical facts in our society, 'the notion of numerical facts as somehow having an objective and value-free status'. On the other hand, however, it is widely accepted in social sciences and humanities that terms like 'truth' or 'facts' are at least somewhat awkward in a philosophical sense. Facts – numerical or non-numerical – are primarily seen as the result of a process of social construction, rather than as objective units of knowledge. In other words, they are seen as products of social activity and the outcome of human choices and efforts. So, journalists tend to worship statistics not necessarily because of their truthfulness but because the intrinsic and ostensible authority of numbers as objective facts well serves their professional ideologies and principles. In a recent piece, Tom Harford (2016) quotes an editorial in the *Guardian* which states that numerical claims would settle no arguments and persuade no voters. Harford points out that this 'statistical capitulation' of the *Guardian* is 'a natural response to the rise of statistical bullshit – the casual slinging around of numbers, not because they are true or false, but to sell a message'. It is somewhat bluntly formulated, but the idea that 'numbers sell' is widespread. And it is at this point that rhetoric comes into the story.

Why numbers sell can be explained by the fact that numbers are – in the eyes of a naive news public – associated with exactness and details. Joel Best (2003, 43) calls statistics 'little nuggets of truth'. And he concludes that 'the public tends to agree: we usually treat statistics as facts'. Teun van Dijk (1988, 88) introduces the notion of a number game in the news. News represents a special kind of discourse where 'rhetorical structures regulate effective comprehension and especially opinion formation and change' (88). Just like metaphors and irony, they serve as rhetorical devices to attract attention. They help readers memorize the author's point of view and ultimately they change people's minds. The use of numbers can be seen in the same perspective as a rhetorical device. Van Dijk (1988, 88) emphasizes that the point with numbers is not precision or the content that numbers add to the text that does the job, but the fact that numbers *are given at all.*

In classical rhetorical terms the use of numbers in news articles can be interpreted as a function of ethos, as a way of establishing the author's reliability or trustworthiness and therefore the trustworthiness of the information and the source of the information. In line with this idea, Roeh and Feldman (1984) see numbers in news articles as 'agents of a rhetoric of objectivity; that is, they contribute to an impression of nothing-but-the-facts-journalism' (Roeh and Feldman 1984, 347). 'They are mobilized to serve the journalism-as-information-model – they contribute to credibility, to facticity' (Roeh and Feldman 1984, 350). Using numbers to transfer information or for quantification is, according to Porter (1996), also a 'technology of distance'. It takes the content away from the author and transforms it into objective facts, unconnected to a writer or sender. In that way, they become the neutral, transparent signs for objectivity or reality that Roeh and Feldman (1984) refer to.

A comparable idea is found by Kilyeni (2009, 2013). In her paper on cosmetic advertisements she shows how numerals are used 'not so much to indicate precision, but to catch the attention, to enhance the credibility of the advertising message and/ or to make the message more emotionally loaded, in order to persuade women to buy the advertised products' (2013, 17). Another phenomenon she refers to is the fact that numbers *connote scientific knowledge* and the prestige and authority that comes with science (Kilyeni 2009). Using numbers, copywriters *mimic scientific discourse*. And in her view, it is not so much the exact or accurate quantification itself that makes numbers so valuable in advertisements: the fact that they are there and look like science increase persuasiveness. What Kilyeni says about advertisements might well be true for news too.

Numbers and the perceived credibility of news

These ideas about the rhetorical role of numbers in news fit perfectly in Kahneman's system-1 thinking. Of course, one cannot really decide whether the information in every news article is reliable. In addition to simple system-1 heuristics about, for example, the source of the news ('It is reliable, because it is in the *New York Times*'), another rule of thumb exists that says: 'It is reliable because there are numbers or statistics involved.' And that would activate a frame that research has been conducted, that someone has carefully done the counting, and that it looks like science.

If this is what numbers do, then it should show in people's judgement about news story credibility. Koetsenruijter (2008 and 2011) reports a series of experiments that test the idea of whether numbers indeed function as rhetorical devices to enhance credibility.[1] The overall design of the experiments is a split-plot with two sets of news articles derived from regular newspapers. The articles deal with typical numerical subjects, are all more or less of the same length (90–115 words) and contain different kinds of numerical information. Each set of ten articles consists of five original articles taken from actual newspapers and five that are rewritten from the actual articles in a way that replaces numerical information with common quantifiers such as 'a lot' 'more', 'less' and so on. Besides these ten test items, each set contains another five articles of the same length, with softer, non-factual, non-numerical news articles. These five items function as fillers to distract subjects from the goal of the experiment. The articles are distributed among the subjects as follows:

	With numbers	**Without number**	**Fillers**
Group 1	Articles 1–5	Articles 6–10	Articles 11–15
Group 2	Articles 6–10	Articles 1–5	Articles 11–15

Here is an example of a (translated) article in the two conditions, with (1) and without (2) numbers:

1. Managers smoke more pot than students

    Brussels. It is not the alternative lefty or the disappointed nihilist but the young, stressed manager that is the greatest cannabis smoker. European research shows that among young adults between fifteen and twenty-four years, the use of cannabis is the highest among successful businessmen and managers. Of them 23.3 per cent

---

[1] Koetsenruijter (2008) reports two experiments with respectively fifty-two and sixty-five bachelor arts students as subjects. In the first one the news articles are presented on paper; in the second experiment they are presented as a video for this purpose from a recorded news show in co-operation with a small local television station. Credibility is measured on a two-dimensional scale. In Koetsenruijter 2011 two series of experiments are reported. The first series consists of three experiments with 52, 57 and 65 bachelor arts students. Credibility is measured on a dichotomous scale. The three experiments are a replication of the 2008 article; in the second and third experiment a new set of ten news articles was used. In the second series another ten news articles were constructed in three versions: without numbers, with four numbers and with eight numbers. These articles were presented in a vignette design in different booklets with two-point, four-point, five-point scales and one with a semantic differential. The data derived from the various scales were recoded in such a way as to make the average for each scale 3. Data in this last experiment were collected from 324 subjects – bachelor students from a diversity of fields and train commuters, evenly divided over the four options.

regularly smokes a joint. Among students that percentage is only 10.7. The so-called Euro-barometer Evaluation of some 7,600 European Young people indicates that that the use of cannabis is higher in France than in the Netherlands, while France strongly disapproves of the Dutch drugs policy. In France 44.9 per cent of the youth population has experience with cannabis; in the Netherlands that percentage is 35.3.

2. Managers smoke more pot than students

Brussels. It is not the alternative lefty or the disappointed nihilist but the young, stressed manager that is the greatest cannabis smoker. European research shows that among young adults the use of cannabis is the highest among successful businessmen and managers. More often than students these managers regularly smoke a joint. The so-called Euro-barometer Evaluation indicates that that the use of cannabis is higher in France than in the Netherlands, while France strongly disapproves of the Dutch drugs policy. In France, more young people has experience with cannabis than in the Netherlands.

In Koetsenruijter 2008 as well in 2011 the medium was varied. In two of the experiments, the examples were given as a video of a news show by a professional news presenter as in a real television news bulletin. We created two versions of the news show, one with the news presenter reading version one of the experiment and another with her reading version two (see table above). The two versions of the television show corresponded to the two paper versions.

In each experiment, half of the subjects (Dutch undergraduate arts students in a classroom setting) read/watched news stories 1–5 in original versions (with numbers) and articles 6–10 in rewritten versions (without numbers). The other half read/watched news articles 1–5 in the rewritten versions (without numbers) and articles 6–10 in the original versions (with numbers). The fillers were for both groups the same. Subjects were randomly assigned to one of the two sets and asked to rate the articles on a credibility scale. The concept of credibility was explained by the test leader as the extent to which subjects *believe* the article and think it is *true*. In general, four principal components of media credibility can be distinguished: source, medium, organization and message characteristics (Saleh 2016). In the experiments, the first three components are not varied. For each item, subjects have no knowledge about the source. The medium and organization are the same, and only the message characteristics, that is, the specific content (news subject) and the presence of numbers, are varied.

The results show that news articles with numbers were indeed more often judged as credible. The difference in the two conditions was large enough to qualify as statistically significant. To check whether subjects were aware of the test design and of the use of numbers as a variable, they were asked what was decisive for their choice. Only a very small number mentioned statistics. The rather vague notion of 'content' was mentioned most often. Despite the considerable varieties across the experiments – for example three different sets of news articles, three different credibility scales, and a semantic difference – the principal outcome did not change: news articles with numbers were more often judged as credible and as more credible than the same articles without the statistics. The effects proved to be robust across the experiments and there were no significant differences in print as well as in oral video news presentation.

The above experiments, however, had some disadvantages. First, there could have been some sort of bias introduced during rewriting the news texts without numbers, causing a somewhat lower validity of the experiment. These were typical news articles where numbers are expected, so the rewritten versions – not by a professional journalist or editor – could cause a feeling of awkwardness and thus influence the judgement of their credibility. Second, the relatively small number of only ten different news articles, each with their specific content, could have influenced the judgement.

With the above critiques in mind, I recently repeated the experiments in an adapted setting. In order to use more news items and at the same time keep the duration of the experiment at an acceptable level, headlines were used instead of complete articles. The headlines were collected during one week from professional (Dutch) news websites, in two variants: with or without numbers. All the material used was unmodified and copied directly from actual news sites. For fifty different news facts a version with and a version without numbers was collected. Thus, in contrast to the earlier experiments, the material was not rewritten or manipulated, but originated from actual professional news sites where professional journalists or editors had made a choice to produce a headline with or without numbers. Some translated examples of the same news fact in the two versions are:

Suicide attacks Yemen

- Lots of deaths at suicide attack in Yemen (Veel doden bij zelfmoordaanslag in Jemen. 29/08/16, Volkskrant)
- At least forty-five deaths at suicide attack in training camp Yemen (Zeker 45 doden bij zelfmoordaanslag trainingskamp Jemen; 29/08/16, NRC).

Coral found

- Large quantity of illegal coral found in barracks in Brabant (Grote hoeveelheid illegaal koraal in Brabantse loodsen aangetroffen; 29/08/16, Nu.nl)
- 2,000 kilograms coral, ivory and peltry of protected animals found in Berghem (2,000 kilo koraal, ivoor en huiden van beschermde dieren gevonden in Berghem, 29/08/16, Omroep Brabant).

Demo record McCartney

- Demo record Paul McCartney under the hammer (Demoplaat Paul McCartney onder de hamer, 29/08/16, showbizzsite.be)
- Demo record Paul McCartney auctioned for 21.000 euro (Demoplaat Paul McCartney geveild voor 21.000 euro, 29/08/16, Volkskrant.nl).

Fifty cases were collected, each in two versions. In this set, twenty extra headlines about less factual news were added as fillers to distract subjects from the goal of the experiment. The material was shown in a lecture room on a screen in an automated Powerpoint presentation. Each headline was visible for six seconds. Undergraduate arts students, who were doing a small course on journalism and new media, were asked to imagine that, as an intern for a news site, they had to pre-select news facts that are worth deepening in a follow-up article. It was emphasized that – of course – the facts they chose must be reliable, serious, true and palpitating. They were asked to choose approximately half of the headlines, 'but ten more or less, is not a problem'. They could give each headline a 1 or a 0 on a form. After the first thirty-five items, there was a break of thirty seconds, the instruction was repeated on screen and the next thirty-five items appeared. Subjects filled in their choice on a paper form. The experiment took 7.5 minutes. Each respondent was also asked to fill in a unique personal code. One week later they did the same test in a mirrored version: items with numbers in the first version were now presented in a version without numbers and vice versa. Cases were presented in a different order than the first time to impede recall. Subjects were again asked to fill in their personal code on the form.

After returning the results back in the same order and (with the personal code) matching the first and second experiment per person, the mean amounts of times an item was chosen in both conditions were compared to each other in a paired t-tests. Each subject was compared with his own choice the first time. In this way, the effect of the specific subject of each news item – an important factor in judging credibility – was ruled out. In the database, sixty-eight students filled in the experiment twice. Students who did not complete both versions were deleted from the database. Outliers who chose fewer than twenty or more than fifty of the seventy articles were deleted from the data.

There was a small but significant difference in the results of the headlines with and without numbers. The fifty headlines with numbers were chosen more frequently than the fifty headlines without numbers ($n = 68$, $t = 2.42$, $p = 0.018$). Comparing, on the other hand, the overall score in both batches, we found a non-significant difference in the overall mean score of all the items ($n = 68$, $t = 1.40 / p = 0.165$). This means that the difference could not be explained by a coincidental variance in the quality of one of the two sets. Afterwards, ten students were asked what they thought the experiment 'was really about'. Some mentioned 'the truth' of the news, the question of whether the news was 'real', whether some of the headlines were in 'the real news' or (in the second experiment) that it was about memory: they recognized some of the headlines from the former experiment. None of them mentioned numbers in the headings. It therefore seems reasonable to assume that they were not aware of what the manipulated variable was.

These results are compatible with the experiments described previously (Koetsenruijter 2008, 2011): they not only show that the effect of numbers on credibility is measurable, but also that the effect is repeatable and that it works for complete articles as well as for headlines. As the students had to make their choice in 6 seconds, the results also show that the effect works especially in situations where system-1 thinking is used. Six seconds is not enough to consider thoroughly whether the news behind a headline is credible. Nevertheless, it is far more time than an average news consumer spends with his finger on the computer mouse before deciding with a mouse click whether to read an article or not. If reliability is one of his/her criteria, we now know that the presence of a number will influence this choice.

Conclusion

In the first part of this chapter it was concluded that there is general evidence to suggest that we cannot be too hopeful about what people understand, remember or are influenced by the numbers in the media. Placing this in the context of the findings from the author's experiments, we have to worry that certain groups of news consumers might be vulnerable to the pervasive use of numbers as a rhetorical device in the news: it can convince people of the credibility of things that they do not understand. As Roeh and Feldman (1984) claim, numbers add to the impression of a nothing-but-the-fact journalism and numbers add to credibility even if the numbers at stake do not mean anything or are wrong or poorly understood. In combination with a bad memory, this provides a toxic cocktail that is a recipe for manipulation. By mimicking scientific language and suggesting preciseness and the availability of underlying research, journalists can reach a situation where uncertain suppositions become hard facts.

Another question at stake is whether these experiments are about news consumers or about journalists. The answer is that they are about a *variable*: the presence of numbers in news articles. We found an effect of this variable in a group of students, but did not structurally investigate whether professional journalists react different than bachelor arts students.[2] Further research can shed light on questions as, for example, whether people with high mathematical skills make different choices than people with low mathematical skills, or whether highly educated people make different choices than lower educated people.

The problem with numbers in news (for journalists *and* consumers) is not a *new* phenomenon, but with an enormous increase in data availability and accessibility, and the ensuing rise of data journalism, it is a growing problem. It presents a serious professional challenge for journalists in the age of big data: their job must be moving from simply bombarding a statistically uneducated public with numbers, to presenting all numbers in a format that is comprehensible and transparent. Such an improvement will enhance the trust in media. Further research on effective ways to present complex statistics for a broader audience should therefore be a crucial task for journalism research. Of course, this includes ways to represent numbers and statistics in graphs and infographics.

---

[2] In one of the experiments we also examined train commuters, but these results could not be used to structurally investigate differences between groups.

References

Angell, R., Gorton, M., Sauer, J., Bottomley, P. and White, J. (2016), 'Don't distract me when I'm media multitasking: Toward a theory for raising advertising recall and recognition.' *Journal of Advertising*, 45(2), 198–210.

Barrio, P. J., Goldstein, D. G. and Hofman, J. M. (2016), 'Improving comprehension of numbers in the news'. In *Association for Computing Machinery Conference on Human Factors in Computing Systems*, San Jose, CA (May 7–12).

Best, J. (2003), 'Audiences evaluate statistics.' In Loseke, D. and Best, J. (eds), *Social Problems: Constructionist Readings*, 43–51.

Chaiken, S. and Trope, Y. (1999), *Dual-process Theories in Social Psychology*. New York: Guilford Press.

Dijk, Teun A. van (1988), *News as Discourse*. Hillsdale, NJ, Hove and London: Lawrence Erlbaum Associates.

European Commission – Directorate-General for Education and Culture Thematic Working Group on Mathematics, Science and Technology (2010 – 2013) *FINAL REPORT Addressing Low Achievement in Mathematics and Science*.

Gorp, B. van (2006), *Framing asiel: Indringers en slachtoffers in de pers*. Leuven: Acco.

Gunter, B. (1987), *Poor Reception: Misunderstanding and Forgetting Broadcast News*. Hillsdale, NJ: Lawrence Erlbaum Associates Routledge.

Gunter, B. (2015), *The Cognitive Impact of Television News*. Houndmills and Basingstoke: Palgrave Macmillan.

Harford, T. (2016), How Politicians poisoned statistics. *Financial Times*, 14 April 2016. Available at: https://next.ft.com/content/2e43b3e8-01c7-11e6-ac98-3c15a1aa2e62.

Hargreaves, I., Lewis, J. and Speers, T. (2003), *Towards a Better Map: Science, the Public and the Media*. Swindon: Economic and Social Research Council.

Hill, R. D., Crook, T. H., Zadek, A., Sheikh, J. and Yesavage, J. (1989), 'The effects of age on recall of information from a simulated television news broadcast.' *Educational Gerontology: An International Quarterly*, 15(6), 607–13.

Hornikx, J. (2005), 'A review of experimental research on the relative persuasiveness of anecdotal, statistical, causal and expert evidence.' *Studies in Communication Sciences*, 5(1), 205–16.

Kahneman, D. (2011), *Thinking, Fast and Slow*. Macmillan.

Kilyeni, A. (2009), '"Hocus-pocusing" the Body. Technology and Femininity in Print Ads.' *Translation Studies. Retrospective and Prospective Views*, 2, 71–9.

Kilyeni, A. (2013), 'The Rhetoric of numbers in print advertisements for cosmetics.' *Buletinul Stiintific al Universitatii Politehnica din Timisoara, Seria Limbi Moderne*, 12, 17–26.

Koetsenruijter, W. (2008). 'How numbers make news reliable.' In L. Dam, L. Holmgreen and J. Strunck (eds), *Rhetorical Aspects of Discourses in Present-Day Society*, 193–205. New Castle upon Tyne: Cambridge Scholars Publishing.

Koetsenruijter, A. W. M. (2011), 'Using numbers in news increases story credibility.' *Newspaper Research Journal*, 32(2), 74.

Lang, A. (2000), 'The limited capacity model of mediated message processing.' *Journal of Communication*, 50(1), 46–70.

Lipkus, I. M., Samsa, G. and Rimer, B. K. (2001), 'General performance on a numeracy scale among highly educated samples.' *Medical Decision Making*, 21, 37–44.

Maier, S. R. (2002), 'Numbers in the news: A mathematics audit of a daily newspaper.' *Journalism Studies*, 3(4), 507–19.

Mérola, V. and Hitt, M. P. (2016), 'Numeracy and the persuasive effect of policy information and party cues.' *Public Opinion Quarterly*, 80(2), 554–62.

Mesbah, H. M. (2005), 'Reading is remembering: The effect of reading vs. watching news on memory and metamemory.' *Speaker and Gavel*, 42(1), 6.

National Center for Education Statistics (2007), *The Nation's Report Card: 12thgrade Reading and mathematics 2005*. Washington, DC: U.S. Department of Education.

National Mathematics Panel (2008), *National Mathematics Advisory Panel: Strengthening Math Education through Research*. Retrieved 12 December 2008 from http://www.ed.gov/about/bdscomm/list/mathpanel/factsheet.html

Olsen, A. L. (2015), 'Human interest or hard numbers? Experiments on citizens' selection, exposure and recall of performance information.'. In *Public Management Research Conference (PMRC)*, Minneapolis, MN, June (pp. 11–13).

O'Keefe, D. (2002), *Persuasion, Theory and Research*. 2nd edn. Thousand Oaks: Sage.

Petty, Richard E. and Cacioppo, John T. (1984), 'Source factors and the elaboration likelihood model of persuasion.' *Advances in Consumer Research* 11, 668–72.

Petty, R. E., Cacioppo, J. T. and Schumann, D. (1983), 'Central and peripheral routes to advertising effectiveness: The moderating role of involvement.' *Journal of Consumer Research*, 10(2), 135–46.

Porter, T. M. (1996), *Trust in Numbers: The Pursuit of Objectivity in Science and Public Life*. Princeton: Princeton University Press.

Price, Vincent and Lauren Feldman (2008), 'News processing and retention.' In W. Donsbach (ed.), *The International Encyclopedia of Communication*. Blackwell Publishing. Blackwell Reference Online. 03 August 2016 http://www.communicationencyclopedia.com/subscriber/tocnode.html?id=g9781405131995_yr2015_chunk_g978140513199519_ss26-1.

Roeh, I. and Feldman, S. (1984), 'The rhetoric of numbers in front-page journalism: How numbers contribute to the melodramatic in the popular press.' *Text. Interdisciplinary Journal for the Study of Discourse*, 4(4), 347–68.

Saleh, H. F. (2016), *Developing New Media Credibility Scale: A Multidimensional Perspective*. World Academy of Science, Engineering and Technology, International Journal of Social, Behavioral, Educational, Economic, Business and Industrial Engineering, 10(4), 1287–1300.

Steen, L. A. (2004), *Achieving Quantitative Literacy: An Urgent Challenge for Higher Education*. Washington, DC: The Mathematical Association of America.

Tal, A. and Wansink, B. (2016), 'Blinded with science: Trivial graphs and formulas increase ad persuasiveness and belief in product efficacy.' *Public Understanding of Science*, 25(1), 117–25.

Wester, F., Pleijter, A. and Hijmans, E. (2006), 'Instrument en codeerformulier; Wetenschap in de krant.' In Wester, F. (ed.), *Inhoudsanalyse: theorie en praktijk*. Nijmegen: Kluwer, 2006, 65–84.

Zillmann, D., Callison, C. and Gibson, R. (2009), 'Quantitative media literacy: Individual differences in dealing with numbers in the news.' *Media Psychology*, 12(4), 394–416.

Zillman, D. and Brosius, H. B. (2000). *Exemplification in Communication. The Influence of Case Reports in the Perception of Issues*. Mahwah, NJ and London: Lawrence Erlbaum.

Zillmann, D. and Brosius, H. B. (2012), *Exemplification in Communication: The Influence of Case Reports on the Perception of Issues*. Routledge.

12

Audience Uses and Evaluations of News Visualizations: When Does an Infographic say More than a Thousand Words?[1]

Yael de Haan, *University of Applied Sciences Utrecht, the Netherlands*
Sanne Kruikemeier, *University of Amsterdam, the Netherlands*
Sophie Lecheler, *University of Vienna, Austria*
Gerard Smit, *University of Applied Sciences Utrecht, the Netherlands*
Renee van der Nat, *University of Applied Sciences Utrecht, the Netherlands*

Introduction

Data visualizations, often referred to as infographics, have become a staple in news outlets around the world. Even though visualizations have been used in some form or other in journalism for several years now, the recent surge in their use represents their emergence as a storytelling genre in their own right, aimed at illustrating complex issues and events. With an audience that is increasingly visually literate, visualizations have also become more popular among journalists today. They are believed to both have the ability to attract news consumers to the news, and to enable the integration of 'big data' into the news (Schroeder 2004; Smit, de Haan and Buijs 2014). Additionally, researchers have suggested that visualizations may help news consumers to actually understand complex news topics (Bakker, de Haan and Kuitenbrouwer 2013), and that they are positively related to political knowledge and news use (Lee and Kim 2016). Yet, others argue that the increased use of visualizations may render the news more confusing to audiences, as visualizations can be too complex for audiences to understand and subsequently process (Brescani and Eppler 2009; Lester 2000).

Nevertheless, so far, a comprehensive empirical view of some of the most basic questions regarding the value of data visualizations in a journalistic context is lacking in the literature. Previous work focuses on how data visualizations are produced in a newsroom (Lowrey 2000, Smit et al. 2014), and the effects on news use and knowledge (e.g. Lee and Kim 2016) but leaves out the basic question regarding how audiences across different news platforms use and appreciate such visualizations. There has been

---

[1] This chapter is a modified and condensed version of the paper 'When Does an Infographic Say More Than a Thousand Words?' published online first in *Journalism Studies* in January 2017 (De Haan et al. 2017).

no empirical evidence to prove intrinsic assumption by journalists that audiences will both use and appreciate visualizations. This is particularly important in current digital media environment where different storytelling platforms might influence the readiness to read it (Adam, Quinn and Edmonds 2007; Bucher and Schumacher 2006). Therefore, this study aims to understand the value of digital news visualizations by determining (a) whether and how news consumers actually *use* such visualizations and (b) the extent to which news consumers appreciate them.

To provide an answer to two fundamental and interconnected questions, we conducted a comprehensive mixed-method study on news consumer evaluations of data visualizations. First, we analysed the extent to which news consumers pay *attention to* (i.e. use) data visualizations by means of an exploratory eye-tracking study. By using an eye movement approach, we are able to directly investigate participants' visual attention, rather than relying on self-reported data (Duchowski 2007). To take into account the changing nature of the news, we test attention to data visualizations in three different news modalities (i.e. print newspaper, e-newspaper on a tablet, and on a news website). Based on the results of this first part of our study, we then analysed news consumers' *appreciation or value* of visualizations by means of triangulation through focus groups and a survey. In focus groups, subscribers of two prominent newspapers and the largest online news website in the Netherlands were asked to provide an understanding of why they use visualizations in print and online news, and what aspects of these visualizations they might appreciate during the news reading process. The findings of this qualitative study are then quantified in an online survey of subscribers of the same media outlets. In adopting this mix-method approach to the expanding genre of news visualization, we also hoped to provide a fundamental empirical building block for future scholarly enquiries into the use and effect of visualizations and other innovative storytelling formats in journalism and mass communication.

Visualizations in the news: An upcoming storytelling genre

Visualizations have seen an unprecedented upsurge both in newspapers and online news media (Bekhit 2009; Smit, de Haan and Buijs 2014; Weber and Rall 2012). The broader transformation of journalism in a digital age including media convergence has created new forms of (visual) storytelling to tailor to the needs of the digitally savvy news consumer (Franklin 2014; Weber and Rall 2012). More specifically, the advent of big data provided a broader scope for visualizations in journalism: Journalists are increasingly confronted with a greater number of data sets that have also increased in size (Segel and Heer 2010). The need to analyse these big data sets – and to make them comprehensible and accessible to a broader range of people – leads to the increasing popularity of data visualizations. This development is also supported by new technologies and accessible design software (Smit, de Haan and Buijs 2014) that make it easier to include visualization into news stories. These range of factors led to the increasing number of designers joining newsrooms (Dick 2014; Giardina and Medina 2013; Schroeder 2004; Segel and Heer 2010) with visualizations increasingly viewed as a tool for storytelling (Segel and Heer 2010; Siricharoen 2013).

Attention to and use of news visualizations across platforms

While the popularity of news visualizations is on the rise, empirical evidence on this new genre of news storytelling remains scarce. Studies in other contexts, such as advertising (Radach et al. 2003) and education (Boucheix and Lowe 2010; Sung and Mayer 2012) have shown that the general use of visual cues in texts may function as an entry point for readers, getting them interested in the content of the article. Based on these findings, a handful of studies have focused on the use of visualizations in print newspapers, stating that visualizations catch readers' attention and extends the reading of the associated text (Holmqvist et al. 2003; Holmqvist and Wartenberg 2005).

Visualizations are considered to be attention-catchers since visual cues can have impact on news consumers' attention and news selection (Bucher and Schumacher 2006; Holsanova 2008). Visualizations are seen as incentives to motivate the user to read news articles and subsequently process information (Garcia and Adam 1991). While data visualizations are shown to be good entry points, they are less looked at comparing to other visuals, such as drawings and pictures (Holmqvist and Wartenberg 2005). Nevertheless, people read texts for a longer time when texts are accompanied by pictures (Holmqvist and Wartenberg 2005). Research examining the use of data visualizations has also found that they take longer than looking at pictures and drawings, and an explanation might be sought in the complexity of the data that is being visualized, as infographics are often used to visualize complex information (Holsanova, Holmberg and Holmqvist 2008). In short, previous research has demonstrated that the inclusion of visual elements into a news story increases the reading of news articles (i.e. visuals attract readers).

However, these studies cannot provide a definitive answer to the question of use, as they neglect the fact that consumers increasingly see visualizations on a computer screen or another electronic device but might not use them substantially (Newman, Levy and Nielsen 2015; Weber and Rall 2012). The modality or form in which a news visualization appears is bound to play a role in how news consumers use visualizations (Adam, Quinn and Edmonds 2007). Research shows that people who read news on websites are more attracted to text than to graphics (or navigation bars and teasers), while printed newspaper readers indicate the opposite (Bucher and Schumacher 2006; Lewenstein et al. 2000). In other words, attention to print media is 'appeal-driven and attention to online media is more content-driven' (Bucher and Schumacher 2006, 365).

Research on the use of visualizations has been conducted through eye-tracking method, as this method can directly show how texts with visuals are read (Garcia and Adam 1991; Holmqvist et al. 2003; Holmqvist and Wartenberg 2005; Holsanova, Holmberg and Holmqvist 2008). In the current study, we build on this work, combining eye-tracking, focus group and survey methods to investigate the extent to which news users devote attention and time to information visuals in the digital storytelling environment. Our first research question reads:

RQ1: Do news consumers actually use news visualizations, and how?

Determinants of a successful news visualization

Assuming that news consumers actually use or see data visualizations, our second question is whether they actually *appreciate* or value them. In a highly fragmented and competitive news market, where the continued existence of newspapers is questioned (Geldens and Majoribanks 2015; Siles and Boczkowski 2012), visualizations are often used to attract news consumers to a particular news outlet (Smit, de Haan and Buijs 2014). Within most news outlets, visualizations are crafted by experienced designers, who make decisions regarding positioning, visual cues, size and colour of the visuals. This is primarily based on their experience on what they think 'works best' on how the news consumer uses and appreciates the visualization (Cairo 2013; Wong 2010). However, naturally, this does not mean that news consumers will look at visualizations in the same way as designers and their editors do.

Hardly any study has tapped into the issue of news consumers' appreciation of visualizations (Lee and Kim 2016). While new technologies have made it possible for journalists to make use of accessible visualization software, and media organizations are hiring designers and data journalists to strengthen the team, we have little knowledge on how visualizations are appreciated by the news consumer and whether they are worth the investment for news organizations. Even though many factors can explain the appreciation of a visualization, including personal preferences, previous knowledge, and news use in the news subject, this study focuses on visualizations as a new storytelling genre, focusing on the design, and more specifically on the relationship between the design of the visual and the text of the news article.

Studies from different fields provide us with several insights regarding this issue. A first factor is how the visualization and the news story that accompanies it are related to each other. A number of studies have concluded that to understand the visualization, it needs to be accompanied by text (Beymer, Russel and Orton 2007; Sung and Mayer 2012). Even though the combination of text and visuals delays the reading time, it does not cause a cognitive overload (Beymer, Russel and Orton 2007). Numerous studies have even shown that the combination of text and visual elements improves the understanding of the story, compared with text only (Butcher 2006; Moreno and Valdez 2005). This is also the case with interactive visualizations (Burmester et al. 2010). More recently, Lee (2015) showed that that the use of visualizations within a textual story stimulates the news consumer to process the story more actively. However, this is primarily the case with news consumers who have less prior knowledge or are less involved in the issue of the news story.

A second factor is the function the visualization serves within the whole story. The comic theorist, McCloud (1994) makes a reference to how text and visual are related, particularly in the field of comics, and what role or function the visual has in relation to the text. Smit, de Haan and Van der Nat (2015) applied the analysis from McCloud when analysing more than 400 news visualizations and introduced three visual functions. The first, the additional function, indicates that the visualization provides additional information to the text. The visualization shows additional information related to the story that was not mentioned in text. The second, the supportive function, indicates that the visualization supports the text by visualizing the same information provided in the text. The text and visual tell the same story with the same information, but in a different form. The last one, the parallel function, shows that text and visualization tell different stories without intersecting or referring to each other. The visualization does not provide information related to the text. The function of the visualization in relation to the text can have influence on how the user reads and appreciates the visualization. An empirical study on visualizations shows that the repetition of information (additional or supportive function), in text and visual forms, facilitates the news consumer in processing the story (Lee and Kim 2016).

A third factor is related to the ease with which a news consumer can navigate through a visualization. This is based on the so-called signalling principle, which states that 'people learn more deeply from a multimedia message when cues are added that highlight the organization of the essential material' (Mayer 2005, 183). When there is no order in how to read the visualization, the reader has to make too many decisions – picking an entry point, choosing a reading order, finding relevant information and connecting the different elements. As a consequence, the material might be more difficult to understand and therefore less appreciated (Holsanova, Rahm and Holmqvist 2006).

A fourth determinant is the presence of cues that help the reader connect the text with the visualization. Textual references, colour coding, arrows or other cues can help the reader to read the text and visualization as one story (De Koning et al. 2007; Lyra et al. 2016). However, decorative elements that do not have a specific function can influence the learning process negatively as they distract the reader from the message (Poynter Institute 2004).

Taken together, studies from different areas provide us with valuable insights on what factors enhance the value of visualizations. Visualizations do not only function as an attention-catcher, but can also help the reader to process information, which in turn leads them to appreciating the information or story that is being read. In this study, we want to advance knowledge about individuals' appreciation of visualizations in both offline and online media outlets, by focusing on news consumers' evaluations of data visualizations through a triangulation of focus groups and a survey. Our second research question is as follows:

RQ2: Under which conditions do news consumers appreciate news visualizations?

Methodology

For this study, we focused on news media platforms that incorporate visualizations tailored at different news consumers in the Netherlands. We included two Dutch quality newspapers (*Trouw* and het *NRC Handelsblad*), a Dutch financial newspaper (*Het Financieele Dagblad),* and the largest online news site in the Netherlands (*NU.nl.*).[2] We chose a mixed-method approach with three complementary methods to be able to

---

[2] The two quality newspapers target highly educated newsreaders, who often have a subscription to the newspaper. The financial newspaper targets a niche market, mainly highly educated people

triangulate data and provide a comprehensive picture on the value of visualizations in the news today (Denzin 1978; Patton 1990).[3]

Eye-tracking

To examine how individuals use news visualizations (RQ1), 122 participants in a laboratory were randomly assigned to either read the quality newspaper *Trouw*, which included a visualization (bar and line graph), to read an e-version of the same newspaper on a tablet or to visit the website of the quality newspaper. Eye-tracking data was collected using *SMI EyeTracking Glasses* (ETG, for the newspaper and e-newspaper condition) and the *SMI RED eye tracker* attached to a 22-inch computer screen (for the website condition). We examined how long participants looked at the text, the visual and data within the visual using *dwell time in seconds*. In total, five variables were constructed, dwell time of the text, first visual (bar graph), second visual (line graph), data within the first visual and data within the second visual. All means and standard deviations are included in Table 12.2 in the results.

Focus groups

To understand how people appreciate news visualizations (RQ2), four focus groups were held with subscribers of the quality newspaper *NRC Handelsblad* and the financial newspaper *Het Financieele Dagblad* and readers of the online news platform *NU.nl*. The focus group sessions consisted of two parts. First, we provided each participant with a selection of news articles including a data visualization stemming from the respective news outlets.[4] The respondents were asked to rank the articles according to which article they would want to read first. Each group then discussed their general news reading patterns, their motivations for their ranking and their choice in relation to topic. Subsequently, participants were asked to read two full articles and visualizations, after which they again discussed their reading pattern, and asked for their motivation on their reading behaviour.

Survey

Based on the qualitative focus group data, we conducted a survey to reach a larger group of readers within the same news outlets as in the focus groups. A total of 503

---

working in the financial sector. The online news site is visited by a much broader group of news consumers, ranging in ages and education as it provides short topical news.

[3] For a more elaborate description of the method, see De Haan et al. (2017).

[4] We chose to show visual types that are often used in news outlets in the Netherlands; data visualizations such as maps and graphs. Moreover, we chose to include a variety of functions these data visualization can have (i.e. additional, supportive, or parallel to the text; Smit, de Haan and Van der Nat 2015) in order to understand the relationship between text and visual influences *and* to gauge the appreciation of the reader. Lastly, we chose articles with a variety of different topics of interest to the reader of the specific outlet.

Table 12.1 Descriptive results from the survey data

	NRC readers (*N* = 209)	*FD* readers (*N* = 83)	*Nu.nl* readers (*N* = 211)
Female (%)	46.90	8.40	50.20
Mean age (years)	45.30	53.44	45.49
Mean educational level (range: 1–7)	5.73	6.40	5.00

respondents participated in the survey. See Table 12.1 for more information on the characteristics.

Participants were first asked to view a news article that was also used in the focus groups. They were asked whether they would 'skip' the visual or whether they would look at the visual. In addition, two questions were asked that measured people's first impression of the visual. In the second part of the survey they were asked to read the article and look at the visual closely. Then questions were asked regarding the evaluation of the text and visuals. They were asked to what extent they understand the text better than the visual, whether they understand the visual and whether text and visual are related.

Results

Do news consumers actually use data visualizations, and how?

The first RQ asks whether news consumers read visualizations across platforms. From the eye-tracking data, we found that visualization is not avoided: both visualizations included in the eye-tracked article (bar and line graph) were noticed (respectively, 95.9 per cent and 89.3 per cent looked at both the text and visualization). However, while the bar and line graphs were noticed, the dwell time was significantly lower than for the text. On average, readers looked at the text for 44.24 seconds, while the visuals received only a limited amount of attention from the readers (respectively 3.74 and 3.86 seconds). When taking a closer look at the results, we observed that people do not always look at highlighted elements in the visual. Only 53.3 per cent and 25.4 per cent of the participants – respectively – noticed the two highlighted elements in the visual. Both elements are published in a much larger font and different colours, but these elements (which highlighted a number in the graphs) were not often noticed. Thus, the reader does not always look at the same entry points that the designer highlights.

Thus, our initial answer to RQ1 is that visualizations as part of an article do not go unnoticed. However, their dwell time is much lower than text, which indicates that the visualization is used to obtain an overall impression and is not looked at in detail. Furthermore, we did not find noticeable differences between modalities (only one highlighted element differed between conditions). It seems that the platform where the news is published does not influence the use of both texts and visuals (see Table 12.2).

Table 12.2 Differences in Dwell Time Between the Modalities in Seconds

	Text M (SD)	Visual (bar graph) M (SD)	Visual (line graph) M (SD)	Percentage M (SD)	Number M (SD)
Dwell time printed newspaper	32.96$_a$	2.85$_a$	3.37$_a$	0.04$_a$	0.31$_a$
	(40.07)	(3.79)	(4.96)	(0.13)	(0.67)
Dwell time e-newspaper	32.62$_a$	4.20$_a$	4.48$_a$	0.09$_a$	0.27$_a$
	(36.40)	(6.70)	(6.58)	(0.20)	(0.39)
Dwell time website	86.48$_b$	4.69$_a$	3.74$_a$	0.37$_b$	0.57$_a$
	(93.72)	(4.89)	(6.63)	(0.48)	(0.84)
Dwell time (overall)	44.24	3.74	3.86	0.13	0.35
	(58.71)	(5.27)	(5.93)	(0.29)	(0.63)

Note. Means with different subscripts within each column are significantly different at the $p < 0.05$ level.

In addition to our eye-tracking data, we used our other two methods to gain some further insights on the use of news visualizations. The focus groups show that interest in a topic correlates positively with the decision to read the article with the visualization. When respondents were asked to prioritize the visualizations according to reading interest, a respondent said: 'I put the visualization on the Dutch economy on the first place as the topic interests me, therefore I would probably read this first.' We also observed that participants' intention to look at the visualizations was related to the design of the visualization. A respondent clearly stated that she chose an article that visualizes trends in education systems across countries because 'this visual uses warm wood colors and I like the design. I usually do not look at graphs because I am afraid that I will not understand them, but this one is designed very nicely'.

The survey confirms that interest and importance of a specific topic and visual design play a role in the intention to use a visualization. For instance, we found that the intention to look at the visualization about politics (e.g. a visual about the Dutch labour party) was positively correlated with participants' perceived importance of politics ($r = 0.287$, $n = 105$, $p = 0.003$). We also demonstrate that this perceived importance of politics correlates with participants' attitudes towards the visualization (e.g. they believe the visual about politics is more beautiful and interesting, respectively $r = 0.371$, $n = 105$, $p < 0.001$ and $r = 0.469$, $n = 105$, $p < 0.001$). Furthermore, using the whole sample, results from a regression analysis (with intention to look at the visual as a dependent variable, see Table 12.3) also show that interest ($\beta = 0.656$, $p < 0.001$) is more important than aesthetics ($\beta = 0.142$, $p = 0.001$). Nevertheless, the survey also shows that aesthetics do play a role. When participants believe the data visualization is more beautiful and interesting, they are also more likely to look at the visualization (see Table 12.4, respectively, $r = 0.627$, $p < 0.001$ and $r = 0.771$, $p < 0.001$).

Table 12.3 Regression Analysis Predicting Intention to Look at the Visual

	Intention to look at the visual		
	B	*(SE)*	*β*
Constant	0.605	(0.368)	
Female	− 0.184*	(0.093)	− 0.057*
Age	0.003	(0.003)	0.026
Education	0.024	(0.037)	0.019
The visual is beautiful	0.168**	(0.049)	0.142**
The visual is interesting	0.724***	(0.047)	0.656***
R^2	0.608		

Note. $*** p < 0.001$, $** p < 0.01$, $* p < 0.05$

Table 12.4 Pairwise correlations between different visual evaluations

Visual evaluations	(1)	(2)	(3)	(4)	(5)
Would look at the visual (1)					
Beautiful (2)	0.627*				
Interesting (3)	0.771*	0.732*			
Helps me to understand the text (4)	0.496*	0.435*	0.535*		
Easy to understand (5)	0.287*	0.249*	0.301*	0.406*	
Visual and text are in concordance (6)	0.351*	0.346*	0.403*	0.546*	0.366*

Note. * Correlation is significant at the $p < 0.01$ level

In sum, our answer to RQ1 is that participants do use news data visualizations and that they do so not only because it is there, but also because they are interested in the topic it covers. Nevertheless, the look and feel of the visual influences its use, and visualizations that are deemed approachable or aesthetically pleasing are used more.

Under which conditions do news consumers appreciate data visualizations?

Knowing that readers look at visualizations, and the topic of interest and aesthetics play a role in the intention to look at it, we now turn to the question on how visualizations are appreciated (RQ2). Whether a news consumer appreciates a visualization is related to both the form and function of the visualization. Relating to the form, when the moderator of the focus groups asked the respondents to read a two-page article on how the stock exchange works, the majority of the respondents agreed that the visual was attractive and that it helps to understand a complex topic.

Whether they feel that they understand the article is not only related to the visualization, but also to the layout of the entire story and the integration of text and visual. The page layout with different text frames, headlines and titles invites the reader to start reading. A respondent explained: 'I like this article, the design is playful and the different pieces of information makes it manageable to read.'

Not only the form but also the function of the visualization influences the appreciation. The function or objective of the visualization in relation to the text needs to be clear. A focus group was invited to read an article about innovation in education. While the design of the visualization and the article layout was found to be attractive, once the participants were asked to read the story, their positive attitude towards the article altered as the text and visual did not tell the same story. A participant said: 'The graphs are unclear, they [the newspaper] focused so much on the form, that the story they want to tell is not clear.' Another group was also confronted with the same dilemma, as one participant explained, referring to an article on wind energy including a map of the Netherlands in which the different wind energy projects were visualized: 'The visualization does not correspond with the text. The visualization is all about the wind farms, but the text is about a particular organization. There is no reference to each other [text and visual].' In other words, form and function do not correspond.

Not only the integration of text and visual is important for the reader to appreciate the story, also the aesthetic elements, such as colour, symbols and other visual cues, need to serve a particular function. The focus groups show that when symbols and metaphors are misinterpreted, it can lead to misunderstanding and less appreciation. Respondents were asked to read a story on economic growth in which a graph was included with a large black wallet in the background of the graph. To the participants, the type of wallet, an old-fashioned design, which is usually only used for coins in the Netherlands, did not correspond with the story. The story was about economic growth, while the wallet symbolized a shrinking economy for the readers.

The findings of our focus groups are supported by our survey data. Participants who found the visualization attractive also believe that the article and the visualization are easier to understand. Participants who believe that the visual is beautiful and interesting also note that the visual is easy to understand (see Table 12.4 for correlations). In other words, the form plays a role in the perceived understanding and therefore appreciation of the visualization. Besides the form, the survey data reaffirm that the function of the visualization influences the appreciation of the visualization, with the survey respondents feeling they understand the text and visual better when the visual and text are in accordance with each other (see Table 12.4).

In sum, we can conclude that attractive visualizations help the reader to take the effort to look into an article that includes a visualization. However, the form should not be merely there to attract, but should have a purpose or function to clarify the text.

Concluding notes: Implications for news industries and journalism scholarship

As an increasing number of media organizations are focusing on making the news appealing by using different forms of news visualizations, this study tests their assumption that visualizations will not only increase their audience, but can also help the news consumer to understand complex information. As a first comprehensive insight in the use and usefulness of visualizations in a journalistic context, this mixed-method study has proved that visualizations are used, but that they are not seen as independent storytelling devices. Use is predicted by a news consumer's interest in the topic of the story, as well as aesthetics. Nevertheless, data visualizations must fulfil a clear function. When visual forms or aesthetics do not serve a purpose, they add no value in the eye of the consumer. In other words, the text and visualization should be interconnected and tell the same story. If they do not, they are perceived as confusing and distracting.

While we focus on news consumer perceptions, our results are also relevant to those that make the news in the first place. At the beginning of the news production process usually stand journalists and designers who need to collaborate to make a textual and visual story. This is often a difficult collaboration where two disciplines without a common background or language are obliged to work together, often with distinct objectives (Lowrey 2000; Smit, de Haan and Buijs 2014; Tufte 1983; Weber and Rall 2012). Our results show that a lack of collaboration, or diverging objectives of text and visuals, might have significant consequences for news organizations. They can lead to visual stories that are not appreciated by the news consumer, and thus loss in business. Our results suggest that an integrated view of design and journalistic content within newsrooms remains crucial.

The study fills a significant gap within that literature by focusing on consumers' evaluation through a mixed-method approach (Kleis Nielsen 2016) and bears some implications for the fast-growing academic literature on journalistic visualizations. In the coming years, news visualizations are likely to become even more important with more advancing interactive possibilities and growing news consumption on online and mobile devices (Fletcher et al. 2015). Studies already show that mobile devices are not only complementing established media, but might even be displacing them (Westlund 2015). While we find no overall difference in how visualizations are used on different devices and on paper, our focus groups with the online news consumers provided us with several indications that visualizations online, and particularly on mobile devices, demand a new analysis on the form and function of visualizations to tell a news story (Burmester et al. 2010). Future research should tap into more advanced forms of visualizations, including interactive possibilities and other engaging facilities.

Naturally, our study design also shows a number of shortcomings, as we are aware of the complexity of the current study. While we did use the same stimuli during our research, we tested these on different groups of individuals. The eye-tracking data is based on a student sample, while the focus group and survey studies are based on a varied sample of news consumers of the same respective media outlets. This divergence is mainly due to logistic limitations of executing an eye-tracking study in a laboratory environment and future studies must validate our findings. Thus the initial use pattern found through the eye-tracking data in this study might need to be examined in other contexts with different groups of users. Secondly, while the experimental design of the eye-tracking study has the advantage of a high level of internal validity, the external validity might be lower as participants were tested in a laboratory setting. Although we asked the respondents to read the news as they would normally do, we cannot make any definite claims that their behaviour in the lab would be the same in a more natural environment. Future work should be conducted to examine our results with studies that are conducted in a more natural setting. Third, our study did not systematically differentiate between different types of visualizations and compare them. Rather, we provide basic knowledge by examining how news consumers generally understand and value visualizations. Future studies should thus examine how different aspects of visualizations affect readers' understanding of the news story, preferably through experimental study designs.

References

Adam, Pegie S., Quinn, Sara and Edmonds, Rick (2007), *Eye-Tracking the News: A Study of Print and Online Reading.* St. Petersburg, FL: Poynter Institute.

Bakker, Piet, de Haan, Yael and Kuitenbrouwer, Carel (2013). 'Wat werkt? Een verkenning van de praktische en wetenschappelijke kennis over informatievisualisaties' [What works? An inventarization of the practical and academic knowledge on information visualizations]. University of Applied Sciences Utrecht.

Behkit, Elsayed (2009), 'Infographics in the United Arab Emirates newspapers.' *Journalism*, 10(4), 492–508.

Beymer, David, Russel, Daniel and Orton, Peter (2007), 'An eye tracking study of how font size, font type, and pictures influence online reading.' BCS-HCI '08 Proceedings of the 22nd British HCI Group Annual Conference on People and Computers: Culture, Creativity, Interaction, 2, 15–18.

Boucheix, Jean-Michel and Lowe, Richard K. 2010, 'An eye tracking comparison of external pointing cues and internal continuous cues in learning with complex animations'. *Learning and Instruction*, 2(20), 123–35.

Brescani, Sabrina and Eppler, J. Martin (2009), 'The risks of visualizations: A classification of disadvantages associated with graphic representations of information.' ICA working paper 1/2. University of Lugano

Bucher, Hans-Jurgen and Schumacher, Peter (2006), 'The relevance of attention for selecting news content. An eye-tracking study on attention patterns in the reception of print and online media.' *Communications*, 31(3), 347–68.

Burmester, M., Mast, Marcus, Tille, Ralph and Weber, Wibke (2010), 'How users perceive and use interactive information graphics: An exploratory study.' *Proceedings of the 14th International Conference Information Visualization (IV 2010)*, London, United Kingdom, July 2010, 361–8.

Butcher, Kirsten R. (2006), 'Learning from text with diagrams: Promoting mental model development and inference generation.' *Journal of Educational Psychology*, 98, 182–97.

Cairo, Albert (2013), *The Functional Art: An Introduction to Information Graphics and Visualization*. Berkeley, CA: New Riders.

De Koning, Bjorn, Tabbers, Huib, Rikers, Remy and Paas, F. (2007). 'Attention cueing as a means to enhance learning from an animation.' *Applied Cognitive Psychology*, 21, 731–46.

Denzin, Norman K. (1978), *Sociological Methods: A Source Book*, 2nd edn. New York: McGraw-Hill.

De Haan, Yael, Kruikemeier, Sanne, Lecheler, Sophie, Smit, Gerard and Van der Nat, Renee (2017). 'When does an infographic say more than a thousand words?', *Journalism Studies*, DOI: 10.1080/1461670X.2016.1267592

Dick, Murray (2014), 'Interactive infographics and news values.' *Digital Journalism*, 2 (4), 490–506.

Duchowski, Andrew (2007), *Eye Tracking Methodology: Theory and Practice*. London: Springer-Vorlag

Fletcher, Rirchard, Radcliffe, Damian, Levy, David, Nielsen, Rasmus and Newman, Nic (2015), *Digital News Report 2015: Supplementary Report*. Reuters Institute for the Study of Journalism. http://reutersinstitute.politics.ox.ac.uk/sites/default/files/Supplementary%20Digital%20News%20Report%202015.pdf.

Franklin, Bob (2014), 'The future of journalism: In an age of digital media and economic uncertainty.' *Journalism Practice*, 8(5), 469–87.

Garcia, Mario and Adam, Pegie S. (1991), *Eyes on the News*. St. Petersburg, FL: The Poynter Institute.

Geldens, P. M. and Marjoribanks, T. (2015), '"A few years ago I thought it was fairly dead in the water": Newspaper printing, new media and job insecurity in Australia.' *Labour & Industry*, 25(2), 134–49.

Giardina, Marco and Medina Pablo (2013), 'Information graphics design challenges and workflow management.' *Online Journal of Communication and Media Technologies*, 3(1), 108–24.

Holmqvist, Kenneth, Holsanova, Jana, Barthelson, Maria and Lundqvist, Daniel (2003). 'Reading or scanning? A study of newspaper and net paper reading'. In Jukka Hyon, Ralph Radach and Heinder Deubel (eds), *The Mind's Eye: Cognitive and Applied Aspects of Eye Movement Research*, 657–70. Amsterdam: Elsevier Science.

Holmqvist, Kenneth and Wartenberg, Constanze (2005), 'The role of local design factors for newspaper reading behaviour: An eye tracking perspective.' Lund University Cognitive Studies 127.

Holsanova, Jana, Rahm, Henrik and Holmqvist, Kenneth (2006), 'Entry points and reading paths on newspaper spreads: Comparing a semiotic analysis with eye-tracking measurements.' *Visual communication*, 5(1), 65–93.

Holsanova, Jana, Holmberg, Nils and Holmqvist, Kenneth (2008), 'Reading information graphics: The role of spatial contiguity and dual attentional guidance.' *Applied Cognitive Psychology*, 22, 1–12.

Kleis Nielsen, Rasmus (2016), 'Folk theories of journalism: The many faces of a local newspaper.' *Journalism Studies*, 17(7), 840–8.

Lee, Eun-Ju and Kim, Ye Weon(2016), 'Effects of infographics on news elaboration, acquisition and evaluation: Prior knowledge and issue involvement as moderators.' *New Media & Society*, 18(8), 1579–98.

Lester, Paul M. (2000), *Visual Communication: Images with Messages*. New York: Whatsworth.

Lewenstein, Marion, Edwards, Greg, Tatar, Deborah and De Vigal, Andrew (2000), The Stanford-Poynter Eye-tracking Study. Retrieved from http://www.poynter.org/eyetrack2000.

Lowrey, Wilson (2000), 'Word people vs picture people: Normative differences and strategies for control over work among newsroom subgroups.' *Mass Communication & Society*, 5(4), 411–32.

Lyra, Kamila, Isotani, Seijim Reis, Rachel, Marques, Leonardo, Pedro, Laiz, Jacques, Patricia and Bitencourt, Ibert (2016), 'Infographics or Graphics+Text: Which Material is Best for Robust Learning?' *Proceedings of the IEEE International Conference on Advanced Learning Technologies* (ICALT).

Mayer, Richard E. (2005), 'Principles for managing essential processing in multimedia learning: Coherence, signaling, redundancy, spatial contiguity and temporal contiguity principles.' Chapter 11 in *Cambridge Handbook of Multimedia Learning*. New York: Cambridge University Press.

McCloud, Scott (1994), *Understanding Comics: The Invisible Art*. New York: Harper Perennial.

Moreno, Roxana and Valdez, Alfred (2005), 'Cognitive load and learning effects of having students organize pictures and words in multimedia environments: The role of student interactivity and feedback.' *Educational Technology Research and Development*, 53: 35–45

Newman, Nic, Levy, David and Klein Nielsen, Rasmus (2015), *Digital News Report 2015*. Oxford: Reuters Institute for the Study of Journalism.

Patton, Michael Q. (1990), *Qualitative Evaluation and Research Methods*. Newbury Park, CA: Sage.

Poynter Institute (2004), Eyetrack III: What news websites look like through readers' eyes. Retrieved from http://www.poynter.org/%20uncategorized/24963/eyetrack-iii-what-news-%20websites-look-like-through-readers-eyes.

Radach, Ralph, Lemmer, Stefanie, Vorstius, Christina, Heller, Dieter and Radach, Karina (2003), 'Eye movements in the processing of print advertisements.' In Jukka Hyon, Ralph Radach and Heinder Deubel (eds), *The Mind's Eye: Cognitive and Applied Aspects of Eye Movement Research*, 609–32. Amsterdam: Elsevier Science.

Schroeder, Roland (2004), 'Interactive Info Graphics in Europe – added value to online mass media: a preliminary survey.' *Journalism Studies*, 5(4), 563–70.

Segel, Edward and Heer, Jeffrey (2010), 'Narrative visualization: Telling stories with data visualization and computer graphics.' *IEEE Transactions*, 16(6), 1139–48.

Siricharoen, Waralak V. (2013), 'Infographics: The New Communication Tools in Digital Age.' Paper presented at The International Conference on E-Technologies and Business on the Web (EBW2013), Thailand. Retrieved from http://sdiwc.net/digital-library/infographics-the-new-communication-tools-in-digital-age.

Siles, Ignacio and Boczkowski, Pablo J. (2012), 'Making sense of the newspaper crisis: A critical assessment of existing research and an agenda for future work.' *New Media & Society*, 14 (8), 1375–94.

Smit, Gerard, de Haan, Yael and Buijs, Laura (2014), 'Working with or next to each other? Boundary crossing in the field of information visualisation.' *The Journal of Media Innovations*, 1(2), 36–51.

Smit, Gerard, de Haan, Yael and Van der Nat, Renee (2015), De Nieuwswaarde van visualisaties [The news values of visualizations]. Final Report of research project. http://www.journalismlab.nl/eindrapportage-de-nieuwswaarde-van-datavisualisaties/

Sung, Eunmo and Mayer, Richard (2012), 'When graphics improve liking but not learning from online lessons.' *Computers in Human Behavior*, 28(5), 1618–25.

Tufte, Edward R. (1983), *The Visual Display of Quantitative Information*. Cheshire, CT: Graphics Press.

Weber, Wibke and Rall, Hannes (2012), 'Data visualization in online journalism and its implications for the production process.' *The 16th International Conference on Information Visualization*. Montpellier, 11–13 July 2012. DOI:10.1109/IV.2012.65.

Westlund, Oscar (2015), 'News consumption in an age of mobile media: patterns, people, place, and participation.' *Mobile Media and Communication*, 3(2), 151–9. DOI:10.1177/2050157914563369

Wong, Dona M. (2010), *The Wall Street Journal Guide to Information Graphics: The Dos and Don'ts of Presenting Data, Facts and Figures*. New York: Norton.

Section Three

Agenda for the Future

13

Towards a Fruitful Relationship Between Statistics and the Media: One Statistician's View[1]

Kevin McConway, *The Open University, UK*

Introduction

Much has been written, by journalists, statisticians, commentators and others, on the role of statistics in the media, and on shortcomings of the presentation of statistical stories and data in traditional media and on the need for journalists to 'do things better' with the numbers. However, one aspect that has been discussed rather little is the role that statisticians can and should play in these matters. For journalists to work effectively with statisticians, it is important for them to understand and take account of the way that statisticians understand (or fail to understand) how journalism works.

At a very high level, what journalists and statisticians do with data is much the same. We all collect, or come upon, data, and use the data to produce insights that lead to a narrative or story. But once one gets below that level, there are major differences in the ways that statisticians and journalists would approach the question of what the important insights are and what the resulting narrative might be. These differences can go well beyond technical issues of data manipulation or calculation. I will mention one example. In the spring of 2014, an introductory Massive Open Online Course (MOOC) was run by the European Journalism Centre.[2] This included teaching by Steve Doig of Arizona State University on the use of Excel spreadsheets to find story ideas. The technical aspects of using Excel to manipulate, sort and summarize data were much the same as one might teach in an introductory course for statisticians. But Doig then pointed out that the news story from a set of data would often lie in its extreme values, because they are potentially more newsworthy than averages. A

---

[1] An earlier version of this chapter appeared as 'Statistics and the media: A statistician's view.' *Journalism* 17(1), 49–65.
[2] At the time of writing, the course is available on the website http://learno.net/, run by the European Journalism Centre.

statistician would usually be much more interested, initially at least, in averages, and descriptions of the shape and spread of the distribution of typical values in the whole of the data set rather than in the extremes – same data, same basic methods, but different orientation and different stories.

There is nothing wrong with these differences of approach – journalists and statisticians do not have the same role. But statisticians and journalists need to understand what is different, if they are to interact fruitfully. This chapter, then, explores how statisticians see news processes. It describes matters principally from a statistical point of view – this is emphatically not meant to imply that journalists should produce only what statisticians would agree with. It is a personal view – hence the phrase '*one* statistician' in the chapter title – so I should explain my position. I am a professor of applied statistics who takes an interest in the use and presentation of statistics in the media. I have experience of working with journalists, particularly in the United Kingdom. Some of this experience comes from the fact that my university co-funds the BBC Radio 4 series *More or Less*, which is about numbers and statistics in the news. I was the university's main academic liaison with the programme for eleven years. I have also from time to time advised and collaborated with other journalists. I work with the United Kingdom's Science Media Centre[3] in providing briefing for journalists on scientific stories with statistical aspects. Finally, I have written a few pieces of print journalism myself. This experience is a basis for the views in the rest of this chapter.

The chapter is structured as follows. The next section briefly outlines the place of statistics in the traditional media, particularly newspapers, with some thoughts and theory on how that position arose. This is followed by an outline of how statisticians (and, arguably, other scientists) see media processes. There is a case study of the media reporting of potential links between mobile phone use and the risk of brain tumours, largely to illustrate what statisticians find strange in the media treatment of such matters. The penultimate section explores issues in helping ordinary readers cope with media stories about health risks. The chapter ends with some reflections and generalizations about statisticians, journalists and the way we interact.

If statistics are so boring, why are the news media so full of them?

Any working statistician must deal with comments from acquaintances that their discipline must be very boring. But this cliché's lack of veracity is demonstrated by a brief look in almost any national newspaper. It should hardly be necessary in a book on journalism to describe the sorts of statistical information and stories one might see in the daily press, but here are some examples:

---

[3]   See http://www.sciencemediacentre.org/

- *Front-page stories*, about economic and financial statistics, or public health data, or, well, almost anything.
- *Infographics* (with the dual aim of getting information across and making the page look more attractive): statisticians can be very critical of these, in cases where the style obscures the substance.
- *Survey results*. There is a huge range here. Stories might be written about major national surveys carried out by Government statisticians or large and respected market research companies. But many survey stories are mainly there as promotion or public relations (PR) for a particular organization such as a commercial company or a charity. Such surveys might be done in-house by the organization in question but are commonly produced, and promoted, by specialist polling and PR agencies.
- *Sport pages*. Many sports stories are full of statistical information of many kinds.
- *Business pages*. Market data and the like are intrinsically numerical, and always have been, but in recent years there has been an increasing tendency to present at least some of the data in a graphical form.
- *Improvement to the page design*. Arguably this is the motivation behind the use of pull quotes involving numbers (often printed in a large typeface and a contrasting colour), within stories, which is a major component of the visual style of several newspapers. Nothing wrong with that, of course, and often the pull quote is really there to draw attention to an important numerical aspect of the story. But sometimes the statistic in the quote does not appear in the actual story and is only loosely connected, and sometimes it adds nothing beyond its visual impact.
- *Fillers*. Some newspapers make considerable use of very short fillers at the foot of columns, and very often these contain a statistic or other number, presumably because with a number one can say a lot in a small space.

Numbers are no less ubiquitous in other media. Cushion, Lewis and Callaghan (2016) report on a major content analysis of BBC television, radio and online news, together with news from some other UK television broadcasters. Over a month, they found 6,916 news items, 22 per cent of which included at least one reference to a statistic. Statistical content was particularly prevalent in news on the economy, politics, taxation and business, though (surprisingly) much less common in stories on sport and crime, for example. See Chapter for more findings and insights from this study.

Why are there so many numbers in the media? The partial taxonomy above does indicate some of the many uses, in the context of a newspaper, for statistical and other numerical information. Numbers in the newspapers may be there for their intrinsic news value, but also for PR purposes, for entertainment and to decorate the page. At a deeper level, various writers have examined the way in which statistics and numbers are used in the media for their rhetorical value rather than for the information they carry (going back at least to Roeh and Feldman 1984), and the extent to which using numbers can increase the credibility of a news story (see, e.g., Koetsenruijter 2011).

I would argue that the relative prominence of numbers has much to do with the special status of numerical facts in our society. Numbers are not construed in the same way as other, equally factual and trustworthy, information. As a fairly trivial example, consider the book *50 Facts that Should Change the World*, by the journalist Jessica Williams (2007). Of the fifty facts, 46 contain numbers (e.g. 'Eighty-two per cent of the world's smokers live in developing countries') and three are comparisons of numbers that are not given directly (e.g. 'Brazil has more Avon ladies than members of its armed services'); leaving just one that is less obviously numerical ('British supermarkets know more about their customers than the British government does').[4]

One author who has taken a broad view of the phenomenon of the predominantly numerical nature of the modern fact is Mary Poovey, a cultural historian and literary critic. She writes, at the start of a major book (Poovey 1998, 1), as follows:

> What are facts? Are they incontrovertible data that simply demonstrate what is true? Or are they bits of evidence marshaled to persuade others of the theory one sets out with? Do facts somehow exist in the world like pebbles, waiting to be picked up? Or are they manufactured and thus informed by all the social and personal factors that go into every act of human creation? Are facts beyond interpretation? Or are they the very stuff of interpretation, its symptomatic incarnation instead of the place where it begins?

Poovey's book explores how the fact became the most favoured unit of knowledge in modern times (up to, roughly, the end of the nineteenth century), and how description (in the shape of 'facts') came to seem separable from theory in the precursors of economics and the social sciences. Much of this exploration is in the context of numerical facts. Poovey starts with the development of double-entry book-keeping in the fifteenth century, and moves on through Francis Bacon, Thomas Hobbes, and onto the 'political arithmetic' of William Petty and others in the late seventeenth century, some of which was explicitly anti-theoretical. Poovey points out that in *The Wealth of Nations*, Adam Smith uses numerical information to lend impartiality (but Smith nevertheless emphasizes the need for theory). It was a strong aspect of early nineteenth century discourse to make a distinction between the facts (data, very often numerical and statistical) and systematic knowledge based on the facts.

The learned society covering my own profession, now the Royal Statistical Society, was founded as the Statistical Society of London in 1834. It is instructive to look at some of its founding documents:

> The Statistical Society of London has been established for the purposes of procuring, arranging and publishing 'Facts calculated to illustrate the Conditions and Prospects of Society.'
>
> The Statistical Society will consider it to be the first and most essential rule of its conduct to exclude carefully all opinions from its transactions and publications –

---

[4] This 'fact' may *look* as if it must be based on numerical data, but according to Williams' book, it comes from an interview with a 'customer forecaster for a British supermarket', who may or may not have based it on specific numerical evidence.

to confine its attention rigorously to facts – and, as far as it may be found possible, to facts which can be stated numerically and arranged in tables. (Statistical Society of London 1834, 492)

The Society's founders were not saying that nobody should have opinions. Instead there was a proposed division of labour, whereby the statisticians were what we might nowadays call technicians, who picked up the pebbles of data and arranged them nicely in tables, and then some rather superior beings, perhaps economists and political scientists and theorists, turned them into theory and justified opinion. This view of statistics was demonstrated in the Society's Latin motto, 'Aliis exterendum', generally translated as 'to be threshed out by/for others'.

This distinction between facts and opinions was not universally accepted even at the time. For instance, a subeditor of the radical *London and Westminster Review* wrote as follows in 1838 (Robertson 1838) in a review of the *Transactions of the Statistical Society of London*:

There is an ambiguity in the word facts which enables the council to pass off a most mischievous fallacy: it either means evidences or it means anything which exists. The fact, the thing as it is without relation to anything else, is a matter of no importance or concern whatever: its relation to what it evinces, the fact viewed as evidence, is alone important. (69)

Views among statisticians have changed since the founding of the Statistical Society of London, and most statisticians would no longer see their role as the provision of some magically value-free set of facts. Indeed, much of the work of many statisticians nowadays consists of the interpretation of numbers produced by others. But the notion of numerical facts as somehow having an objective and value-free status has perhaps survived into present-day public attitudes towards statistics, and hence, arguably, into some of the reasons why statistical information is prominent in the media. The cultural development of this large-scale equating of quantification with objectivity and impersonality has been explored broadly by the historian of science Theodore Porter (1996) and, particularly in the context of governance and power, by the social theorist Nikolas Rose (e.g. Rose 1991).

But these views lead to dilemmas that most people, at some level or other, do recognize. People consider numbers to be 'solid facts' which must be correct. But sometimes they self-evidently are *not* correct. So statistics are absolute truths at the same time as being labelled alongside 'lies, damned lies'. Such dilemmas do little for the credibility of statistical reports and stories to lay readers. Others have put it in a more scholarly manner. Here is another quote from the start of a book:

Unemployment, inflation, growth, poverty, fertility: these objective phenomena, and the statistics that measure them, support descriptions of economic situations, denunciations of social injustices, and justifications for political actions. ... As references these objects must be perceived as indisputable, above the fray. How then should we conceive a debate that turns on precisely these objects? How can we dispute the undisputable? ... Do statistics lie? ... These measurements, which are reference points in the debate, are also subject to debate themselves. (Desrosières 1998 [1993], 1)

Desrosières is a sociologist and historian of science, and most of the rest of his book, in theoretical terms, is taken up with explaining how these different ways of looking at statistics (as the indisputable basis for debate, and as objects of debate themselves) can be reconciled, and indeed, have to be reconciled. The same dilemma about the basis of numerical facts exists just as strongly, of course, in their use in the media, though it is not often considered explicitly. Finally, looking more broadly, a social psychologist writes:

In principle we may say that 'facts speak for themselves'; in practice they do so only when accompanied by a chorus of approval. (Kelvin 1970, 30)[5]

There is always a social context.

Statisticians and media processes

A major difficulty that most statisticians (and, arguably, most scientists) have in dealing with journalists and journalism is that we have very little idea of how stories actually get into the papers or the broadcast news. Looking at it from our end, it is all so complicated! The culture of media production is so alien to us!

A (very) simplified version of what many research scientists and statisticians do is as follows. We bid for funding from research funders. We eventually get some money. We do (roughly) what we said we would do in the research proposal, over a timescale often measured in years, and we find something out. We present our results at a research conference, or publish them in an academic journal. That is what we are trained to do, paid to do, and it is what we understand. This is a particular culture and, to outsiders, one that probably looks strange.

But then, now and then (and quite possibly never, in many perfectly good research careers) someone decides the results of the research have a wider impact, and things get a bit out of hand. People who we may never have heard of begin to be involved – press officers, science correspondents, perhaps the Science Media Centre, and we have to deal with mysterious concepts like embargos and press briefings. Press releases are written, by press officers from one's university, the journal, the funders, who knows who else. We might be consulted about the content but we might not be. In my experience, most statisticians know little or nothing about these processes until and

---

[5] When I was checking the reference for this quote, I looked in the index of this book under Facts. It said 'See Values'.

unless they are personally involved. One suddenly must deal with all sorts of people and processes one knows nothing about. One's work can be reported in a lot of places, which never quite say what one would want them to. We are simply not trained to deal with any of this!

And meanwhile, the conditions arise for growth of a blame culture where the wrong party is usually blamed. Sometimes a media report makes factual errors about what happened in the research study, or about the methodology used, but commonly there is disagreement between scientists and journalists on what the story revealed by the research really is. Scientists feel they have been misreported and blame the journalist whose byline is on the story. That might be because they feel they have a monopoly on the interpretation and presentation of our findings and do not recognize that their culture is not the only one in society. Journalists will understand well that, because of the pressure of time and of other stories, they must often rely to a major extent on the press release, and for many reasons cannot always comply with the counsel of perfection to read the original source paper or report, and talk to those who produced it. The lack of resource in the media that makes these shortcuts increasingly essential may be – *is* – regrettable, but we cannot ignore its existence. Furthermore, some factors that play a key institutional role in the production of media outputs – for instance, news agendas and the notions of news value and newsworthiness – constrain and shape how media outputs are produced. These factors either do not arise, or arise in a very different way, in the production and reporting of statistical research.

The scientists most closely involved may or may not understand all this as well, but in many cases, their scientific or statistical peers do not understand it at all. In the worst case, colleagues blame the scientists for doing poor research work in the first place, while in a slightly better case, the scientist colleagues join the chorus of condemnation of the journalists whose names are on the story, and/or their newspapers or broadcasting stations.[6] All this points to a clear lack, in the training and education of statisticians and scientists, of coverage of how the media operate. Some of us do go through 'media training', but this is very far from universal. Maybe it is not essential that all of us should be exposed to this information, because, depending on one's research field, many of us will never have to engage with the media in our professional capacity. But one never knows when one might be thrown into this melee, and in my view, some basic understanding should be part of the background knowledge we all have.

I am not in a position to judge the extent to which journalists are comparably ignorant of how scientists and statisticians do their work. Science correspondents are generally very aware of our working practices (and indeed many science correspondents have spent time working as research scientists). But I am less confident

---

[6] Issues of the role of press releases have been discussed in this context, for example, by Riesch and Spiegelhalter (2010). We statisticians do not commonly read the kind of journal in which this article was published, though I should mention that Spiegelhalter *is* a statistician. A more recent article, Sumner et al. (2014), *was* in a journal of which more statisticians are aware, and it did lead to more comment and discussion in the statistical community.

of the position in other areas of journalism. Even some data journalists seem to be reasonably ignorant of the way that others who use data as the basis for research, such as academic statisticians, do our jobs.

The strange case of mobile phones and brain tumours

To explore further the potential for the lack of understanding, by statisticians and journalists, of how each other works, I shall use an example that I have previously employed to attempt to show statisticians and scientists some aspects of media processes that they may not have considered.

There has been a certain amount of concern, going back many years, that radiation from mobile telephones (cell phones) may increase the risk of certain kinds of brain tumours. It is true that mobile phones do emit electromagnetic radiation at low levels – namely radio frequency radiation that they use to communicate with their network (and without which they would simply not work). It is also very well known that, in high enough intensity, certain kinds of electromagnetic radiation can initiate tumours of various sorts. So the question of whether the radiation from mobile phones might possibly increase tumour risk does have some kind of basis in science.

The problem is, how to establish whether such a link exists. It is obvious – from previous knowledge and from the fact that early users of mobile phones did not come down with brain tumours in their thousands – that the risk cannot be immense. Several epidemiological studies, that is, studies where populations of people are observed, have been carried out over the years. The fact that all these studies are observational does, in a sense, make their interpretation difficult. In an observational study, if one observes that two things are related, it is generally difficult to establish whether the relationship is causal. In this context, for instance, if a study finds that people whose phones are used a lot are more likely to get certain tumours, then that may be because the radiation from the phones causes the tumours, or it may be because people that use their mobile phones a lot are untypical in some other ways, and it is these other ways (rather than the phone use) that are leading to the increased tumour risk. Epidemiological studies can establish correlations, but as the statistical mantra has it, correlation is not causation. There are ways in epidemiology of increasing one's confidence that an association is in fact causal, for example in the form of the so-called 'Bradford Hill criteria' (Hill 1965), as well as the more recent (and much more complicated and technical) methodology of causal inference (see, e.g., Pearl 2009; Rothman and Greenland 2005), but they are far from foolproof.[7]

Those remarks would apply to almost any epidemiological study – but the position with investigating the possibility of a link between mobile phones and brain tumours is particularly difficult. Brain tumours (despite the number of media stories about them)

---

[7] An extra complication in *writing* about causal or non-causal relationships in the media is that readers' understanding of some statements about cause is inconsistent (Adams et al. 2016).

are thankfully rather rare, so to carry out a study that includes enough people to observe enough cases of brain tumours can be difficult, time-consuming and very expensive. Some brain tumours have a very long latent period (the length of time between their initiation and when they can be detected), running into many years, and we must recall that mobile phones have been in common use for only twenty years or so.

Because of these features, most epidemiological studies of this possible link have been case–control studies – that is, they have taken some people with the brain tumours in question (the cases), and some roughly comparable people without brain tumours (the controls), and looked back at the history of both groups to see if the cases had, in the past, used mobile phones more than the controls had. But it is harder to produce convincing evidence from case–control studies than from some other kinds of epidemiological study, for several reasons. Perhaps a difference between the cases and the controls might have nothing to do with a difference in their phone-using status but be due to something else. Perhaps the two groups differ in the way that they recall previous mobile phone use, years later, rather than in the actual use. Because of these difficulties, the position until at least 2011 was that the conclusions from the epidemiology on mobile phones and tumours were diverse and contested. Some studies showed no evidence of a link (not yet anyway). Others did show evidence of some increased tumour risk in long-term heavy users of mobile phones. Both sets of studies had methodological difficulties, so there was room for criticism of almost any of them.

Mobile phones and cancer produce a classic example of high news value, because the mundane (mobile phones) is combined with the dread (radiation and cancer). Even though most of the evidence is related to non-cancerous brain tumours, the word 'tumour' is bound to raise fears associated with cancer.

I originally became interested in this research area, and the corresponding press stories, because of an example in Blastland and Dilnot's popular book *The Tiger that Isn't*. Blastland and Dilnot (2007, 116) describe the reporting (by the press, and by the president of the British Radiological Protection Board) of the conclusions of a Swedish study (Lönn et al. 2004) that suggested long-term use of mobile phones was associated with a doubling of the risk of acoustic neuroma, a non-cancerous but still very unpleasant type of brain tumour. The press reports concentrated very much on this risk doubling – an example of *relative risk* because it compares one risk with another rather than giving the actual absolute values of the risks being compared. But there was almost no mention in the media of the original baseline risk. In fact, it was very small. The original study on which the reports were based did not itself state the baseline risk. It reports a case–control study, and such studies cannot directly measure baseline risks. But Blastland and Dilnot interviewed one of the original researchers, who told them (based on other studies) that the baseline risk was very small – roughly one case in 100,000 people per year – because acoustic neuromas are not at all common. The study found that this risk was increased, in people with ten years of regular mobile phone use, to about two cases per 100,000 people per year. Twice not very much is still not very much. Given also the methodological problems of knowing whether the increased risk was caused by the mobile phone use, arguably this study should not have provided much cause for alarm – but the press reports treated it differently.

Because I used this example in teaching, I was particularly interested when, in June 2011, the International Agency for Research on Cancer (IARC), part of the World Health Organization (WHO), classified the radiation from mobile phones as 'possibly carcinogenic to humans'. IARC classifies the potentially hazardous agents that it has investigated into five different groups. The brief descriptions of the groups, and the numbers of agents in each group at the time of writing this chapter, are shown in Table 13.1 (IARC, 2016).]

The classification is based on the likelihood that there is a non-zero risk each agent causes cancer. It is not necessarily the case that agents in Group 1, for instance, present a greater risk than those in Group 2A – the distinction is that for agents in Group 1, the evidence that they increase cancer risk at all is very secure, whereas it is less secure for agents in Group 2A.

Predictably, the press had a field day after IARC released their classification of mobile phones as a possible carcinogen. Some of the headlines were as follows. The *Sun* said, '"Cancer" mobiles: World health chiefs in U-turn on phone fears'. (It was not actually a U-turn – IARC had not previously categorized this potential risk.) The *Daily Express* went with

'SHOCK WARNING: MOBILE PHONES CAN GIVE YOU CANCER'. Some newspapers showed more restraint. The *Guardian* wrote, 'Mobile phone radiation is a possible cancer risk, warns WHO.'

The details of the stories were also interesting. IARC's Group 2B has almost 300 agents that are 'Possibly carcinogenic to humans', but most of them are chemicals that most people have never heard of. Thus, quite reasonably, newspapers picked out a few agents from the list that their readers would know, to show the company that mobile phone radiation was now keeping. These differed between newspapers. The *Sun* and the *Daily Express* chose the same comparators: petrol exhaust fumes and DDT. Both are generally considered in the public awareness to be harmful to health – though not because of their potential (or otherwise) to cause cancer. Other newspapers chose comparators that would be more likely to be seen as benign: talcum powder and working in dry cleaners (*Guardian*) and coffee (*The Times*).[8]

One point of interest here to a statistician is that no newspaper seems to have picked up clearly that the IARC classification is not based on the size of the risk but on the

Table 13.1 Groups used by IARC to classify the potentially hazardous agents they have investigated, with numbers of agents in each group as of November 2016.

Group 1	Carcinogenic to humans	118 agents
Group 2A	Probably carcinogenic to humans	81
Group 2B	Possibly carcinogenic to humans	292
Group 3	Not classifiable as to its carcinogenicity to humans	505
Group 4	Probably not carcinogenic to humans	1

Source: International Agency for Research on Cancer (IARC 2016).

[8]   DDT has since been moved to Group 2A, and coffee to Group 3.

strength of evidence. That confusion has arisen on many other occasions, and IARC seem to have recognized that it causes difficulty by publishing 'Questions and Answers' (IARC 2015) that provide a relatively accessible description of the way they classify.[9] Another aspect is that in choosing comparators from the available possibilities, the newspapers put a slant on the way the IARC statement about mobile phones should be taken. Exhaust fumes and DDT are, to some extent, matters of dread, whereas talcum powder and coffee are mundane. No statistician (I hope) is naive enough to think that media stories are all entirely objective, or even that they should be so, but the interest here is that the slant was very largely introduced by the choice of comparators rather than other aspects of the way the stories were written.

The IARC announcement was not the only published item about mobile phones and brain tumours that summer. On 1 July 2011, the International Commission for Non-Ionizing Radiation Protection Standing Committee on Epidemiology published a statement:

> Although there remains some uncertainty, the trend in the accumulating evidence is increasingly against the hypothesis that mobile phone use can cause brain tumours in adults. (Swerdlow et al. 2011, 1534)

This was reported in all the 'serious' UK media. It says nothing about children (who had been picked out for particular concern in much previous reporting), but otherwise it is negative about the possibilities of an increased risk from mobile phone use. Since it was fairly heavily reported, this goes against the view that the press selectively reports bad news in relation to health risks.

The next I heard about mobile phones and brain tumours was on 15 July 2011, when the popular web comic strip XKCD published a strip referring to the matter (Munroe 2011). One of the characters in the strip says, 'Another huge study found no evidence that cell phones cause cancer. What was the W.H.O. thinking?', and then the other character pokes fun at inferring causality from observational data. Apart from my delight at the fun, the other thing that struck me was that 'another huge study' appeared to have been published, but I had seen no trace of it in the press.

The study in question seems to have been conducted by Schüz et al. (2011) – another study looking at a potential link between mobile phone use and acoustic neuroma (under its alternative name of vestibular schwannoma), with an arguably stronger type of epidemiological design (a cohort study rather than a case–control study). As the comic strip said, this study reported that 'overall, no evidence was found that mobile phone use is related to the risk of vestibular schwannoma'. The Schüz et al. paper was published online on 28 June 2011 and appears to have been met by press silence, in the United Kingdom and the United States at least. Reuters put out a story on it on 14 July 2011. This received heavy coverage in the United States, but nothing in the UK national press. At the time, I personally hypothesized that the lack of UK coverage was because

[9]   Even this Q & A did not prevent IARC classifications from being wrongly reported again in some subsequent news.

the UK press was far too full of the News International phone hacking scandal – a rather different angle on mobile phones.

But I was (almost certainly) wrong. On 27 July 2011, the work of Aydin et al. (2011) was published online. This study of a possible link between mobile phone use and brain tumours *in children and adolescents* reached a generally negative conclusion – that is, it found essentially no evidence of risk. It was reported rather little in the United States, though there were some press reports of criticism of its epidemiological methods, indicating that there was at least knowledge of it in some quarters of the US press. But again there was no mention in the UK national press. This time I could not plausibly attribute the lack of UK coverage to competing news. In fact, it was rather a slow news week in the United Kingdom, between the Breivik massacres in Norway and Amy Winehouse's death (the previous week) and the UK riots. Furthermore, the following month (23 August 2011) saw the online publication of what was the biggest and most authoritative and international study on the matter so far (INTERPHONE Study Group, 2011). This, again, came to a generally negative conclusion on the existence of any increased risk. There was very little reporting in the United States, and again none in the UK national media.

This pattern of publication raises more questions than I can answer. Why were Schüz et al. (2011), Aydin et al. (2011) and INTERPHONE Study Group (2011) not reported in the UK nationals? It seems improbable that it was merely because of their negative conclusions on the existence of increased risk, because Swerdlow et al. (2011) also had a negative conclusion, and was fairly heavily reported. It seems not entirely to have been because of competing pressures from other news. So what caused this to stop being newsworthy, when it had been just a few weeks earlier? The Science Media Centre had circulated documentation and also held a face-to-face briefing on some of these study reports, in which it was made clear that there was no real evidence of increased risk – but the timing of when the stories stopped does not seem to match the dates of these briefings very precisely.

Maybe the clearest conclusion from this is that I do not entirely understand how news values and newsworthiness work. But it is clear (and hardly novel) that these processes are complicated, and at least to statisticians concerned about balanced and appropriate media reporting, they are worth studying further.

How should the public read statistical news stories?

It is clearly possible to be cynical about the nature and purpose of statistical media stories. Yes, journalists, press officers (and statisticians) make mistakes. Yes, statistics can be used in media reports for rhetorical purposes and to create illusions of objectivity, credibility and authority. But that does not imply that the stories have no value to their readers. The public cannot possibly be expected to get their information on economic trends or on public and personal health issues by finding and reading the original reports and papers.

There is a need for continuing improvement in the standard of reporting of stories involving statistics. But the detail of how to make this improvement is largely beyond the scope of this chapter. New approaches in training journalists in the relevant skills continue to emerge – the data journalism MOOC that I referred to in the 'Introduction' section is just one part of this trend. There is no shortage of advice and checklists, produced by statisticians and others, for journalists working in relevant fields, and I cannot summarize them all here. One excellent example, though it relates to the broader field of science and health reporting and not specifically to statistics, is the set of ten recommendations submitted to the Leveson Inquiry by the Science Media Centre (Fox 2012), and indeed the section in the Leveson Report on Science Reporting (paragraphs 9.57 to 9.75 of Volume II in Leveson 2012) provides much food for thought for anyone writing about science.

However, I believe there is also a need to help puzzled readers of the resulting stories – and indeed material serving this need does exist. An excellent example in relation to health stories is White (2014). Another is McConway and Spiegelhalter (2012), which, despite its title, is not solely or even principally about radio journalism. In McConway and Spiegelhalter (2012), when producing a (not entirely serious) checklist for the public, we went beyond much of the other available advice and chose to concentrate as much on the way that a health risk story is reported as on the study itself and its applicability to the reader. This is partly because some of the issues of study quality and applicability need technical statistical or epidemiological understanding that an everyday reader cannot be expected to have. Journalists have access (if they look for it and can find it) to support and advice on such matters (see, for instance, Senn 2009), but one cannot reasonably expect readers of media stories to acquire the necessary skills. However, our choices for the checklist also reflect our experience that the process of news production is not well understood and that (perhaps because of the particular status of numerical facts) many readers would not approach statistical stories with even the degree of critical awareness they would apply to other media content.

Our checklist therefore has four items about the quality of the study being reported on (e.g. 'Is it just an isolated study, rather than building on a broader context or putting together a range of studies?'), and two on the extent to which the study's findings apply to the reader (e.g. 'Might the study simply not be relevant to you – e.g. animal studies, untypical people, study population not like you?'). But there are six items on the nature of the reporting, some of which are merely the sort of things that should be asked about any media story:

1. Is there enough information to allow you to find the original source?
2. Is the headline exaggerated? (We point out that, if it is, that does not necessarily mean it is a poor story; someone else probably wrote the headline.)
3. Is there any independent comment?
4. Does the story give only relative risks (because twice next to nothing is still next to nothing)?

5. Is there really anything substantial to the story, or is it (for instance) irrelevant PR? In other words, why do the original study authors and/or the journalist want you to know this?
6. What are you not being told?

These may not, in fact, be the most relevant points. They do deal with one key aspect of statistical thinking – that of taking a suitably sceptical attitude – but they largely do not deal with another key aspect of statistics, that of describing and accounting for uncertainty. Advice from journalists on how to produce a better list would be welcome. Perhaps some of the points from Chivers (2014) could usefully be added. Furthermore, it is obvious that the way to change public perceptions and attitudes towards statistical news is not by statisticians producing a few checklists – journalists and journalism can contribute much more effectively. The checklist is presented to its readers primarily as a way of deciding which health risk stories are ignorable – perhaps it will also help those producing the stories to make them less ignorable in the first place.

Advice for statisticians on working with journalists

In conclusion, though, I return to the relationships between statisticians and journalists, and outline the advice that I routinely give to statisticians on working with journalists. This may seem contrary in a chapter aimed more at journalists than statisticians, but I do hesitate, from my position, to give too much direct advice to journalists about their own trade – and I believe that sharing this advice more widely will help further to demonstrate to journalists what we statisticians do not understand! A list of the key differences in the way journalists and statisticians think about each other is presented in Table 13.2.

This list contains some major stereotyping, and certainly does not apply to all statisticians or all journalists. But as often with stereotypes, there is some truth in them. I also try to make it clearer to my statistical colleagues that they need to recognize that

Table 13.2 Some stereotypical views from both sides

Statisticians think journalists ...	Journalists think statisticians ...
Are innumerate	Are illiterate
Do not understand quantitative reasoning (or indeed any kind of logic)	Are pedantic
Distort and oversimplify statistical information and conclusions	Make the story so boring that it does not get across
	Concentrate so much on the ifs and buts that the overall message disappears
Won't listen	Won't listen

journalists are different from statisticians in training, in attitudes, and in the context in which they work, in at least the following ways:

- **Timescales**. Deadlines count for a journalist and are typically much shorter than those that most statisticians are used to (because much of the work of statisticians is on projects that last weeks or months). Journalists will want to get something out by the deadline even if they know it is not perfect, because it is better to get something slightly wrong out there than to get nothing out at all because the deadline has been missed. If you (the statistician) cannot keep up, they will not wait for you, because they cannot.
- **Agenda.** A journalist will be reporting on statistical work not because it is an excellent piece of statistical reporting, but because it has some news value. That news value will depend on the context in which the journalist is working; it might be explicitly political in some way, it might be to do with the editorial policy of the publication involved, or it might be something else. But wherever it comes from, if you (the statistician) want to work effectively with the journalist, you must understand what his or her agenda is, and understand that there is little or nothing you can do to change it.
- **The pressure to be personal**. It is an exaggerated cliché of journalism that every story needs to refer to a named individual; the new drug might decrease everyone's risk of dying of the disease, on average at any rate, but the story will be that Mrs Bloggs, aged fifty-two, was cured by the drug. Getting a balance between this personalization and the overall statistical message is crucial in getting a good story out of statistically based conclusions, but most statisticians have difficulty in thinking in this way because we are trained not to. In many cases, the statistical culture of looking at entire, variable, populations is so ingrained in us that we may not even realize that others do not necessarily share it.

So, my take-home message for statisticians is that journalists have strengths that (most) statisticians do not have – for example, knowing their audience, getting things across in a short space, and, generally, *telling stories*. We statisticians should not simply blame journalists for getting things wrong: *we must help them to get things right*. We need to be proactive in making known to journalists what we do, and why and how we do it. In my experience, journalists will respond, and positively. It is this experience that makes me broadly optimistic about the reporting and use of statistics in the media. Although appalling mistakes and misrepresentations do still occur, the ability to deal with the numbers is spreading, much more widely than it used to, outside the boundaries of specialist science and economics correspondents, and I am increasingly impressed with journalists' willingness to listen and learn, despite all the pressures on their time and resource.

I hope my approach to statisticians helps us move forward together with the media in a direction both sides are happy with. But there needs to be give and take. We statisticians do need to understand something of the journalist's world, but journalists also need to understand something of our world. Perhaps this chapter will help a little.

References

Adams, R. C., Sumner, P., Vivian-Griffiths, S., et al. (2016), 'How readers understand causal and correlational expressions used in news headlines', *Journal of Experimental Psychology: Applied*. Advance online publication, available at http://dx.doi.org/10.1037/xap0000100 (accessed November 2016)

Aydin, D., Feychting, M., Schüz, J., et al. (2011), 'Mobile phone use and brain tumors in children and adolescents: A multicenter case-control study.' *Journal of the National Cancer Institute*, 103, 1264–76.

Blastland, M. and Dilnot, A. (2007), *The Tiger that Isn't*, London: Profile Books.

Chivers, T. (2014), 'How not to be taken for a mug by misleading health stories this New Year.' *Spectator blogs*, 31 December. Available at: http://blogs.spectator.co.uk/spectator-surgery/2014/12/how-not-to-be-taken-for-a-mug-by-misleading-health-stories-this-new-year/ (accessed November 2016).

Cushion, S., Lewis, J. and Callaghan, R. (2016), 'Data journalism, impartiality and statistical claims.' *Journalism Practice*. Advance online publication, available at http://www.tandfonline.com/doi/full/10.1080/17512786.2016.1256789 (accessed November 2016).

Desrosières, A. (1998 [1993]), *The Politics of Large Numbers: A History of Statistical Reasoning*, Cambridge, MA: Harvard University Press.

Fox, F. (2012), *10 best practice guidelines for reporting science & health stories*, Science Media Centre. Available at: http://www.sciencemediacentre.org/wp-content/uploads/2012/09/10-best-practice-guidelines-for-science-and-health-reporting.pdf (accessed November 2016).

Hill, A. B. (1965), 'The environment and disease: Association or causation?.' *Proceedings of the Royal Society of Medicine*, 58, 295–300.

IARC (2015), *IARC Monographs Questions and Answers*. Available at: https://monographs.iarc.fr/ENG/News/Q&A_ENG.pdf (accessed November 2016).

IARC (2016), *Agents Classified by the IARC Monographs, vols 1–117*. Available at: http://monographs.iarc.fr/ENG/Classification/ (accessed November 2016).

INTERPHONE Study Group (2011), 'Acoustic neuroma risk in relation to mobile telephone use: Results of the INTERPHONE international case-control study.' *Cancer Epidemiology*, 35, 453–64.

Kelvin, P. (1970), *The Bases of Social Behaviour*. New York: Holt, Rinehart and Winston.

Koetsenruijter, A. W. M. (2011), 'Using numbers in news increases story credibility.' *Newspaper Research Journal*, 32, 74–82.

Leveson, L. J. (2012), *An Inquiry into the Culture, Practices and Ethics of the Press: Report*, London: The Stationery Office. Available at: http://webarchive.nationalarchives.gov.uk/20140122145147/http://www.official-documents.gov.uk/document/hc1213/hc07/0780/0780.asp (accessed November 2016).

Lönn, S., Ahlbom, A., Hall, P., et al. (2004), 'Mobile phone use and the risk of acoustic neuroma.' *Epidemiology*, 15, 653–9.

McConway, K. and Spiegelhalter, D. (2012), 'Score and ignore: A radio listener's guide to ignoring health stories.' *Significance*, 9, 45–8.

Munroe, R. (2011), 'Cell phones'. Available at: http://xkcd.com/925/ (accessed November 2016).

Pearl, J. (2009), 'Causal inference in statistics: An overview.' *Statistics Surveys*, 3, 96–146.

Poovey, M. (1998), *A History of the Modern Fact*. Chicago, IL: University of Chicago Press.

Porter, T. M. (1996), *Trust in Numbers: The Pursuit of Objectivity in Science and Public Life*, Princeton, NJ: Princeton University Press.

Riesch, H. and Spiegelhalter, D. (2010), '"Careless pork costs lives": Risk stories from science to press release to media.' *Health, Risk and Society*, 13, 47–64.

Robertson, G. (1838), 'Review of the transactions of the statistical society of London, vol. I, Part I.' *Westminster Review*, 31, 45–72.

Roeh, I. and Feldman, S. (1984), 'The rhetoric of numbers in front-page journalism: How numbers contribute to the melodramatic in the popular press.' *Text*, 4, 347–68.

Rose, N. (1991), 'Governing by numbers: Figuring out democracy.' *Accounting, Organizations and Society*, 16, 673–92.

Rothman, K. J. and Greenland, S. (2005), 'Causation and causal inference in epidemiology.' *American Journal of Public Health*, 95(Suppl.), S144–50.

Schüz, J., Steding-Jessen, M., Hansen, S., et al. (2011), 'Long-term mobile phone use and the risk of vestibular schwannoma: A Danish nationwide cohort study.' *American Journal of Epidemiology*, 174, 416–22.

Senn, S. (2009), 'Three things that every medical writer should know about statistics.' *The Write Stuff*, 18, 159–62.

Statistical Society of London (1834), 'Prospectus of objects and plan of the Statistical Society of London.' In: *Report of the Third Meeting of the British Association for the Advancement of Science*, 492–5. London: John Murray.

Sumner, P., Vivian-Griffiths, S., Boivin, J., et al. (2014), 'The association between exaggeration in health related science news and academic press releases: Retrospective observational study.' *British Medical Journal*, 349, g7015.

Swerdlow, A. J., Feychting, M., Green, A. C., et al. (2011), 'Mobile phones, brain tumors, and the interphone study: Where are we now?' *Environmental Health Perspectives*, 119, 1534–8.

White, A. (2014). 'How to read health news.' *NHS Choices*. Available at: http://www.nhs.uk/news/Pages/Howtoreadarticlesabouthealthandhealthcare.aspx (accessed November 2016).

Williams, J. (2007). *50 Facts That Should Change the World*, Rev. edn. London: Icon Books.

14

Mind the Statistics Gap: Science Journalism as a Bridge Between Data and Journalism

Holger Wormer, *TU Dortmund University, Germany*

Introduction

The media seem to love nothing more than figures, surveys and opinion polls which are regarded appropriate to transport (apparently) simple messages. This kind of 'love affair' (cf. Wormer 2007[1]) obviously has been amplified by the tendency to tell stories in 'charticles' by combining articles with charts which, according to the American Journalism Review, was already en vogue in 2008 (Stickney 2008) and has been growing with colourful and interactive graphics in online media. At the same time, there is no real evidence that journalists in general have significantly improved their knowledge on data and statistics in recent years. However, the emerging field of data journalism or data-driven journalism (formerly also known as computer-assisted reporting) promises some improvement in the complex relationship between figures, statistics and journalists. But are most data journalism projects really meaningful from a statistician's point of view? Are they, at least, better than any 'he-said-she-said-journalism'? Or do they rather give journalism products just a more trustful appearance often without a significant conclusion as some scientists criticize? This chapter seeks to explore these issues and argues that one possible bridge to connect both sides could be systematically educated science journalists who are familiar with journalistic needs and skills (between news selection, classical investigation and storytelling) but also with basic principles of statistics and scientific methods. From this perspective science journalists and science journalism education strategies may be not only one key to improve the strange relationship between journalists and statistical data in the newsroom but also predestined for a 'second generation data journalism'.

---

[1] Some basic ideas of this article correspond with this publication of the author from 2007.

Numbers and figures: News or just nice?

Forty thousand is a story,[2] 341 is another one,[3] so is 120, not to forget 250,000,000 000[4]: numbers are so attractive that media (the German broadsheet Süddeutsche Zeitung in these examples) sometimes just put them as a stand-alone headline. And they have also created a regular format called 'The figure' appearing nearly every day.

Other figures appear in countless surveys and rankings as a popular and very simple form of statistics. Rankings and numbers often seem to give just simple messages fitting in short headlines (although the given data and messages often do not seem very reliable for experts in statistics). Regarding the media circus of numbers and opinion polls in the popular press the daily German daily newspaper TAZ already claimed a couple of years ago: 'It is astonishing that there is no house of editors yet who has launched a special magazine on surveys and studies on the market.'

The attraction of numbers and simple data presentation is an international phenomenon – perhaps, let's say, a special variation of the very successful international Guinness Book of World Records principle. Some authors have tried to quantify the significance of quantification in special sections of the news. For example, Maier (2002) analysed in a one-month content review more than 500 news stories in order to estimate how often mathematical calculation is involved in local news reporting. He found that nearly half of the stories examined have required some sort of mathematical calculation or numerical point of comparison.

The digital age of journalism has probably further increased the significance of figures, numbers and data in the media. First, many formats on the internet need databases to generate the interactive and colourful forms of journalism which promise higher page impressions, which in turn are monitored with live web metrics in the newsroom (e.g. see Nguyen and Lugo-Ocando 2016). Second, data generated and presented in the context of crowdsourcing projects shall improve the ties between media houses and their users. Third, databases from everywhere are much more accessible for investigation than before and often deliver automatically tools and statistical material that can easily be used for reporting. For example, football statistics is an emerging field of numbers in the media and it can be generated rather simply by computers (which is regarded also as one of the first fields of robot journalism). It is especially attractive for every user and supporter: even if your favourite team has lost a game you may even prove with the delivered material that, regardless of the

---

[2] '40 000 people already rejected as refugees in Italy and Greece will be housed in Germany ...' (translated from Süddeutsche Zeitung Online, 11 November 2016: www.sueddeutsche.de/muenchen/erding/die-zahl--1.3245931).

[3] 'A human being has been 341 times forced to take over the steering of one of the self-driving cars of Google. The company reported this number of events to the road traffic authorities of California' (translated from Süddeutsche Zeitung, 14 January 2016, 16).

[4] '14 most interesting aspects of vasting; '120 … e-mails are managers receiving everyday'; '250,000,000,000 cubic meters of water are consumed worldwide for the cultivation of cotton' (all translated from Süddeutsche Zeitung, No. 301, 31 December 2015, 19).

result counted in goals, your team has also won to a certain extent, by measures of, for instance, the number of corner balls, number of shots, or percentage of ball possession during the game.

Data and databases as 'automatic news value generator'

The example of football statistics illustrates one possible answer to the question why (at least simple) figures and databases are so attractive for the media and media users. Putting it into a colourful metaphor, we may compare the situation with a public map of trails installed at the frontier of a forest or a National Park: obviously, the most important element of the map is the 'You're here' point (which is, for natural reasons, very often not clearly readable any more, after having been touched by thousands of fingers on the map). Similarly, simple numbers, figures and rankings give everybody the possibility to check where her or his personal 'I am here' point is located in the world and the different contexts of daily life. Furthermore, following a simple explanation for the love for number proposed by statistician Kevin McConway (2005), we can check where the others are at the same time: 'We do like knowing what our fellow citizens are getting up to. And, if the information is in the form of impersonal statistics, we don't have to feel guilty about spying on our neighbours. I became a statistician partly because I like poking my nose into other people's business – most people don't go quite to this extreme, but we're still interested.'

In light of basic theories such as news values – see, for example, Galtung and Ruge (1965) for foreign news factors and Badenschier and Wormer (2012) for science news – databases (especially the interactive ones) can be perhaps even regarded as a kind of automatic 'news value generator' for journalists and their audience. Whereas visualization is already one news factor, other important news factors such as proximity or relevance can be constructed by the media user himself. So, every football supporter can choose the figures of his favourite team from the league statistics. Every inhabitant of one city or region can construct its personal proximity and find his/her personal 'I am here' point on the map, then compare it with other regions in the country, with other countries in the world or with the characteristics of other people in relation to his/her physical properties, opinions, preferences and so on. The interactive database seems to deliver clear and simple answers. Probably it would be worthwhile to empirically further analyse how media users use such databases and to see how these pattern fits with classical theoretical frameworks.[5]

---

[5]　For such an approach, of course, existing research on users has to be included as the theory of news values describes primarily the action of journalists and not of media users. However, journalists do anticipate preferences and needs of their users and their choices are therefore a helpful construction regarding this perspective.

The limits: Daunting data and the statistical skills of everyday journalists

Besides their described attraction, numbers, databases and statistics include also strong elements of 'negative news factors': beyond just simple figures indicating clarity and simplicity, everything that has to do with mathematics sticks with a tag of complexity which is difficult to understand. Attractive figures can easily change into daunting data.

The dimension of the problem can be illustrated by an anecdote of Steven Hawking who explains in his bestseller, *A Brief History of Time*, why he has avoided all formulas except E= mc2 in the book: he was told that every formula in a popular book would divide the number of readers by two. Obviously, there exist also psychological reasons for the difficulties of the average (not only of the average journalist) to estimate proper results and messages by given figures as well as for the perception of risks and probabilities (e.g. Gigerenzer 2003). Classical news journalism is accustomed to providing facts about things that have already happened – for example the number of injured people and the estimated damage of a traffic accident. So, for journalists, reporting on things that have not yet happened but may happen in the future with a certain uncertainty (such as the risk of cancer or the dimension of climate change) is a big challenge. Therefore, risk communication which is closely related to statistics can be regarded as a 'natural' problem zone for journalism.

One part of the problem cannot be influenced easily because media users' preferences and statistical literacy will not change overnight. Meanwhile, statistics and data presented according to the ideals of a statistician will often remain 'too complicated' for a general audience. Somebody reading a data story on the small screen of his smartphone in an overcrowded subway is not able (or not willing) to read and decode a sophisticated data analysis. In the classical sense of 'The Medium is the message' (McLuhan 1964), some more sophisticated messages may be – at least to some extent – incompatible with the perception of the general media user.

In contrast to the media user, the statistical literacy of the journalist is much easier to address. The literature does not yield much (statistically reliable!) information on the real statistical literacy of journalists in different countries (e.g. Griffin and Dunwoody 2016), but there are many indicators for limited knowledge concerning these issues among this group. Although 'the stereotype of the statistically incompetent journalist is not entirely correct', Nguyen and Lugo-Ocando (2016, 3) report that 'it is still quite hard to find statistic course in university journalism programmes' (ibid., 2). So what does it mean when we assume, on the one hand, a more or less constantly limited knowledge about figures and statistics among journalists and, on the other, an increasing presence of figures, tables and data-based graphics in the news, especially in digital journalism?

First of all, it has to be underlined that the journalistic tendency to look more for figures and figure-based facts should be first and foremost regarded as an evolution going beyond the evidence level of an opinion-based 'he-said-she-said' reporting, which is still a common weakness in journalism, especially in political journalism.[6] However, figures and tables strongly suggest a level of clarity, precision and sometimes simplicity which is often not the case in reality, at least not from a scientific/statistical perspective. The downside of many media presentations of numbers and statistics begins with the choice of data and the basic population (as already illustrated in the football example, a selective choice of numbers can help to prove nearly everything).

Furthermore, statisticians would probably claim that the data source, validity, reliability, accuracy limitations and confidence intervals should be presented, at least as 'error bars' (see also Chapter 1). But on the other hand, most of such additional information will probably reduce the news value of a simple figure which has been, at the first glance, simple, nice and clear. Even if the journalist had been able to evaluate, calculate and include such information, probably most of the users would not understand or would not really care about the limitations and framework. However, the note that media users would not understand the additional information anyway must not be used as a general excuse. As many media data journalism projects suggest, quality standards that are a compromise between scientific and journalistic standards, are needed to ensure a flair of scientific precision. In other words, something that looks like science or has the appearance of it in the media – such as data journalism – should fulfil at least some basic scientific standards.

Media sections of hope: Science and economy

The good news of the whole story is that the basic problem addressed above is not new: at least some sections of the media are used to dealing with such difficulties for a long time. There are at least two sections that are routinely dealing with numbers, statistics and probabilities more than any other: the science and, to a certain extent, the business/economy section. Journalists working in these areas very often have at least a basic scientific education which includes mathematical and statistical skills. I focus here on science journalism, since some of its sub-areas, such as medical reporting or reporting on environmental sciences, have been prescribed with quality standards that are a synthesis of both scientific and journalistic standards (see Schwitzer 2008; Wilson 2009; Wormer 2011; Rögener and Wormer 2015). There is also some evidence that media houses with specialized science journalists fulfil such criteria better than those without such specialized journalists (Wormer and Anhäuser 2014; Wilson et al. 2010). Science journalists are also used to crosschecking scientific data published even in prestigious journals or provided by medical doctors and they were traditionally forced to develop strategies on how to communicate numbers and probabilities (e.g.

---

[6] To a certain extent this argument refers to the field of Evidence-Based Medicine where different evidence levels are defined according to the used study design and the information source. According to this classification, the lowest evidence level is an expert opinion, even lower than the evidence from a single case of observation.

by avoiding percentages and use rather absolute numbers; see Gigerenzer, Mata and Frank 2009; Weymayr 2011 and the reporting guidelines on the German site of Media Doctor: http://www.medien-doktor.de/medizin/sprechstunde/die-krux-mit-den-relativen-und-absoluten-grosenangaben-teil-1-2/).[7]

Ideally the skills of science journalists in understanding statistics and reading research studies should enable them to fulfil their role as a critical observer of science in a similar way as it is claimed for political journalism for politics: science journalists should have moved, as Rensberger (2009) claims, from 'cheerleaders to watchdogs'. As many other fields of reporting are more and more connected with studies, or to put it in other words, as data and statistics have become 'part of the fabric of the contemporary world' (Nguyen and Lugo-Ocando 2016, 1), it seems to be rather logical to bring the different journalistic sections closer together in order to bridge the gap between data, science and journalism and to enrich the emerging field of data journalism (discussed in the next part). Furthermore, some existing quality criteria in science journalism could help to develop a similar matrix for data journalism and for reporting on numbers and statistics in non-science news beats.

The common roots of data-driven journalism and science journalism

Regardless what terms are used – 'Date Driven Journalism' (DDJ), 'Data Journalism' (DJ) or, a decade before, 'Computer Assisted Reporting' (CAR) – all of these forms may be regarded, to a certain extent, as a further development of 'Precision Journalism', which Philip Meyer (1973) proposed as an inquiry method that uses 'quantitative social science research methods … to gather news' (Demer and Nichols 1987, 10–11). According to these authors, there is even room to date back its origin some decades earlier: 'Precision Journalism is already described for some journalistic investigations in the 1930s.' Simon Rogers (2013), data journalism pioneer at the *Guardian*, traces one of the first data stories even back to 1858, when Florence Nightingale was reporting in texts and graphics on the situation of British soldiers. However, only the combination of powerful computers and the World Wide Web has offered the possibility of accessing (as well as scraping or generating by web robots) big databases, storing and analysing them in the way as we know it today.

---

[7] One classic example is the information given about the effect of a regular breast cancer screening in Germany (cf. Wormer 2007). 'Within 10 years the breast cancer screening reduces mortality by 25%' was an oft-cited statement of medical doctors. Only some media outlets succeeded in uncovering this misleading information by using absolute instead of relative numbers. That is, 'without breast cancer screening 4 out of 1000 women die of breast cancer within ten years. With the use of screening there are 3 out of 1000.' The weekly newspaper that reported this, Die ZEIT (No. 23, 18th of June. Hamburg. pp 27–8), even found a way not only to explain these facts but also to give their readers a feeling of the figures – by printing hundreds of symbols of women who were supposed to have been tested to identify and to heal one woman more than without screening (Koch and Weymayr 2003).

Temporarily CAR was strongly located in the field of investigative reporting and its scientific roots seemed to be a little forgotten. The first steps of CAR are usually dated in the second part of the 1970s with the foundation of the Society of Investigative Reporters and Editors (IRE) in the United States (www.ire.org), followed by the National Institute of Computer Assisted Reporting (NICAR, founded by IRE and supported by the Missouri School of Journalism). Since then NICAR has documented, stored and distributed databases to journalists systematically as well as offered help for analysing data (www.ire.org/nicar/). Already between 1989 and 2000 there was at least one Pulitzer-Prize story every year generated by CAR techniques (Redelfs 2001). According to the former Acting Director of IRE between 2003 and 2007, more than 700 investigations that involved significant CAR work had been posted on the IRE website. These represent a small percentage of the total stories that used CAR in some way for a fact, context or graphics (Houston 2007). Examples of database driven stories can be still searched on the IRE site (http://www.ire.org/resource-center/stories/) but a reliable total number of such stories could not be estimated anymore.

Different journalism cultures and media structures have influenced the dynamics and the kind of data journalism practised in different countries. CAR became popular much earlier in the United States than in Europe. Since there is no evidence that the average US journalist was better qualified to deal with numbers and statistics than their colleagues in Europe there must be other structural reasons for the earlier emergence of the field. One reason may be an earlier take-up of intensive internet use and development of computer-based tools in the United States. Another could be frameworks such as the Freedom of Information Act which forces US government agencies to be transparent in providing facts and figures and to make databases more easily accessible than in Europe. A third reason probably is the traditionally higher degree of specialization of US journalists on certain forms that make it easier for a sub-species of journalists as 'number crunchers' to emerge and establish itself. This new group started to analyse databases to generate stories at a time when such behaviour was more or less unpopular among journalists in many other countries.

Nevertheless, there are also some parallels between Europe and the United States. In Germany it was also the national association of investigative reporters and editors, Netzwerk Recherche (www.netzwerk-recherche.de), that has started the first notable CAR efforts to make the field more popular in Germany. Since 2006 the preconditions for CAR/data journalism have improved also in Germany by the first nationwide 'Informationsfreiheitsgesetz' which – although not as strict as the US Freedom of Information Act – promises a better access to governmental databases for statistical (and journalistic) analysis.

If official sources cannot deliver adequate datasets, the digital world around the internet offers many possibilities to generate the necessary data by data journalists themselves. Approaches such as crowdsourcing or scraping, for example, have also become more popular in recent years. But the highest popularity of the field was probably achieved by whistleblowing investigative stories. As is also observed by

Hewett (2016, 119): 'Data journalism has evolved rapidly since some pioneers gathered in Europe in 2010 ... . This was the year in which data journalism came of age, with the WikiLeaks war logs from Iraq and Afghanistan, and US embassy cables.' And citing other sources from different countries Hewett (2016, 120) comes to the conclusion that 2014 was the year when data journalism went 'mainstream'.

However, it must remain open what 'mainstream' really means. It might mean that many media would like to do 'something with data' rather than that they are really able to do so with their existing staff and resources. In 2013, Weinacht and Spiller (2014) could identify not more than thirty-five data journalists in Germany; in one of our own students' projects in 2015, we did not exceed this magnitude (see www. datenjournalismus-dortmund.de/who-is-who). According to Weinacht and Spiller's survey among the identified data journalists 'their role and activity profile differs in various dimensions compared to the average journalist in Germany'. They found a highly interdisciplinary field including graduates especially from social sciences and computer sciences (Weinacht and Spiller 2014, 424–5). Again, this survey indicates that most data journalists are still strongly connected to the investigative department and see their main function in control and criticism of politics, economy and society. On the other hand, everyday journalists seem to have still a rather diffuse picture of data journalism and often their idea of the field seems to be primarily focused on the aspect of visualization.[8] However, a detailed analysis of broad data journalism projects (documented on the IRE website or for Germany on sites such as http://katalog. datenjournalismus.net/#/) is still missing.

Mainstream data journalism of today: Doing some statistics or just visualization?

Beyond visualization, which is well noted as an important aspect of data as well as of classical journalism (e.g. Tufte 2001), statisticians criticize that many data journalism projects published lack of a real statistical data analysis or use only very simple forms of statistical tools. To check this assumption, we have examined sixty-two best practice data-driven projects which were awarded either a Data Journalism Award of the Global Editors Network, a Pulitzer Prize or a German Grimme Online Award between 2011 and 2015 (Marty et al. unpublished). Only a minority was based on self-collected or in-house created data; 92 per cent of the projects used existing data from public institutions, companies, NGOs, scientific institutions and so on. Even more interesting were the results of a content analysis conducted in cooperation with our statistics faculty in order to classify the used statistical methods in the award-winning data journalism projects: In 15 of 62 examined projects, no statistical methods could be identified; the vast majority

---

[8] This observation is not based on empirical data but among others on a debate at the annual meeting of the members of the Association of German Science Journalists ('Wissenschaftspresse-Konferenz').

(39) projects were based on descriptive statistics, more sophisticated methods such as multivariate regression (2), inductive statistics (1), unsupervised learning (2) and network analysis (3) were the exception. Comparing these results with data journalism projects that students could regularly realize in the teaching context of tandem seminars at our university (consisting of at least one student of statistics and one from the journalism department in each project), we see a much broader variance of applied statistics tools and methods. Reimer and Loosen (Chapter 6, this volume) found similar and other issues from their detailed content analysis of DJA-nominated projects during 2013–15.

Although we can only postulate an added value that the examined award-winning projects would have been carried out following an approach more oriented towards statistical methods,[9] our findings seem to be in line with the general observation that many data journalism projects do not include even a basic framework of analysis and information that a statistician would expect. Also, popular and freely available teaching material from practitioners for practitioners, such as *The Data Journalism Handbook* (Gray, Bounegru and Chambers 2015), seem to lack of contributions by academics from the field of statistics. In any case, the connection and network between data journalism, investigative journalism and possible scientific contributions (especially from the field of statistics and computer science) could be further developed.

Towards a next generation data (analysis) journalism

Besides all obstacles and constraints concerning data literacy in the broader public as well as among many journalists there are reasons for optimism. One reason is the general tendency of an increasing presence of data in the life of the average Joe: people of today are not only routinely buying data (e.g. for their smart phone) by themselves but also cannot escape anymore the reporting on data manipulation, data sharing and data security in their environment. Before the 2016 US Election, topics such as 'filter bubbles', 'echo chambers', the distribution of 'fake news' and 'roboter tweets' were rather limited to an expert or at least especially interested audience. Within a short time after that, these issues had become temporarily omnipresent, spreading from talk shows to political debates. At least, a certain 'I do not really understand it but it seems to be important' status concerning data issues should have been reached in the broader public. Big data leak projects, such as the Panama Papers recently initiated by the German nationwide newspaper Süddeutsche Zeitung and realized in a broad international cooperation with many journalists from all over the world, have also underlined the significance of the

---

[9] An interesting approach would be to replicate the projects in cooperation with statisticians and comparing the results with the results gathered by the purely data-journalistic analysis. In our ongoing research, we are envisioning such a two-arm investigation for a current project.

field. Such big data projects lead to a second reason for optimism: their dimension and complexity has let many data journalists realizing that such amounts of (mostly rather unstructured) data could not be handled anymore without approved scientific methods and knowledge. Although Panama Papers was again a data journalism project clearly connected to investigative journalism, it has underlined the need to build bridges with scientific expertise. To put it in other words: it brought data journalism a little bit back to the roots of 'precision journalism' and thereby to science and statistics.

Such broadly visible projects are a good starting point and together with the described increasing need of interactive (data-based) reporting and the increasing presence of web metrics as a marketing tool in the newsroom, there are new options to establish data literacy and better statistical skills among journalists and to implement them on a broader scale in journalism education. Until today, most of the journalists have probably become journalists because they liked to deal with language, pictures and telling stories (if they had preferred to deal with mathematics, they probably would have become physicists, statisticians, data or computer scientists). Therefore, their potential to check statistical data by themselves will be limited – especially in the extremely limited time slots of daily news and deadlines. Statisticians may have days up to weeks for their analysis of a database; journalists usually have minutes to hours. However, journalists should at least be able to check for plausibility and to communicate with data experts in a team (see also Chapter 6).

But who could ideally bring the two worlds together? If we assume that specialist science journalists (not the general reporters assigned to cover science stories) have the closest relationship to figures and formulas and that they usually have developed strategies to deal with abstract data in journalistic reporting, they may play a leading role in the media's endeavours to harness the vast potential of data journalism. They may encourage journalists from other fields to approach this 'strange' field. And they understand the language of journalists as well as of computer scientists and statisticians, and that makes them very valuable as translators or even leaders to direct interdisciplinary teams and bring people with different qualification together.

One strategy for an appropriate education of data journalists to reach such goals is to offer tailored courses combining skills of journalism and science, following the model of existing science journalism courses, such as the one we have established at Dortmund University since 2003 (www.science-journalism.org). The data journalism courses that we offer are part of the science journalism programme: science journalism students can choose between a specialization in life sciences and medicine, physics, engineering sciences, and data journalism. Establishing data journalism as a specialization in science journalism should also help to attract students who feel from the very beginning comfortable both with language and other purely journalistic skills and with figures, datasets and statistics – that is, nearly the same profile of people that we have always tried to attract in the classical science journalism programme. To cut it short, such students should neither fear words nor numbers. With this strategy, we avoid just from the beginning the problem pointed out by Hewett (2016, 132) that 'despite promising job prospects, data journalism risks being rejected as unappealing by potential students with an arts or humanities background'.

In addition to teaching methodological skills (purely statistics courses shared with students in the faculty of statistics in the first year) and journalistic skills, our programme[10] includes elements of investigation and presentation with a special focus on data aspects, from data visualization and infographics, data protection law and copyright law. Also, questions such as about the generation and the origin of data or why some data is missing are especially addressed, following the tradition and recommendation of CAR experts like Brant Houston (2007), who claims that it has always to be 'assumed that every database is in some way incomplete or imperfect and it is important to determine what its weaknesses are'.

Another characteristic of the programme is that most of the basic courses are open for other students from the department of journalism (not studying data journalism or science journalism) and particularly for students of statistics, too. The deep cooperation and integration of data journalism education in the faculty of statistics also has a strong positive side effect: it immediately strengthens the possibilities of cooperation between journalists and statisticians already at a student's level and, at the same time, attracts students from statistics to think about the media world as a possible working field. Some courses also help to improve the skills of the students in statistics. Interestingly, in the above-mentioned tandem seminars with students from both fields, the limitation factor for a successful data journalism project by far comes not just the journalistic part with its difficulties to deal with figures and statistics. A clear limitation for a better treatment of numbers in the news comes from the statistics side itself: statisticians (as well as the students in this field) rarely recognize by themselves the 'big media story' behind an interesting data analysis that they have done. They have much more difficulties to present results in an understandable way; they are not used to think in terms of storytelling frames and news factors; they do not know about other investigation tools that help professional journalism to find from some statistical findings a political or economic angle that is of high interest for the public. Very often the journalistic students are even better in developing the right (research) questions and have more awareness of possible pitfalls and biases in a given dataset (e.g. because of conflict of interests of the source). To put the whole matter in a nutshell: the programme does not only promise a new generation of data journalists with a much better knowledge in statistics than most of the data journalists in the market (who often learned such skills by themselves in a rather autodidactic manner). At the same time, the programme shows that journalistic thinking can also be helpful for scientific work of young statisticians.

---

[10] An overview of our program can be found at www.datenjournalismus-dortmund.de .

Conclusions and perspectives

In a programmatic article on the future of journalism and journalism education, Wolfgang Donsbach (2014) has described journalism as a new knowledge profession with the two core societal functions of validation and shared reality. For such reasons, journalism programmes had even a responsibility to guide their students 'choices outside their major': 'Journalist also need deeper knowledge and understanding of the subjects they are covering. … While the level of this knowledge journalists have will rarely compare to the level that the experts in the respective field have, it has to be sufficiently deep so that the structure of the field is understood and the main actors are known' (Ibid., 668). Interestingly, Donsbach continues by referring to Philip Meyer when he recommends that 'journalism students receive instruction in the process of knowledge-tested reporting. Such training would educate journalism students to be truth-seekers in a scientific sense and provide evidence that is always tested against alternative explanations. Philip Meyer has long argued that journalists need to apply the logic of the scientific method to their work.'

Notably Donsbach was talking not about science journalism but about journalism education in general[11] and, if we take his (and Meyer's) approach seriously, the fields of journalism and science have indeed need to be brought closer together (see Foreword and Introduction for Walter Lippmann for similar views in the 1920s). Interdisciplinary courses in data journalism as presented here, as well as other double-branched study approaches in science journalism or economical or financial journalism, could be a bridge and a helpful blueprint for journalism educators to tailor programmes to the different needs in different media environments in different countries. Instead of continuing to keep complaining for another decade about the limited statistical knowledge of journalists, a triple strategy for action to improve treatment of numbers, data and statistics in the media is recommended:

1. Improving general journalism education by including more basic modules from science journalism, data journalism or similar courses that are, in Donsbach's demand, 'sufficiently deep so that the structure of the field is understood'. In any case, journalists have to be enabled to work efficiently in teams with data experts and computer experts in the newsroom.
2. Professionalizing journalists' data literacy and statistical skills in predestined fields (such as science journalism or economics/business/finance journalism) which enable them to do a broader spectrum of statistical analysis by their own and which could be profiled and labelled as 'next generation data journalism' or 'data-analysis journalism' (DAJ).

---

[11] This is even more striking if we compare Donsbach's proposal with Rensberger's (2009, 1056) claims for science journalism only five years before: 'If science journalists are to regain relevance to society … they must learn enough science to analyse and interpret the findings – including the motives of the funders.'

3. Attracting and directing interested statisticians, data analysts and computer scientists in the context of interdisciplinary teaching and real-life cooperation projects towards the journalistic mass media world as a potential working field.

All in all, journalism has to become much more scientific in the future, particularly in the field of investigation, without abandoning journalistic virtues and time-tested (e.g. narrative) forms of presentation or losing its ability to use the attractive aspects of numbers as news factors. Substantial journalist–scientist collaborations seem to be an important strategy. Such cooperation can help to ensure quality (e.g. with verification tools to detect 'fake news') and, in general, to generate the needed added value of a journalistic product that might persuade media users to pay for news in the future. Hopefully such journalism will be recognized as more valuable than what the user can just get from data machines (where he/she pays with his/her own personal data).

Foundations such as the Volkswagen Foundation have already identified the potential of such journalist–scientist collaborations (Volkswagen Foundation 2015). Consulted by some members of our institute, the foundation has launched a funding line in order 'to initiate joint research and reporting projects which enable both sides science/scholarship and journalism to learn from each other and to generate new impulses for their respective activities'. As a part of that programme an initial conference brought scientists, data journalists and other journalists together[12] and permission for following-up conferences for 2017–19 have been granted by the foundation. Parallel to the first conference, a survey on proposals for possible quality standards among the participating scientists and journalists is in progress.

Finally, besides all the practical and technical similarities or win–win constellations for scientists and journalists, there is another normative argument to bring both professions stronger together: their shared responsibility for truth-seeking, fact-checking and strengthening debates and opinion formation in a modern democracy. This common role has recently underlined in a recommendation and position paper by all German academies of Sciences (Acatech 2014, 2):

Science and journalism are among the essential pillars of a democratic society. This is why Article 5 of the German constitution (Grundgesetz) guarantees freedom of the press and scientific freedom. Despite their necessary mutual independence and their often divergent purposes, both freedoms also fulfil similar functions. They supply policymakers and society with a diverse array of information that is as reliable as possible, reinforcing the education and knowledge of the population and stimulating democratic discourse. They should also provide a basis for reasoned political, economic and technological decisions.

---

[12] Datenlabor – Exkursion an die Schnittstellen von Datenjournalismus und Wissenschaften: https://netzwerkrecherche.org/termine/konferenzen/fachkonferenzen/daten-labor15/ and www.scicar.de/index.php?article_id=1&clang=1

Acknowledgements

The author would like to thank Christoph Marty and Claus Weihs (Department of Statistics of TU Dortmund University) for a fruitful discussion and especially for the analysis and further development of data journalism courses at TU Dortmund University.

References

Acatech, German National Academy of Sciences Leopoldina, Union of the German Academies of Sciences (2014), *On Designing Communication between the Scientific Community, the Public and the Media*. Position Paper. www.acatech.de/fileadmin/ user_upload/Baumstruktur_nach_Website/Acatech/root/de/Publikationen/ Kooperationspublikationen/3Akad_Stellungnahme_Kommunikation_2014_EN_final. pdf

Badenschier, F. and Wormer, H. (2012), 'Issue selection in science journalism: Towards a special theory of news values for science news?' In Rödder, S., Franzen, M. &Weingart, P. (Hrsg.), *The Sciences' Media Connection & Communication to the Public and its Repercussions. Sociology of the Sciences Yearbook* (28), 59–85. Dordrecht: Springer.

Demers, David P. and Nichols, Suzanne (1987), *Precision Journalism. A Practical Guide*. Newbury Park: Sage.

Donsbach, W. (2014), 'Journalism as the new knowledge profession and consequences for journalism education'. *Journalism*, 15(6), 661–77.

Galtung, Johan and Ruge, Mari Holmboe (1965), 'The structure of foreign news.' *Journal of Peace Research*, 2, 64–91.

Gigerenzer, Gerd (2003), *Reckoning with Risk. Learning to Live with Uncertainity*. London: Penguin Press.

Gigerenzer, G., Mata, J. and Frank, R. (2009), Public Knowledge of Benefits of Breast and Prostate Cancer Screening in Europe. *Journal of the National Cancer Institute*, 101(17), 1216–20.

Gray, J., Bounegru, L., Chambers, L. (2015), *The Data Journalism Handbook*. Retrieved from http://datajournalismhandbook.org/1.0/en/index.html.

Griffin, R., Dunwoody, S. (2016), 'Chair support, faculty entrepeneurship, and the teaching of statistical reasoning to journalism undergraduates in the United States'. *Journalism*, 17(1), 97–118.

Hewett, J. (2016), 'Learning to teach data journalism: Innovation, influence and constraints'. *Journalism* 17(1), 119–37.

Houston, Brant (1998), *Computer Assisted Reporting. A Practical Guide*. New York: St. Martin's Press.

Houston, Brant (2007), *Personal Communication* as cited from Wormer (2007: 394).

Koch, Klaus and Weymayr, Christian (2003), Vom Segen des Nichtwissens. Bislang gilt die Früherkennung als die wirksamste Waffe im Kampf gegen den Krebs. Doch ist sie das wirklich? *Die ZEIT*, 23, 18th June, Hamburg, 27–8.

Maier, Scott R. (2002), 'Numbers in the news: a mathematics audit of a daily newspaper'. *Journalism Studies,* 3(4), 507.

McConway, Kevin (2005), Statistics and the media. *More or Less* Radio Programme, *The Open University/BBC.* Available at www.open.edu/openlearn/science-maths-technology/mathematics-and-statistics/statistics/statistics-and-the-media.

McLuhan, Marshall (1964). *Understanding Media: The Extension of Man.* London: Routledge.

Meyer, P. (1973), *Precision Journalism.*Bloomington: Indiana University Press.

Nguyen, An and Lugo-Ocando, Jairo (2016), 'The state of data and statistics in journalism and journalism education – issues and debates'. *Journalism,* 17(1), 3–17.

Rensberger, Boyce (2009), 'Science journalism: Too close for comfort'. *Nature,* 459, 1055–6.

Redelfs, Manfred (2001), Computer Assisted Reporting als neue Form der Recherche. Kleinsteuber, Hans. *Aktuelle Medientrends in den USA.* Wiesbaden: Westdeutscher Verlag.

Rögener, W. and Wormer, H. (2015), Defining criteria for good environmental journalism and testing their applicability: An environmental news review as a first step to more evidence based environmental science reporting. *Public Understanding of Science,* August, 1–16.

Rogers, Simon (2013), *Facts are Sacred: The Power of Data.* London: Faber and Faber/ Guardian Books.

Schwitzer, G. (2008). How do US journalists cover treatments, tests, products, and procedures? An evaluation of 500 stories. *PLoS Medicine,* 5, e95.

Stickney, Dane (2008), Charticle fever. *American Journalism Review,* October/November 2008. Available online: http://ajrarchive.org/Article.asp?id=4608

Tufte, Edward R. (2001), *The Visual Display of Quantitative Information,* 2nd edn. Cheshire, CT: Graphics Press.

Volkswagen Foundation (2015), Science and data-driven journalism. Available online: https://www.volkswagenstiftung.de/en/funding/completed-initiatives/science-and-data-driven-journalism-completed.html

Weinacht, S. and Spiller, R. (2014), 'Datenjournalismus in Deutschland. Eine explorative Untersuchung zu Rollenbildern von Datenjournalisten'. *Publizistik* 59 (4), 411–33.

Weymayr 'Die Krux mit den relativen und absoluten Größenangaben (Teil 1&2),' published on 8 March 2011 on 'http://www.medien-doktor.de.'

Wilson, A., Bonevski, B., Jones, A. and Henry, D. (2009). 'Media reporting of health interventions: Signs of improvement, but major problems persist'. *PLoS ONE,* 4, e4831.

Wilson, A., Robertson, J., McElduff, P., Jones, A. and Henry, D. (2010), Does it matter who writes medical news stories? *PLoS Medicine,* 7, e1000323.

Wormer, H. (2007), 'Figures, statistics and the journalist: An affair between love and fear. Some perspectives of statistical consulting in journalism'. *AStA Advances in Statistical Analysis,* 91(4), 391–7.

Wormer, H. (2011), 'Improving health care journalism'. In G. Gigerenzer and J. A. M. Gray (eds), *Better Doctors, Better patients, Better decisions: Envisioning Health Care 2020.* Strüngmann Forum Report, Band 6 (S. 317–337). Cambridge, MA: MIT Press.

Wormer, H. and Anhäuser, M. (2014), Gute Besserung! – und wie man diese erreichen könnte. Erfahrungen aus 3 Jahren Qualitätsmonitoring Medizinjournalismus auf medien-doktor.de und Konsequenzen für die journalistische Praxis, Ausbildung sowie Wissenschafts-PR. In V. Lilienthal, D. Reineck and T. Schnedler (eds), *Qualität im Gesundheitsjournalismus. Perspektiven aus Wissenschaft und Praxis.* Wiesbaden: Springer VS.

15

Teaching Statistical Reasoning (Or Not) in Journalism Education: Findings and Implications from Surveys with US J-Chairs

Robert J. Griffin, *Marquette University, USA*
Sharon Dunwoody, *University of Wisconsin-Madison, USA*

Introduction

Technology has given all of us access to a tsunami of data in this twenty-first century, and journalists the world over are scrambling to take advantage of it. In an occupation characterized more by math anxiety than by a passion for numbers (Maier 2003; Harrison 2014), reporters and journalists-in-training are confronting statistical training as a required – not an optional – part of their toolkit. Responses to that need vary widely, from one-day workshops and instructional websites to professional degrees and even to initiatives that pair journalists and statisticians. As Lewis notes in a recent special issue of the journal *Digital Journalism* titled 'Journalism in an Era of Big Data', 'these data-centric phenomena, by some accounts, are poised to greatly influence, if not transform over time, some of the most fundamental aspects of news' (Lewis 2015, 321).

But even as university journalism programmes worldwide begin building courses and specializations in (take your pick) computer-assisted reporting, data journalism and computational journalism,[1] it is important to ask about the extent to which statistical reasoning is being embedded in these pedagogical efforts. We posed this question to journalism school directors in the United States in two national surveys over the course of a decade. That decade preceded today's data journalism explosion, but at the end of this chapter we will suggest that, even in the midst of a determined curricular focus on 'big data', there remains the danger that the fundamentals of statistical reasoning are being given short shrift.

---

[1] For a conceptual effort to distinguish among these labels, see Coddington (2015).

Statistical reasoning

Statistical reasoning and numeracy (or mathematical literacy, a term utilized in some countries), although clearly related, are conceptually distinct phenomena. The Organization for Economic Co-operation and Development (OECD), which conducts the periodic International Student Assessment (PISA) exam among teenagers worldwide, defines numeracy as

> an individual's capacity to identify and understand the role mathematics plays in the world, to make well-founded judgments, and to use and engage with mathematics in ways that meet the needs of that individual's life as a constructive, concerned and reflective citizen. (OECD 2004, 15)

Coben, in turn, argues that

> To be numerate means to be competent, confident, and comfortable with one's judgments on whether to use mathematics in a particular situation and, if so, what mathematics to use, how to do it, what degree of accuracy is appropriate, and what the answer means in relation to the context. (Coben 2000, 35).

Statistical reasoning, in contrast, becomes relevant in the face of uncertainty. Garfield explains:

> Much of statistical reasoning combines ideas about data and chance, which leads to making inferences and interpreting statistical results. Underlying this reasoning is a conceptual understanding of the important ideas, such as distribution, center, spread, association, uncertainty, randomness, and sampling. (Garfield 2002)

Nguyen and Lugo-Ocando, in an introduction to a recent issue of *Journalism* dedicated to statistical reasoning in the field, create a relatively bright line between numeracy and statistical literacy in journalism:

> The job of handling numbers for the news is often wrongly perceived as that of measuring, calculating and analysing things with eye-numbing formulae. Statistics and mathematics are two different things: it is not necessary to be adept at mathematics to be able to use statistics effectively. However frightening they might look, statistical analyses are about the application of valid reasoning, not calculation. (Nguyen and Lugo-Ocando 2016, 4–5)

To illustrate the comparison, let's take a situation where a state health department issues a report about a cancer cluster in a community in the state. A journalism student will exercise numeracy skills in the course of writing that story by describing the size of the cluster relative to the neighbourhood, taking care to distinguish between raw numbers and percentages, reflecting on the prevalence (nationally or statewide) of the particular cancer etc. A student with statistical reasoning skills will do all of that as well but will also, importantly, explore the extent to which the cluster may have occurred by chance, taking care to explain that people typically underestimate the likelihood that rare things will co-occur.

Statistical reasoning is apparently uncommon. One can find ample evidence that individuals, confronted with decision-making in the face of uncertainty, tend to employ reasoning strategies that can lead them astray. Studies of the ubiquity of heuristic processing, for example, suggest that we default to a wide array of non-statistical frameworks, such as using information easily brought to mind (the availability heuristic), slotting new stimuli into established categories (the representativeness heuristic), and interpreting subsequent information based on one's initial starting point (anchoring).[2] And although applicable data are scarce, Stocking and Gross (1989) make a strong argument that journalists employ the same kind of cognitive biases as they carry out their professional responsibilities.

Can formal education turn these deficits around? Most of the available data, which focus principally on numeracy, are equivocal at best. Although there are exceptions, young people in many countries struggle to acquire reasonable levels of mathematical literacy. In its latest global survey of fifteen-year-old students, conducted in 2015, the Organisation for Economic Co-operation and Development (OECD) found that, while more than 1 in 4 students in countries such as China, Singapore and Taiwan did well on numeracy tests, proving they could handle tasks that 'require the ability to formulate complex situations mathematically, using symbolic representations',[3] students in many countries, including the United States, scored well below the global average on mathematical questions.[4]

Historically, journalism schools have not offered much help in numeracy training for their students. One 2009 survey of department chairs in the United States found that respondents classified their students' math skills as 'poor' but were unlikely to offer any pedagogical recourse (Cusatis and Martin-Kratzer 2010). And Harrison (2014) reports a dearth of numeracy instruction in British journalism education as well. Efforts to design classroom modules that teach mathematical skills have been shown to be effective (see, e.g., Ranney et al. 2008), but such efforts appear to be uncommon.

When it comes to statistical reasoning per se in journalism education, data are even more limited. One can find descriptions of occasional local initiatives,[5] but systematic data are elusive. We, thus, turn to our decadal surveys of journalism department directors in the United States to look for some patterns.

---

[2] For a detailed discussion of these and other heuristics, see Kahneman, Slovic and Tversky (1982).
[3] https://www.oecd.org/pisa/pisa-2015-results-in-focus.pdf
[4] https://nces.ed.gov/surveys/pisa/pisa2015/pisa2015highlights_5.asp
[5] See, for example, http://www.worldofstatistics.org/2015/10/28/statistics-meets-journalism-in-statsstories-webcast/.

Research questions

In 1997 and 2008, we sent questionnaires to a sample of the head administrators of college-level journalism programmes in the United States (we will refer to these individuals as chairs, regardless of their actual job titles). Our purpose was to explore their views of various challenges and potential benefits of teaching statistical reasoning to their undergraduate students. Included in the two identical sets of questionnaires were items that asked chairs to reflect on their students' abilities and willingness to learn statistical reasoning, the readiness of their faculty to teach statistical reasoning, the integration of statistical reasoning into the curriculum, and their perceptions of the use of statistical reasoning by working journalists.

Detailed results can be found elsewhere (Dunwoody and Griffin 2013; Griffin and Dunwoody 2013; Griffin and Dunwoody 2016). Our purpose in this chapter is to summarize and highlight some of our key findings, specifically along these lines:

1. Were there any noteworthy differences in journalism chairs' perceptions of statistical reasoning instruction between the two survey waves over a decade-plus?
2. If statistical reasoning skills are valuable to journalists, why are they not being offered in the major?
3. What structural factors (e.g. offering a graduate degree) seem to facilitate the teaching of statistical reasoning to journalism undergraduates?
4. How willing are the chairs to support the teaching of statistical reasoning to their undergraduates?
5. What key benefits and constraints do chairs see in teaching statistical reasoning to journalism students, and how might such factors translate into the way they reward 'entrepreneurial' faculty who try to bring this type of instruction into their classrooms?
6. What is the role of such faculty entrepreneurs?

Method

Sampling

For the 1997 survey, we constructed a probability sample of 219 out of the 430 journalism undergraduate programmes in the United States as listed in the latest editions of the *Journalism and Mass Communication Directory* and the Dow Jones *Journalism Career and Scholarship Guide*. For the 2008 survey, we used the same sample so as to make the results from the two waves as comparable as possible.[6] Four of the programmes from the 1997 responses had gone out of existence.

---

[6] The American Statistical Association provided funding for the 1997 survey. The 2008 survey was funded by an internal grant from the Diederich College of Communication at Marquette University, Milwaukee, WI, USA.

In both waves, personalized invitations to participate in the survey were sent in advance to the chairs associated with each sampled programme, and several reminders were sent to those who had not as yet submitted their questionnaires. Ultimately, 75 per cent of the chairs in the 1997 sample responded with usable questionnaires (164/219) as did 63 per cent of the chairs in the 2008 sample (135/215). We considered these fairly high response rates to be indicative of the chairs' interest in the topic of statistical reasoning instruction for journalism undergraduates. All but twenty programmes had changed chairs in the decade-plus between the two survey waves. Thus, for statistical purposes, we considered the respondents from the two waves as comprising two independent groups.

In 1997, the entire survey – for example, invitations, questionnaires and follow-up reminders – was conducted by surface mail. In 2008, the survey employed a combination of electronic systems and surface mail. Invitations and follow-up reminders were sent by both surface mail and email, and, while the questionnaire was online, chairs could request a hard copy instead to fill out (29 per cent of the returned questionnaires in 2008 were hard copies). In both waves, we took steps to make sure that the identities of the chairs were physically or electronically disconnected from their completed questionnaires. The Marquette University Institutional Review Board approved the surveys.

Questionnaire

To help the chairs understand what we meant by the term 'statistical reasoning', we began the questionnaires in both waves with the following (boldfaced emphases in original):

In this survey we are interested in your ideas about the extent to which your undergraduate journalism students should be introduced to statistics and especially to **statistical reasoning**. By 'statistical reasoning' we don't mean their ability to compute statistical tests. Instead, we mean their **ability to think systematically and reason using numerical data,** for example:

- to assess critically the quality of data;
- to apply data appropriately to problem solving;
- to understand the limits to generalizability;
- to understand probability and risk;
- to recognize when better data and information are needed for decision-making (e.g. when the data provided are incomplete or not comparable), and to diagnose what information is missing.

The questionnaire then posed fifteen statements about statistical reasoning as applied to journalism education and the journalism profession, and asked the chairs to respond to each on a five-point, Likert-type, agree–disagree scale. Towards the end, the questionnaire also asked the chairs: (1) how they would prefer statistical reasoning to be taught to journalism students, if at all; (2) how many, if any, of their faculty were making efforts to teach statistical reasoning in their classes; (3) how well integrated into their curriculum any statistical reasoning instruction was at the time; (4) the highest degree their programme offers; and (5) the size of the programme in terms of the number of students and number of full-time faculty. Both of the latter variables were transformed to reduce positive skew, were standardized, and then summed to produce the variable representing programme size (alpha = 0.90).

Data analysis

Data were analysed using the IBM Statistical Package for the Social Sciences (SPSS) and its 'AMOS' structural equation modelling (SEM) programme. To clarify the narrative that follows, we have combined the 'strongly agree' and 'agree' percentages from the Likert-scaled items in the Appendix into a single category of 'agreement', and similarly condensed the 'disagree' and 'strongly disagree' responses. In the 1997 survey, the margin of error is ±6.0% for percentages around 50 per cent, and ±4.8 per cent for percentages around 20 per cent or 80 per cent (95 per cent confidence interval). For the 2008 survey, the 95 per cent CI is ±7.0 per cent for percentages around 50 per cent and ±5.5 per cent for percentages around 20 per cent or 80 per cent. Detailed results can be found in the Appendix.

Results

Did J-chairs' attitudes about statistical reasoning instruction change over the decade? Journalism administrators across the decade endorsed the value of statistical reasoning for their students. In 1997 some 92 per cent agreed that such training was important and in 2008 91 per cent agreed. About three quarters (74 per cent in 1997 and 77 per cent in 2008) preferred this instruction to be included somewhere within their journalism programmes. Yet actual provision of such training remained scant. In fact, as the reader will see, the first decade of the twenty-first century seems to have produced little variance in administrator attitudes or behaviour. Indeed, the survey results in the Appendix show that the chairs in 2008 perceived the teaching of statistical reasoning in the programmes in essentially the same way as did the chairs in 1997 across a variety of attitudinal measures.

Challenged to describe the best venue for such training, a plurality of administrators (41 per cent in 1997 and 47 per cent in 2008) felt it should be embedded in courses across the journalism curriculum. The second most popular choice was requiring their students to take a statistics course offered elsewhere in the university (30 per cent chose this option in 1997 and 25 per cent in 2008). Finally, one in five administrators in both surveys preferred a statistical reasoning course designed for journalism majors.

Regardless of administrators' preferred venue, however, it is clear that journalism students were rarely exposed to this kind of material. Only a quarter of the chairs agreed that their students 'receive adequate instruction in the application of statistics to everyday problems' (23 per cent in 1997 and 25 per cent in 2008). Yet many administrators (56 per cent in 1997 and 41 per cent in 2008) left inclusion of statistical reasoning in the journalism curriculum up to individual professors, virtually guaranteeing its erratic appearance across courses. And only somewhat more than a third (35 per cent in 1997 and 36 per cent in 2008) said they required some or all of their majors to encounter such instruction within the major, for example as part of a research method or a computer-assisted reporting course. Approximately half of the administrators did note that their students were receiving statistical reasoning instruction elsewhere in the university, but the nature and volume of that contribution remained unclear.

The bottom line is that, over the course of a decade, journalism administrators in our surveys maintained a belief that statistical reasoning training was a valuable component of journalism training but did little to make that a reality in their own programmes even in the face of accrediting standards that require that majors learn to 'apply basic numerical and statistical concepts' (Accrediting Council on Education in Journalism and Mass Communication 2013). The stability of attitudes over time and across administrators is quite pronounced.

Why are statistical reasoning skills not being offered in the major? Why did we encounter such a chasm between beliefs and behaviour? We will suggest a number of factors in ensuing sections. But important among them, we think, were chair perceptions that, when it comes to statistical reasoning, both instructors and students would rate themselves low in self-efficacy. Put another way, chairs felt that neither instructors nor students would feel competent handling statistical reasoning concepts and processes.

For example, these chairs indicated that most of their faculty members would have difficulty offering statistical reasoning material in their classrooms; 53 per cent felt this way both in 1997 and in 2008. Additionally, more than three quarters (78 per cent in 1997 and 79 per cent in 2008) indicated that most of their student majors would avoid such material rather than embrace it. As one chair noted by way of illustration:

> We worked with local professionals in print journalism and advertising to develop an information gathering course required of all majors. It included applied statistics. After one year of listening to student complaints the faculty voted to eliminate it.

More than 40 per cent of chairs in each wave (48 per cent in 1997 and 42 per cent in 2008) believed that most of their majors did not have the mathematical skills to do well in the world of numbers and uncertainty, a strong disincentive to including statistical reasoning in coursework. Although journalism students may well agree with these views, data do not support such perceptions. At the time of our surveys, two analyses of SAT scores of high school students intending to major in journalism indicated that the students possessed quantitative skills sufficient for a general university education that includes basic math and statistics (Becker and Graf 1994; Dunwoody and Griffin 2013).

What structural factors facilitate the teaching of statistical reasoning? Despite such frustrations and misperceptions, some key structural elements – specifically the size of the programme, the highest degree it offers, and the presence of some flexibility in the curriculum – seem to make statistical reasoning instruction in the major more likely. To represent how much the department was active in such efforts, we examined the chairs' reports of the extent to which statistical reasoning instruction was integrated into the curriculum and the number of faculty who have tried, successfully or not, to teach statistical reasoning to undergraduate students. We termed such professors statistical reasoning entrepreneurs. (Analyses involving the number of entrepreneurs on the faculty are adjusted by the size of the faculty.) Then we compared these structural elements to the two indicators of department activity in teaching statistical reasoning across both waves of the study, as shown in Table 15.1.

Table 15.1 Correlation of structural variables with Integration of Statistical Reasoning (SR) Instruction, and with Faculty S.R. Entrepreneurship, in Journalism undergraduate programmes

	S.R Integration[1]			S.R. Entrepreneurship[2]	
	Programme size[3]	Highest degree offered[4]	Curricular flexibility[5]	Highest degree offered[4]	Curricular flexibility[5]
Both waves combined (N = 299)	0.20*** [6]	0.18*** [6]	0.23*** [6]	0.18** [7]	0.14* [7]
First wave, 1997 (n = 164)	0.23**	0.17*	0.26***	0.17* [8]	0.21** [8]
Second wave, 2008 (n = 135)	0.18*	0.20*	0.20*	0.19* [8]	0.09 [8]

Significance key: *p = 0.05 **p = 0.01 ***p = 0.001

[1] Whether department offers a course featuring statistical reasoning instruction: (0) no; (1) elective; (2) required for most or all journalism students.

[2] The number professors who have 'done anything creative (whether successful or not) to teach statistical reasoning to journalism students', as reported by the chair. The variable was log-transformed (see article text) and controlled via partial correlation by the number of full-time faculty. Number of students did not correlate with entrepreneurship with number of faculty controlled, the lack of correlation probably due to necessarily using faculty size as the covariate.

[3] Number of full-time students and number of full-time faculty in department (each standardized). Alpha = 0.87.

[4] (1) Bachelor's; (2) Master's; (3) Ph.D.

[5] Chair's disagreement with the questionnaire item: 'The journalism curriculum is too tight to offer in-house instruction in statistical methods and their applications.' See Appendix.

[6] First-order partial correlation, controlled by wave of study.

[7] Second-order partial correlation, controlled by wave of study and number of full-time faculty.

[8] First-order partial correlation, controlled by number of full-time faculty.

Table (amended) from Griffin and Dunwoody (2016), 105.

The most noteworthy result? Consistently, the higher the degree offered by the programme, the more likely that faculty entrepreneurs were to be found there, striving to teach statistical reasoning to journalism undergrads. While the reason for this relationship is not clear, programmes with graduate degrees – especially those that offer doctorates – are typically staffed by faculty with strong research interests. Those professors who have quantitative acumen may be actively incorporating that instruction into their undergraduate classes. Similarly, programmes with graduate degree offerings are also more likely to integrate statistical reasoning instruction into their curriculum, that is, to offer this instruction as a specific feature of elective or required courses.

In both waves, larger programmes – those with more students and faculty – were more likely to include statistical reasoning in their home-grown instruction. Similarly, programmes that have more room or flexibility in course offerings were more likely to have entrepreneurs and to incorporate statistical reasoning instruction. This suggests that programme size and flexibility might allow more experimentation (e.g. via special topics courses), thus providing interested faculty avenues to develop ways to teach statistical reasoning.

Next we will explore further the roles of chairs and entrepreneurs in infusing statistical reasoning instruction into the journalism curriculum.

How willing are the chairs to support the teaching of statistical reasoning to their undergraduates? Arguably, support from one's departmental head can help shape and implement curricular policy and encourage classroom innovation. To what extent were the chairs willing to actively support teaching statistical reasoning to undergraduates? When the chairs responded to the questionnaire statement, *'To the extent I can, I reward faculty who bring statistical reasoning into their classes,'* their most common answer was noncommittal (53 per cent in 1997 and 42 per cent in 2008 answered 'feel neutral' or 'don't know'). A little over a fourth agreed (26 per cent in 1997 and 29 per cent in 2008), and roughly as many disagreed (21 per cent in 1997 and 29 per cent in 2008).

So, to learn more about what motivates chairs to reward faculty for these efforts (or not), and what outcomes that might have on faculty entrepreneurship and the curriculum, we constructed and tested a model using the IBM SPSS 'AMOS' programme. The bottom line of that analysis: chair rewards help to encourage faculty entrepreneurship, which in turn promotes the integration of statistical reasoning instruction into the curriculum. Next we explore these dynamics further.

How do J-chairs' perceived benefits and constraints of statistical reasoning instruction translate into the way they reward 'entrepreneurial' faculty who try to bring it into their classroom? We wanted to understand more about the factors that might lead chairs to reward their faculty for statistical reasoning instruction. So we chose from the questionnaire three items that represented the chairs' perceptions of significant constraints on that instruction, thinking that perhaps some chairs might want to reward faculty who try to teach statistical reasoning despite these roadblocks. These items, as described previously, all concerned the chairs' perceptions of student and faculty lack of efficacy, specifically that: (1) most of their students would have difficulty learning statistics, (2) most would really not want to be exposed to statistical reasoning, anyway, and (3) most of their faculty would have difficulty teaching it in their classes.

To balance these perceived constraints, we chose three items that represented the chairs' views concerning benefits from teaching statistical reasoning. Might these perceived advantages prompt chairs to reward faculty who try to be entrepreneurial? Overwhelmingly, chairs agreed that 'it is important for our journalism students to be able to reason statistically' (92 per cent in 1997 and 91 per cent in 2008). In very practical terms, at least two-thirds believed that 'statistical reasoning skills give students a competitive edge in the journalism job market' (67 per cent in 1997 and 72 per cent in 2008). And what role might consistency with university policy have? There were highly mixed responses when the chairs were queried: 'Our university's goal is to integrate statistical reasoning into the curriculum.' Some agreed (23 per cent in 1997 and 32 per cent in 2008). Many others, however, disagreed (45 per cent in 1997 and 42 per cent in 2008) or were noncommittal (32 per cent in 1997 and 27 per cent in 2008), the latter perhaps not knowing whether their universities had relevant policies.

To examine the relationship of these benefits and constraints with the chairs' rewarding of faculty for entrepreneurship, we created a structural equation model that proposed that perceived benefits and constraints would positively affect the chairs' rewarding of faculty who bring statistical reasoning instruction into their classes. In turn, those rewards could increase the probability of professors being entrepreneurial and thus enhance the integration of statistical reasoning instruction into the curriculum. (We adjusted the results to control for the structural factors and relevant survey design features; see Griffin and Dunwoody 2016 for analysis details.)

As shown in Figure 15.1, which combines both waves of the surveys, perceived benefits motivate chairs to reward faculty for bringing statistical reasoning instruction into their classes (beta = 0.44, p≤ 0.05) more than do perceived constraints, which have no significant effects on reward decisions (beta = 0.10, ns). This pattern is duplicated in

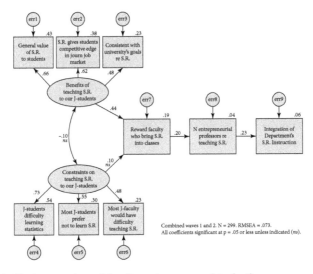

Figure 15.1 Chair rewards and faculty entrepreneurship, both waves.

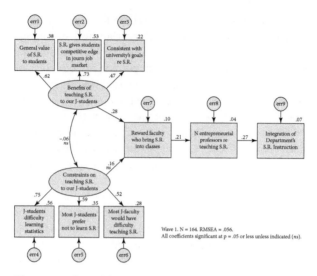

Figure 15.2 Chair rewards and faculty entrepreneurship, first wave.

both 1997 (Figure 15.2) and 2008 (Figure 15.3). In 2008, however, perceived benefits seem to have played a bigger role in the chairs' reward decisions than they did in 1997 (beta = 0.28 in 1997, 0.58 in 2008, both p ≤ 0.05).[7] Apparently the most impactful perceived benefits are that statistical reasoning has a general educational value to the students and that it gives them a competitive edge in the journalism job market. The most salient perceived constraint is the perception – albeit inaccurate in general – that journalism students have difficulty learning statistics.

What is the role of faculty entrepreneurs? As noted, chair rewards seem to encourage statistical reasoning entrepreneurship (e.g. beta = 0.20, p ≤ 0.05, Figure 15.1). In turn, the more faculty who engage in entrepreneurial behaviour, the more likely that statistical reasoning is integrated into the curriculum (e.g. beta = 0.23, p ≤ 0.05, Figure 15.1). Put another way, while a 'lone wolf' entrepreneur may do little to integrate statistical reasoning in her/his unit, once a department has several entrepreneurs on board, integration picks up.

These results are encouraging, but variables other than structural factors and chair rewards, unmeasured in our brief survey, must be accounting for substantial portions of variance in entrepreneurship and the integration of statistical reasoning instruction. For example, chairs may have used means other than rewards to facilitate entrepreneurship among their faculty or might have hired faculty with interests in teaching statistical reasoning to journalism undergraduates. Professional organizations and peers elsewhere might have had some direct impact as well on faculty teaching

---

[7] Although the reason for this difference is not clear, in 2008 a recent nationwide recession had weakened the job market for journalism and mass communication graduates (Becker et al. 2009). Chairs in 2008 may thus have been more likely than chairs a decade earlier to consider whether statistical reasoning ability might enhance their students' employability in deciding whether to reward faculty entrepreneurship.

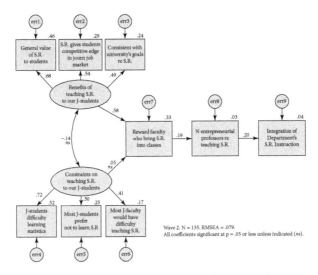

Figure 15.3 Chair rewards and faculty entrepreneurship, second wave.

interests and abilities in this area, and could even have inspired the local development of courses (e.g. survey methods, precision journalism, 'big data', data visualization) that could readily include statistical reasoning content.

Implications for the future of statistical reasoning in journalism curricula

Although these surveys date from, respectively, one and two decades ago, they still have implications for today and the future. Whether a similar survey conducted now would find any progress in the teaching of statistical reasoning to journalism undergraduates is unknown, and would be a worthy research undertaking. However, the results we currently have yield some useful and generalizable points.

First, structural factors facilitated or constrained the offering of statistical reasoning instruction in these surveys. And given the glacial pace of change in such components, we suspect they still hold in many of today's journalism departments. In particular, journalism programmes that offer research-oriented graduate degrees will still be more likely to have entrepreneurial faculty whose quantitative research interests will be incorporated into their undergraduate teaching. Larger units will continue to be more likely to devote resources to statistical reasoning instruction than will smaller units, which typically have fewer means. And programmes that lack the kind of course flexibility that encourages innovative curricular strategies – especially those with credit-hour restrictions based on ACEJMC accreditation – will still be pressed to find room in their curricula for instruction in math (Cusatis and Martin-Kratzer 2010) or statistical reasoning.

Secondly, the chairs' views of statistical reasoning instruction and their rewarding of it were remarkably stable over the course of the decade. This result is especially noteworthy because, by and large, the chairs were different individuals in 2008 than in 1997 even though their academic units were the same in both waves. That suggests that the chairs were standard bearers for relatively unchanging cultures within the various journalism programmes they directed across those times, and it also leads us to speculate that change since 2008 is likely to have been slow.

In short, when it comes to statistical reasoning instruction for journalism undergraduates, chairs overwhelmingly support it in principle and see its benefits for their students' intellectual development and future job prospects. But only a little over a quarter of these administrators thus try to reward their faculty for their attempts to teach statistical reasoning. Such rewards seem to encourage faculty entrepreneurship, which in turn appears to help integrate statistical reasoning into the curriculum. But commonly, chairs seem unable or unwilling to make this instruction an educational reality.

All that said, it is clear that the last decade or so has brought rapid changes in the journalism business itself. As the occupation struggles to develop viable, revenue-generating niches, it is increasingly focusing on analytical products, many of which require statistical reasoning savvy. Additionally, the new world of 'fake news' calls on journalists to develop evaluative capacities in order to align truth claims with evidence (see, e.g., Graves 2016). Journalists themselves are quite sensitive to these changes. In a recent national survey, Willnat and Weaver (2014) found that the percentage of respondents who identified the importance of 'analysing complex problems in society' (68.8 per cent) had jumped a dramatic eighteen points over results from a similar survey in 2002, and more training in data journalism was one of the top three areas for desired further training (28.1 per cent wanted that).

But access to data alone is insufficient; in fact, data can be dangerously misleading if journalists do not know how to reason with such resources. There is some cause for concern here, as individuals in learning mode are often more focused on the pragmatic than on the conceptual aspects of the task at hand. For example, in a recent series of interviews with data journalists working in American newsrooms, Fink and Anderson (2015) found these individuals to be quite articulate about describing the ideal skills of a data journalist; however, those descriptions were dominated by pragmatic, statistical and visual capabilities, not foundational aspects of statistical reasoning. A number of journalists in this study had taken advantage of 'big data' workshops offered by such organizations as Investigative Reporters and Editors (IRE) and the National Institute for Computer-Assisted Reporting (NICAR), so one could detect a genuine focus on knowledge gain. But the extent to which such organizations build foundational learning into their efforts is unknown.

Provision of conceptual information in the real world is often in short supply. Permit one other example: science communication scholars John Besley, Anthony Dudo and their colleagues wondered if scientists who receive public communication training from consultants or other trainers were getting anything more than skills training. In other words, do individuals who provide skill-based instruction also delve into such conceptual issues as understanding differences between communication and engagement processes, understanding audiences, developing communication goals, and so on? The answer, alas, was no (Besley et al. 2016).

It is in these conceptual arenas that university instruction comes to the fore. Statistical reasoning training for journalism students in journalism schools should not ignore the practical, of course, but its *raison d'etre* must be the conceptual. In a world dominated by uncertainty, how does a journalist make judgements about what is most likely to be true? What kinds of evidence does she need, and what does she do with that evidence? When is a pattern well grounded and when is it on shaky ground? What role does chance play in what looks like cause and effect?

In the future, the ease of lassoing big data and taming it with computers will make dealing with these epistemological questions ever more essential for journalists. For example, although correlation does not prove causation, nonetheless it is the starting point most researchers use to look critically for further evidence of cause and effect. Feed the software a huge number of cases, a large number of variables, and chance correlations flourish. 'With such large data sets, it is all too easy to find rare statistical anomalies and confuse them with real phenomena,' Steven Salzberg, professor of medicine and biostatistics at Johns Hopkins University, told author Sherri Rose (2012, 47). So, the task of filtering the signal from the noise will itself become more demanding. Nonetheless, large data sets could also offer researchers the opportunity to control for confounding variables, for reverse causality, and for other competing explanations of the results found in the data (see, e.g., Stray 2016).

Why should such considerations matter to reporters? Arguably, journalists using data in their reporting – especially big data sets – are often looking for verifiable causal patterns. Some of the stories may investigate broad-ranging national social, economic, or political issues (e.g. what effect might new statewide voter identification laws have on voter turnout in American elections?), but other stories can be local and just as impactful. Ben Poston, Keegan Kyle and Grant Smith (2008) of the *Milwaukee Journal Sentinel*, for example, used existing city databases to determine that the prevalence of racial minorities in certain sections of the city was the primary factor delaying street repairs in those areas. Not satisfied with mere correlation, they gathered as much other evidence as possible and used the SPSS multiple regression to control for some alternate explanations of their results. Poston at the time was the newspaper's computer-assisted reporting specialist. Kyle and Smith were journalism student interns.

Indeed, understanding probability, learning to think analytically about uncertainty, and practicing judgment calls under such conditions will be worth its weight in gold for journalism majors now and in the future. But getting there, our decadal surveys tell us, will be challenging. Since structural forces will morph at only a glacial pace, chairs will have to become more proactive in supporting the teaching of statistical reasoning, in whatever way makes sense in their particular institutions. Professional and academic J&MC organizations are clearly already in the 'big data' game, but we suspect they may be giving short shrift to the conceptual issues that we identify as 'statistical reasoning' fundamentals. Those organizations that are offshoots of academic environments, in particular, need to be sensitive to these conceptual topics.

And finally, journalism scholars need to continue to monitor provision of this training through surveys and other data-gathering efforts. The two surveys we discuss in this chapter desperately need a modern update, ideally through ongoing research into curricular issues in journalism training in general. Important as well will be pedagogical research that examines the effectiveness of specific training strategies. For example, experiential learning is something of a gold standard in university training these days. Is it being employed in statistical reasoning classroom environments? Can an instructor highlight the conceptual in an experiential process, where the practical often overwhelms other components?[8]

University journalism training in the United States has remained determinedly profession-centric; many schools still maintain such occupational divisions as 'newspaper', 'magazine', 'television', 'public relations'. The rigidity of such structures might lead one to despair, but the rapid and dramatic changes in journalism itself are likely to prod journalism schools into increased instructional fluidity and innovation. We hope that, in the midst of such upheaval, statistical reasoning will find its place in the curriculum.

References

Accrediting Council on Education in Journalism and Mass Communication (2013), ACEJMC accrediting standards. Retrieved from http://www2.ku.edu/~acejmc/ PROGRAM/STANDARDS.SHTML

Becker, L. B. and Graf, J. D. (1994), *Myths and Trends: What the Real Numbers Say about Journalism Education*. Arlington, VA: The Freedom Forum.

Becker, L. B., Vlad, T., Olin, D., et al. (2009), *2008 Annual Survey of Journalism & Mass Communication Graduates*. Athens, GA: Grady College of Journalism and Mass Communication, University of Georgia.

Besley, J. C., Dudo, A. D., Yuan, S. and Abi Ghannam, N. (2016), 'Qualitative interviews with science communication trainers about communication objectives and goals'. *Science Communication*, 38(3), 356–81.

Coben, D. (2000), 'Numeracy, mathematics, and adult learning'. In I. Gal (ed.), *Adult Numeracy Development: Theory, Research, Practice*, 33–50. Cresskill, NJ: Hampton Press.

Coddington, M. (2015), 'Clarifying Journalism's Quantitative Turn: A typology for evaluating data journalism, computational journalism, and computer-assisted reporting'. *Digital Journalism*, 3(3), 331–48.

Cusatis, C. and Martin-Kratzer, R. (2010), 'Assessing the state of math education in ACEJMC-accredited and non-accredited undergraduate journalism programs'. *Journalism & Mass Communication Educator*, 64(4), 356–77.

Dunwoody, S. and Griffin, R. J. (2013), 'Statistical reasoning in journalism education'. *Science Communication*, 35(4), 528–38.

---

[8] Lawson et al. (2003) in their experiment aimed at improving undergraduate students' statistical reasoning found that 'combining training in basic statistical concepts with training in everyday applications of those concepts – as we did in our statistics-reasoning group – will result in far greater gains' in statistical reasoning abilities (109).

Fink, K. and Anderson, C. W. (2015), 'Data journalism in the United States: Beyond the "usual suspects"'. *Journalism Studies*, 16(4), 467–81.

Garfield, J. (2002), 'The challenge of developing statistical reasoning'. *Journal of Statistics Education*, 10(3). Retrieved from www.amstat.org/publications/jse/v10n3/garfield.html

Graves, L. (2016), 'Anatomy of a fact check: Objective practice and the contested epistemology of fact checking'. *Communication, Culture & Critique*. doi:10.1111/cccr.12163.

Griffin, R. J. and Dunwoody, S. (2013), 'Promises and challenges of teaching statistical reasoning to journalism undergraduates: Twin surveys of department heads 1997 and 2008'. In Brennen, B. (ed.), *Assessing Evidence in a Postmodern World*, 169–96. Milwaukee, WI: Marquette University Press.

Griffin, R. J. and Dunwoody, S. (2016), 'Chair support, faculty entrepreneurship, and the teaching of statistical reasoning to journalism undergraduates in the US'. *Journalism*, 17(1), 97–118.

Harrison, Steven (2014), 'History of Numeracy Education and Training for Print Journalists in England'. *Numeracy*, 7(2). DOI:http://dx.doi.org/10.5038/1936-4660.7.2.2

Kahneman, D., Slovic, P. and Tversky, A. (1982), *Judgement Under Uncertainty: Heuristics and Biases*. Cambridge, UK: Cambridge University Press.

Ranney, M. A., Rinne, L. F., Yarnall, L., Munnich, E., Miratrix, L. and Schank, P. (2008), 'Designing and assessing numeracy training for journalists: Toward improving quantitative reasoning among media consumers'. In P. A. Kirschner, F. Prins, V. Jonker and Kanselaar (eds), *International Perspectives in the Learning Sciences: Proceedings of the Eighth International Conference for the Learning Sciences*, 2. International Society of the Learning Sciences, Inc, 2-246-2-253.

Rose, S. (August 2012), 'Big data and the future'. *Significance*, 9(4), 47–8.

Lawson, T. J., Schwiers, M., Doellman, M., Grady, G. and Kelnhofer, R. (2003), 'Enhancing students' ability to use statistical reasoning with everyday problems'. *Teaching of Psychology*, 30(2), 107–10.

Lewis, S. C. (2015), 'Journalism in an era of Big Data: Cases, concepts, and critiques'. *Digital Journalism*, 3(3), 321–30.

Maier, S. R. (2003), 'Numeracy in the newsroom: A case study of mathematical competence and confidence'. *Journalism & Mass Communication Quarterly*, 80(4), 921–36.

Nguyen, A. and Lugo-Ocando, J. (2016), 'The state of data and statistics in journalism and journalism education–issues and debates'. *Journalism*, 17(1), 3–17.

OECD (2004), Learning for tomorrow's world: First results from PISA 2003. Paris: OECD.

Poston, B., Kyle, K. and Smith, G. (September–October 2008), 'Disparity of repair: Race gap found in pothole patching'. *Investigative Reporters and Editors (IRE) Journal* 31(5), 8–9.

Stocking, S. H. and Gross, P. H. (1989), *How do journalists think? A proposal for the study of cognitive bias in newsmaking*. ERIC Clearinghouse on Reading and Communication Skills, 2805 E. 10th St., Smith Research Center, Suite 150, Bloomington, IN 47405.

Stray, J. (24 March 2016), *The Curious Journalist's Guide to Data*. New York: Tow Center for Digital Journalism. http://www.cjr.org/tow_center_reports/the_curious_journalists_guide_to_data.php

Willnat, L. and Weaver, D. (2014), *The American Journalist in the Digital Age: Key Findings*. Bloomington, IN: School of Journalism, Indiana University. Accessed at http://news.indiana.edu/releases/iu/2014/05/2013-american-journalist-key-findings.pdf.

Appendix: Percentages and adjusted means for fifteen items addressing chairs' attitudes towards statistical reasoning instruction by survey wave[1]

The following are statements that some professors and administrators have made about the teaching of statistical methods and statistical reasoning to undergraduates in journalism. Please indicate the strength of your agreement or disagreement by checking one response to the right of each one.

	Wave	SD (%)	D (%)	FN/DK (%)	A (%)	SA (%)	Adjusted Means[2]	$F_{1,293}$
1. It is important for our journalism students to be able to reason statistically.	1997	0.6	2.4	5.5	44.5	47.0	4.35	0.05 *ns*
	2008	0.7	1.5	6.7	41.5	49.6	4.38	
2. It is important for our journalism students to take a course in basic statistics.	1997	1.2	6.1	26.8	42.1	23.8	3.80	0.51 *ns*
	2008	2.2	15.6	18.5	42.2	21.5	3.67	
3. Statistical reasoning skills give students a competitive edge in the journalism job market.	1997	1.2	6.7	25.0	45.7	21.3	3.71	3.17 *ns*
	2008	0.7	7.4	20.0	43.7	28.1	4.00	
4. Most of our journalism students lack the mathematical aptitude required to do well in the basic statistics course at our university.	1997	2.4	26.8	23.2	29.3	18.3	3.28	0.00 *ns*
	2008	3.0	26.7	28.1	30.4	11.9	3.29	
5. Some instruction in statistical reasoning should be a part of our journalism courses.	1997	1.8	9.8	14.6	49.4	24.4	3.86	0.30 *ns*
	2008	1.5	3.7	17.8	51.9	25.2	3.95	
6. Most of our journalism students would rather not learn statistical reasoning.	1997	0.0	4.9	17.7	47	30.5	3.99	0.04 *ns*
	2008	1.5	8.1	11.9	54.8	23.7	3.96	
7. The statistics courses at our university meet the needs of our journalism students.	1997	6.1	19.5	35.4	32.9	6.1	3.14	0.00 *ns*
	2008	6.7	21.5	30.4	34.8	6.7	3.13	

Statement	Year	1	2	3	4	5	Mean	Stat
8. Most of my faculty would have difficulty teaching statistical reasoning as part of their journalism classes.	1997	1.8	26.8	18.3	35.4	17.7	3.31	0.60 *ns*
	2008	5.2	23.0	18.5	37.0	16.3	3.47	
9. The journalism profession does not reward statistical reasoning by journalists.	1997	1.2	25.0	32.3	34.8	6.7	3.21	1.51 *ns*
	2008	5.9	29.6	31.9	25.2	7.4	2.99	
10. The inclusion of statistical reasoning in a reporting course should be up to the faculty member teaching the course.	1997	1.2	23.8	18.9	45.1	11.0	3.34	1.58 *ns*
	2008	3.0	37.8	18.5	34.8	5.9	3.11	
11. Our journalism students receive adequate instruction in the application of statistics to everyday problems.	1997	5.5	43.9	28.0	17.7	4.9	2.69	0.05 *ns*
	2008	8.9	48.9	17.0	23.7	1.5	2.65	
12. The journalism curriculum is too tight to offer in-house instruction in statistical methods and their applications.	1997	4.3	32.9	14.0	34.8	14.0	3.17	0.39 *ns*
	2008	7.4	35.6	17.0	31.1	8.9	3.04	
13. To the extent I can, I reward faculty who bring statistical reasoning into their classes.	1997	3.0	18.3	53.0	22.6	3.0	2.99	0.00 *ns*
	2008	10.4	18.5	42.2	25.9	3.0	3.00	
14. Our university's goal is to integrate statistical reasoning into the curriculum.	1997	8.5	36.6	31.7	19.5	3.7	2.72	0.77 *ns*
	2008	8.9	32.6	26.7	24.4	7.4	2.89	
15. In general, the news media do a good job of interpreting statistically based information [e.g., polls, health risks] for their audiences.	1997	11.0	47.0	23.2	17.1	1.8	2.48	0.20 *ns*
	2008	15.6	41.5	20.0	23.0	0.0	2.56	

[1] Scale responses: 1 = strongly disagree, 2 = disagree, 3 = feel neutral (or don't know), 4 = agree, 5 = strongly agree / Wave 1 (1997): n = 164; Wave 2, 2008: n = 135)

[2] Means in this Appendix are adjusted by the following control variables: size of the programme, highest degree offered, and (for 2008 data) whether the questionnaire was completed online or on hard copy. Percentages are not adjusted by the control variables.

Adapted from Griffin and Dunwoody (2013).

16

Four Conceptual Lenses for Journalism Amidst Big Data: Towards an Emphasis on Epistemological Challenges

Oscar Westlund, *Volda University College, Norway*
and *University of Gothenburg, Sweden*
Seth C. Lewis, *University of Oregon, USA*

Introduction

The social world is increasingly quantified, rendered as digital trace information – geolocation, web metrics, self-tracking, social graphs, likes and shares, and much more. Such data may be collected and analysed at ever larger scale amid the growing ubiquity of mobile devices, always-on sensors, 'smart' homes, algorithms and automated systems, digital repositories and archives, and the many fragments of social activity represented by ones and zeroes (Kitchin and McArdle 2016; Mayer-Schönberger and Cukier 2013).[1] Harnessing, combining, manipulating and visualizing such data – techniques that once required supercomputers – increasingly can be accomplished via standardized software or cloud computing (Manovich 2012). In academia, for instance, such contexts have facilitated the rise of computational social science (Shah, Cappella and Neuman 2015). Importantly, big-data developments are not merely a technological transition but truly a *sociotechnical* phenomenon – a complex amalgamation of digital data abundance, emerging analytic techniques, mythology about data-driven insights, and growing critique about the overall consequences of big-data practices for democracy and society (boyd and Crawford 2012). As its opportunities and challenges come into focus, big data promises enormous potential for yielding various types of value – social, cultural and monetary – for the institutions that manage it effectively, while simultaneously posing troubling questions about privacy, accuracy and ethics (Howard 2015; O'Neil 2016; Tufekci 2014).

Like the corporate, scientific and government sectors, institutions of journalism increasingly need to work out how they are to approach big data. News organizations have long been familiar with data and databases – through computer-assisted reporting (CAR) and other forms of information visualizations – both as a form of journalistic evidence and as

---

[1] This chapter draws in part on material previously published in Lewis and Westlund (2015a).

a type of news product (Anderson 2013a, 2015). But, as Usher (2016) describes, journalism is at a constitutive moment in which changes in technology, culture and economics have enabled the rise of 'interactive journalism', which, at its core, is about the application of data, code and software to novel forms of news presentation. Moving forward, the open question 'is not whether data, computers, and algorithms can be used by journalists in the public interest, but rather how, when, where, why, and by whom' (Howard 2014, 4).

In a conceptually oriented research review article on the intersection of big data and journalism in a special issue in *Digital Journalism* guest-edited by Lewis, we argued that big data, in the context of journalism, embodies emerging ideas about, activities for, and norms connected with datasets, algorithms, computational methods, and related processes and perspectives tied to quantification as a key paradigm of information work (Lewis and Westlund 2015a). Importantly, we think of big data as neither good nor bad for journalism. Rather, it is freighted with equal parts potential and pitfall, depending on how it is imagined and implemented – and, crucially, towards what purposes and in whose interests. Scholars have begun to examine and conceptualize big data in the broad context of media and public life, but less has been accomplished with regard to journalism. Just as journalism is trying to make sense of big data, journalism studies must develop conceptual and theoretical toolkits to understand what it means for how news is perceived and practiced.

In that context, our *Digital Journalism* article presents four conceptual lenses for understanding the nexus of big data and journalism: epistemology, expertise, economics and ethics. Each of these addresses relevant questions for news media, helping both to illuminate existing research and to guide future work. Briefly summarized, the concept of epistemology addresses the legitimization of new claims about knowledge and truth. Expertise is concerned with the negotiation of occupational status, authority and skill sets as new specializations are developed and deployed. Economics encompasses the potential for and challenges with new efficiencies, resources, innovations, value creations and revenue opportunities. And, finally, ethics deal with the issues raised by these developments for the norms and values that guide human decision-making and technological systems design. In this chapter, at the editor's invitation, we reprise our discussion of the four concepts, but this time place particular emphasis on *epistemology*, addressing current and future research on big data's implications for journalism and the production of news knowledge. As such, the chapter will first revisit the three conceptual lenses of expertise, economics and ethics before synthesizing current research literature on the epistemology of big data and journalism and, finally, discussing future directions for such research.

The expertise, economics and ethics of big data and journalism

In reviewing expertise, economics and ethics, we first offer a brief concept introduction, followed by a discussion of the concept's meaning for big data and journalism and finally a series of examples or possible applications in the context of news production and distribution. This systematic approach will be used also in the succeeding section that goes into further details about epistemology.

Expertise

The term 'expertise' comes from the Latin root *experiri*, meaning 'to try', and generally refers to 'the know-how, the capacity to get a task accomplished better and faster because one is more experienced' – hence *expertus*, or 'tried' (Eyal 2013, 869). The sociology of professions field has since long focused on how *expert* (the social actor) is connected with *expertise* (the specialized know-how). There are diverse studies into how occupations (and the experts who constitute them) work to forge and maintain 'jurisdictional control' over the boundaries around a body of abstract knowledge and the application of that knowledge through work practices. Consequently, this allows professionals to claim autonomy, authority and other benefits (Abbott 1988, 60). Expertise thus distinguishes those possessing specialized knowledge *and* experience from those who do not.

The expertise of journalists is being contested by increasingly active and prolific 'produsers', people who both produce and consume media content (Bruns, 2012), as well as new forms of digital aggregation and curation (Anderson 2013b). Moreover, as media users spend more and more time on social media platforms, and as those platforms – Facebook especially – assume greater control over information distribution writ large (Bell 2016), the role of journalist-as-expert-gatekeeper falls into question (cf. Braun 2015). And, unlike law and medicine, journalism has a more tenuous link to formalized expertise, lacking the protective trappings of credentials bestowed via examination. Altogether, journalism's claim to social expertise through its 'professional logic' (Lewis 2012) – a bargain to control the production and distribution of news on society's behalf – is rather beset by challenges to authority (Carlson 2017). Against that backdrop, and alongside Collins and Evans' (2007) typology of expertise that presumes expertise is a 'real' feature of socialization and experience, Reich (2012) has argued that journalists can be understood as 'interactional' experts. Journalists' expertise lies in their ability to work with and among *other* types of experts, ultimately synthesizing and translating others' specialized knowledge for non-experts. Through their interactions and engagement with lay audiences, Reich argues, journalists also develop a *bipolar* form of interactional expertise. Finally, within the sociology of knowledge, there is a greater recognition for a competence-based approach focusing on what people can do, rather than what they can learn (Collins and Evans 2007; cf. Reich 2012). Skills are thus recognized as integral to expert distinction. Next, we revisit these constructivist (Eyal 2013) and normative (Collins and Evans 2007) frameworks for a brief discussion of three concepts important for understanding journalistic expertise amid big data (Lewis and Westlund 2015a).

1. *Social interactions.* As noted by Reich (2012), some journalists have primarily source-interactional expertise (e.g. long-time beat reporters) while others primarily have audience-interactional expertise (e.g. editors who hear from readers), and others some combination (e.g. columnists and commentators).

The social, cultural and technological nature of big data may affect the character of these interactions with sources and audiences. Audience metrics can facilitate decisions for news reporting and news distribution (e.g. Tandoc 2014; see also Nguyen and Vu, this volume). News companies are increasingly hiring data scientists to make sense of data – both *data as source material* for journalistic storytelling as well as *data on audiences* for business purposes.

2. **Networked interactions.** Networked interactions involve the more socio-technical interactions of expertise that might occur between journalists (as actors) and machines (as actants – see Lewis and Westlund 2015b). From Eyal's (2013) approach, the growing deployment of algorithms and automation in journalism might entail new arrangements of 'networked expertise', altering how we imagine what it is that journalists know and how they represent that knowledge to the world. In this vein, Anderson (2013b) has shown how the dividing lines of expertise between 'original' reporting and 'parasitic' news aggregation are hardly clear-cut. In fact, networks of social actors and technological actants, when viewed holistically, yield rather complicated renderings of journalistic expertise under different conditions of digitization. Ultimately, as human expertise is increasingly inscribed into technical systems used for news production and news distribution, it challenges what is 'human' and what is 'machine' about such expertise (Lewis and Westlund 2016).

3. **Skill sets.** Expertise is manifest in actual, practical skills (Collins and Evans 2007). Professional approaches to big data prioritize certain skills, such as data analysis, computer programming and visualization, drawn from disciplinary origins such as computer science, mathematics and statistics (Mayer-Schönberger and Cukier 2013). Consequently, news organizations striving to succeed with big data increasingly need expertise in areas such as computer programming and sophisticated back-end databases to comprehend data, and to publish it in ways that allow users to explore the data for themselves (Parasie and Dagiral 2013). *Data* and *code* thus constitute skills-based forms of expertise that news organizations are working to cultivate. Yet bridging the skills gap between journalists and technologists, or helping journalists develop such data-and-code skills, is neither easy nor broadly institutionalized as yet (Howard 2014; Lewis and Usher 2014, Westlund 2011). Journalistic expertise may evolve by having journalists learn to write basic software, and having external 'algorithmists' – expert reviewers of big-data analysis and predictions (Mayer-Schönberger and Cukier 2013) – critique computational journalism in a similar fashion as ombudsmen (cf. O'Neil 2016). Incorporation of such skills into the news production and distribution process may alter the notion of what truly *counts* as expertise in journalism.

Economics

Economics, with its etymology in the 'management of household', is a discipline that studies the behaviours of agents in households and organizations, focusing on how resources are managed to achieve certain ends. Typically, these agents are assumed to act rationally, making choices about how to use limited resources towards desired outcomes and strategic goals. Applied to the context of communication, *media economics* is defined as 'the study of how media industries use scarce resources to produce content that is distributed among consumers in a society to satisfy various wants and needs' (Albarran 2002, 5), and thus includes questions of media management and media innovations.

Many legacy news media companies around the world – especially local newspapers – have historically benefited financially from their standing as oligopolies or monopolies in a distinct geographical market (Picard 2010). In the twenty-first century, such firms have been facing a shrinking advertising base, fragmenting audiences, and rising competition from mobile, social and digital media (Anderson, Bell and Shirky 2012). Amid a general call for media companies to innovate (Storsul and Krumsvik 2013; cf. Westlund and Lewis 2014), big data represents an opportunity for value creation through revised business processes as well as new products and services. Big data has obvious relevance for business-side revenue opportunities, allowing media companies to better understand and serve particular audiences and advertisers. Nevertheless, these developments come with concerns about the ultimate social and political outcomes, including how targeted advertising based on data mining leads to pressures on (news) media companies to personalize content in response (Couldry and Turow 2014). The march towards big-data personalization, in this view, threatens the very ecology of common knowledge upon which representative democracy depends. To date, and after the 2016 American presidential election, academics and pundits alike have expressed alarm about the pronounced individualization of media behaviour – whether in the form of selective exposure or in unintentional filter bubbles created by algorithm-controlled social platforms (Lewis and Carlson 2016).

Importantly, can big data afford new value creation without undermining the church–state divide that, for many journalists, is central to professional autonomy? Big data promises economic efficiency by enabling 'more observation at less cost' (Crawford, Miltner and Gray 2014, 1666) – as in the case of labour-saving forms of automated journalism (Carlson 2015; Westlund 2013). Moreover, it may be associated with augmenting, rather than displacing, human labour by catalysing new types of technologically enabled forms of news work (Powers 2012), or by allowing journalists to function more like 'knowledge managers' who better gather, organize and analyse disparate information flows in a community (Lewis and Usher 2013).

We hereby suggest two types of applications for envisioning the value creation opportunities for a journalism leveraging big data. The first involves social actors – especially journalists but also technologists – manually drawing upon large datasets to report and present news in ways that differentiate their work from the traditional storytelling paradigm, thereby creating value for distinct publics interested in new types of news and creating a competitive advantage vis-à-vis commodity news in the marketplace. This shift has been called 'method journalism' (Madrigal 2014), moving from an *area of coverage* (a topic, beat or location of interest) to focus instead on the *method of coverage*. Several news startups and initiatives are emblematic of this change, built around method-oriented objectives, such as FiveThirtyEight and Vox (Madrigal 2014).

The second approach involves journalists and technologists employing technological capacities – in the form of technological actants – to algorithmically gather, link, compare and act upon big data of interest to audiences (cf. Lewis and Westlund 2015b). These algorithms can be tailored to fit with the personalized preferences and behaviours of individuals, promoting specific news articles and news categories to specific individuals (building upon processes described by Thurman 2011). Importantly, social actors within news media firms need to actively assess what technological actants they are to use for such purposes (if acquired from external providers), or how they are to be developed and configured (if developed internally). Diverse social actors need to collaborate in inscribing the technological actants with logics and news values for their operation, ensuring that journalistic values and ambitions are built into automation in a way that suits audiences (cf. Westlund 2012). Such actant-focused automation, as Anderson, Bell and Shirky (2012) have argued, offers an important yet underexplored avenue for news media to cut expenses (e.g. by no longer wasting resources on stories that a robot could write just as well) and simultaneously create value (e.g. by redeploying humans towards projects where they uniquely can contribute).

Ethics

The term 'ethics' derives from the Greek *ethos*, meaning 'character' or 'personal disposition' on the part of the individual, and relates to the Greek *mores* (or 'morals') and its emphasis on the customs of a group. As such, ethics is internally concerned with personal decision-making and externally situated in relation to the rules of society (Ward 2010). Ethics is thus concerned with appropriate practice within a framework of moral principles.

Ethical standards serve an essential function in orienting journalists, especially, to work in ways that promote honesty, accuracy, transparency and public service (Ward 2010). For journalists, ethical codes and conduct serve not only to guide their choices but also to define who they are as professionals. The big-data phenomenon is freighted with its own set of ethical quandaries – about issues such as user privacy, information security and data manipulation (Crawford et al. 2014) – that deserve scrutiny and reflection as journalists determine how to adapt to and along with the series of innovations associated with it. Next we briefly discuss three important areas of concern: (1) data transparency and quality, (2) social science research ethics, and (3) inscription of values into technological systems.

1. ***Data transparency and quality.*** Publishing data and making large datasets publicly available online is linked to transparency. An ethos of openness is shared among many data journalists (Howard 2014), even as journalists broadly have struggled to embrace such openness as a professional norm

(Lewis 2012). Data journalists often seek to make complete datasets and programming code open to public examination and collaboration. Lewis and Usher (2013) have suggested this may lead to journalism reinventing itself, integrating norms like iteration, tinkering, transparency and participation that are connected with the social, cultural and technological framework of digital technologies (cf. Lewis and Usher 2016). There are, however, often underlying problems with public data provided by diverse stakeholders, problems that may go unnoticed either because of the size of the data involved or because of the attractiveness of making it freely available. Crawford (2013), for instance, mentions deep structural signal problems when collecting big data from Twitter, as such data has little or no representation of less-connected communities. Moreover, big-data analysis may come with ethical challenges regarding disclosing private or sensitive information (Howard 2014). Altogether, even well-intentioned efforts to revise journalistic norms through big-data run up against ethical questions embedded in the organization, analysis and dissemination of such data.

2. *Social science research ethics.* Journalists have long drawn on social science methods, but they are doing so to a greater extent in this data-rich environment (Howard 2014). Meanwhile, at the same time that journalists are embracing such techniques, 'social scientists are undergoing a fundamental shift in the ethical structure that has defined the moral use of these techniques', rethinking what it means to protect individuals from harm and to allow for informed consent in a world of big-data research methods involving millions of human subjects (Fairfield and Shtein 2014, 38). Journalists, of course, are not subject to institutional review boards (IRB), and yet they should be cautious: just because certain content is publicly accessible does not mean that it was intended to be *made public* to everyone (boyd and Crawford 2012).

3. *Inscription of values into technological systems.* Finally, journalists should consider the ethics of technological systems design, or the inscription of values into technological systems (Nissenbaum 2001). To the extent there is a turn towards technology-led practices in journalism (see Westlund 2013), what happens as humans embed technological actants like algorithms with *some* assumptions, norms and values, and not others? To what extent (and how) should technological actants be designed to manage thorny issues such as fake news and manipulation, or otherwise be 'taught' to act ethically? Is there an ethics of algorithms (Kraemer et al. 2011)? Such an ethics will need to unpack various factors of selection, interpretation and anticipation, revealing 'how algorithms structure how we can *see* a concern, why we think it *probably matters*, and *when* we might act on it' (Ananny 2013, 6, original emphasis), all issues of deep relevance for journalistic decision-making. Altogether, this attention to code and structure behind technology encourages a study of the 'black boxes' of big data, in order to uncover matters such as biases, influences and power structures (Diakopoulos 2015).

The epistemology of big data and journalism: Current research[2]

Finally, we address the fourth 'E' of big data: epistemology. Epistemology, as a theory of knowledge, differs from ontology. In the philosophy of science, ontology refers to fundamental inquiries into the nature of existence, that is, the 'science of what is' (Smith 2001, 79). While the world undoubtedly exists in the form of nature, people or events, any attempt to represent the world will, in fact, turn into some sort of re-presentation, with inherent limitations. Epistemology thus points to the nature and boundaries of human knowledge about the world and the determination of truth in that process of re-presentation. The term derives from the Greek *episteme*, which means knowledge, and was developed on the basis of *epistanai*, which means to understand. A fundamental issue in epistemology concerns the work of legitimizing certain types of information as knowledge relative to others. The academy, like other knowledge-producing fields of practice, long has developed epistemologies that shape what counts in this regard (Schon 1995).

Journalism has become, and remains to be, one of the most influential knowledge-producing institutions in society. The concept of epistemology in journalism refers to the norms, standards and methods that determine how journalists know what they know, as well as how they develop and display that knowledge through their production practices and news products (Ekström 2002; Ettema and Glasser 1998). Ekström (2002) has shown how the epistemology of journalism involves the rules, routines and institutionalized procedures that are used in a specific social setting, and which impact how knowledge is produced and how claims about knowledge are explicitly or implicitly expressed. Indeed, how journalists justify their truth claims is just as important as the claims themselves (Ettema and Glasser 1998). Research into news production has shown how journalists develop methods for adjudicating knowledge claims in a routinized fashion (Tuchman 1978), such as by adhering to ideals such as objectivity and practices such as multiple-sourcing (Wiik 2010). Ultimately, epistemology concerns how journalists – and other agents of news production, such as news bots – make assessments of diverse information and sources, determining that the information gathered is sufficiently correct and reliable in order to be published as news (Ekström 2002; Ettema and Glasser 1998; cf. Godler and Reich 2012).

While there are many important studies of digital journalism, few focus on epistemology. Exceptions include Matheson's (2004) early work on the hosting of blogging by the *Guardian*. His approach to epistemology characterized a kind of

---

2   This section builds on our journal article (Lewis and Westlund 2015a) as well as a grant proposal ('The epistemologies of digital news production') that was successfully funded by the Swedish Foundation for Humanities and Social Sciences in October 2016. The grant proposal was developed together with Professor Mats Ekström at the University of Gothenburg; his important reflections and contributions to our joint grant proposal are reflected in this discussion of epistemology and journalism.

journalism built around online connections more than 'fact' involving interpersonal relations and a different kind of authority, as the public was invited to participate in the construction of knowledge. More recently, Wahl-Jorgensen has analysed matters of epistemology in the study of journalism and emotion (see, e.g., 2015; 2016). A key point in her works is that in relation to the rise of citizen journalism, social media and user-generated content, journalism has developed new conventions of storytelling. These challenge the objectivity ideal and established ways of knowing, and involve new forms of truth claims, giving room for journalism in which emotion and personal experience are key ingredients in the narratives. Exemplifying this, a citizen sharing a personal experience through amateur footage with a mobile device may be seen as truly authentic, perhaps more so than the truth claims of a story produced by a professional journalist (Wahl-Jorgensen 2016). And, other recent research has focused on the intersection of data journalism and epistemology. Borges-Rey (2017) employed three conceptual lenses – materiality, performativity and reflexivity – for an interview study, concluding that while data journalism reinforces norms and rituals, it also displays a distinct character.

Aside from these and a few other studies, relatively little is known about epistemology in contemporary journalism, even while there is great concern among scholars, pundits and practitioners about the direction of journalism in the digital era. There have indeed been substantial shifts in the structural and technological conditions of journalism (see, e.g., Anderson, Bell and Shirky 2012; Meikle and Redden 2010; Westlund 2011, 2012), shifts that relate closely to weakening the financial base of legacy news media and their capacity to provide quality news reporting (Picard 2010; cf. McChesney 2012). In much of the Western world, journalism is under stress: legacy providers are shrinking, some failing altogether, leaving a range of stakeholders – media managers, policymakers and citizens – to wonder about the future of news. In countries such as Denmark, Norway and Sweden, where press subsidy and publicly financed media have long played an important part in the media system, politicians have begun redesigning media policies, launching media inquiries that investigate the situation and propose directions for future media policy.

Against that backdrop, it is fair to ask how journalists are coping with such pressures: is news quality worsening, maintaining status quo or improving? This is an important question because the democratic significance and future legitimacy of journalism is inexorably connected to the extent that journalists succeed in publishing relevant and 'accurate' news (Ekström 2002; Karlsson 2011).

While this chapter does not address the concept of quality per se, it focuses on how journalistic quality relates to epistemology through news production processes. To date, the literature on journalism and epistemology mostly consists of studies of news production processes at newspapers or news broadcasters before the rise of digital journalism (see, e.g., Ekström and Nohrstedt 1996; Ettema and Glasser 1987, 1998). Throughout the 2000s and 2010s there has most likely been not only continuity but also change in journalistic epistemology – that is, how journalists think about and act towards knowledge production and knowledge expression – particularly amid increasingly technologically oriented forms of news production.

How journalists justify the stories and facts they publish appears to vary with different forms of news, as seen in the case of investigative journalism (Ekström 2002; Ettema and Glasser 1987) and more specific sub-genres of reporting (Ekström and Nohrstedt 1996). It matters, therefore, to understand if and how the fundamental principles and work processes of what might be called 'traditional news journalism' have changed – and, in turn, how such changes may be manifest in emerging forms of journalism as well. The introduction of diverse technologies into news work has indeed raised questions about the relative knowledge value that is associated with specific forms of news work such as photojournalism and programmer-journalism (Powers 2012). Here, we will focus on the literature that takes up the intersection of epistemology and (big) data journalism.

The development of computer-assisted reporting (CAR), nearly a half-century ago, pointed to a hope for 'precision journalism' – the potential for achieving greater accuracy through a combination of computer and social science (Meyer 1973). While computers have improved significantly since then, journalists have maintained their established epistemology, which assumes that 'data have no journalistic value on their own' and therefore journalists must work to find the story 'hidden' in the data (Parasie and Dagiral 2013, 859). Such a view fits the normative paradigm of journalists as essential knowledge-producers for society (Lewis 2012). However, questions have emerged about the role that technology might play in developing the capacities for and practices of knowledge production in journalism – for example, in the form of augmented reality for digital storytelling (Pavlik and Bridges 2013) or technological systems for customizing diverse types of news for diverse types of audiences (Westlund 2013). Through it all, the point remains: technology has uncertain consequences for journalistic epistemology.

The extraction, combination and analysis of big data offer improved possibilities for specific forms of journalism, such as investigative journalism (cf. Parasie 2015), and with it a stronger link between the production of social facts in journalism and matters of 'science' and 'precision'. Importantly though, data should never be taken as a proxy for the 'science of what is', in the ontological sense, but rather as one form of epistemological knowledge in which numbers carry great significance. There is no such thing as 'raw data' (Gitelman 2013), and thus (big) data does not represent an objective truth. Figures yielded by big data – even if enormous, robust and highly correlated – require interpretation.

There is a need to study the implications of big data for the epistemologies of news production, on the one hand, and news distribution, on the other hand. In our previous research (Lewis and Westlund 2015b, 2016; Westlund and Lewis 2017), we have discussed the role of diverse agents – social actors and technological actants – in connection to the five stages of news work described by Domingo et al. (2008): access/observation, selection/filtering, processing/editing, distribution and interpretation. The first three stages, especially, are closely connected with journalistic epistemology, even when humans are not tightly involved. For example, a programmer-journalist at *The Los Angeles Times* developed an algorithm to record earthquake notifications, process such alerts into epistemological facts, and facilitate easy editing and rapid publication.

Altogether, technology and datasets may be implicated in each of the five stages, in each case helping to determine if something is worthy and credible enough for publishing – a key element of epistemology. The access/observation phase may involve 'watchdogging in code' (Stavelin 2014), through which journalists use algorithms to continuously and automatically monitor what politicians are doing. Relevant both to access/observation and selection/filtering, journalists use different data-driven means to check 'facts'. For example, the Truth Teller prototype at the *Washington Post* is used by journalists to fact-check political speech in real-time by applying a speech-to-text algorithm to fact-oriented databases.

Turning to processing/editing, big data and related processes – for example algorithms and automation – complicate traditional notions of journalistic judgement, insofar as the technology adjudicates findings in the data (Carlson 2015). For news distribution, meanwhile, big data is connected with emerging representations of digital journalism such as infographics, interactive data visualizations and customizable probability models (Howard 2014; Usher 2016). These news products, in turn, carry certain epistemological assumptions about how audiences might acquire knowledge, as users are encouraged to 'play' with the data to comprehend a particular and personalized version of the news narrative. For instance, some news organizations have sought to make datasets more accessible, transparent and exploratory for users, in line with the ethos of open-source software and open-government advocacy (Lewis and Usher 2013; Parasie and Dagiral 2013). Processes of news delivery and audience engagement in this big-data context thus present new questions about the legitimation of knowledge in and through such data-driven participation.

The epistemology of big data and journalism: Directions for future research

This chapter has reprised our discussion of epistemology, expertise, economics and ethics as four key conceptual lenses for the intersection of journalism and big data. While representing distinct concepts, these lenses are interrelated when it comes to journalism practice and future research. For instance, research into the epistemology of data journalism is linked to research questions about the skill sets and authority of journalists vis-à-vis other social actors in the organization (i.e. expertise). This, in turn, may connect both to ethics (such as the norms and values that guide what actors and technological systems do) as well as to economics (such as innovation, efficiency and value creation). Having said this, we now turn our focus more exclusively to the nexus of epistemology, data and journalism. Ultimately, a more holistic approach to this area is missing in contemporary research, with a number of key questions waiting to be addressed.

In his analysis of television news, Ekström (2002) conceptualized journalistic epistemology in three parts: form of knowledge (i.e. medium-specific concerns – in his case those associated with television as a media form); production of knowledge (i.e. professional norms and routines); and public acceptance of knowledge claims (i.e. the conditions for social legitimacy). Each of these perspectives, brought into conversation with our data-oriented emphasis, leads to various questions for future research. For instance, how might data journalism be associated with particular types of knowledge claims? What are the institutionalized routines and procedures that social actors adopt to guide the production of data-backed knowledge claims? And, how are such routines conditioned to ensure that claims are legitimate and justified? Ultimately, what is the audience's role in the legitimation of knowledge in and through datasets?

More specifically, as an example, we can outline here directions for a study of epistemological similarities and differences between data journalism and 'traditional news journalism'. Such study can provide insights into changing as well as constant approaches to epistemology, yielding insights about determinations of journalistic quality and thus indications about the future legitimacy of journalism. This means examining, from a sociological perspective, how journalists know what they know; how various knowledge claims are articulated; and how facts and news stories are justified.

The epistemology of what one may call 'traditional journalism' corresponds to long-standing professional practice, yet one that most likely has transformed over time. The emergence of data journalism quite naturally will relate to such established professional practice, remediating old practices and values when approaching the new, while at the same time developing new approaches oriented to the affordances of computer programming, software development, algorithms and datasets – the stuff of big data, in a sense. Importantly, there likely are processes of cross-fertilization between data journalism and traditional journalism, though with the former being seen as 'the new' phenomenon being institutionalized into established practices within 'the old'. For instance, in what ways does data journalism – and the journalists making knowledge claims through datasets, large and small – apply epistemological values and practices from traditional journalism, and vice versa?

Future research should adopt a systematic approach that ensures clarity and coherence –for instance, by analysing journalistic epistemology in the context of news production frameworks like those described above. There is a need for theoretical concepts and questions that may guide the empirical data collection in relation to the stages outlined by Domingo et al. (2008). Following the lead of Ekström and Nohrstedt (1996), we suggest addressing three sets of questions for each stage and form of news journalism. First, what norms, standards and methods are applied when processing information, guiding decisions on whether the information is credible enough for publication or not? Second, what network of sources do journalists turn to and rely on, and how are these assessed? Third, what characterizes the knowledge claims articulated?

Answering such questions generally involves research methods that are qualitative and ethnographic, in line with classic and contemporary newsroom studies (Willig 2013). There are, of course, growing challenges to conducting such studies, particularly as news work becomes harder to pin down methodologically: It is more embedded in machine systems that may be black-boxed to researchers; it is increasingly technologically dependent, obscuring the converging roles of human and machine; and, as news organizations increasingly rely on freelance labour, news work is less connected to a single site for ethnographic study (see discussion in Deuze 2007; Lewis and Westlund 2016). Altogether, a combination of observations at key sites, interviews with key informants on-site/off-site, and studies of the systems that mediate journalistic judgements is critical for capturing the nature of epistemology in various forms of journalism.

Ultimately, journalism matters to democratic society as a form of knowledge production about public affairs. As such, research into the epistemology of news production matters not simply for what it may contribute to the academic study of journalism and communication. It also speaks to this 'post-truth' moment for media, politics and society: an opportunity to understand how information is legitimated as 'real' or 'fake', to make sense of journalism's co-production of knowledge with publics, to identify dynamics of information production and circulation that lead to various interpretations on the part of audiences. In the end, to study the epistemology of journalism in an era of big data is to assess how technologies and their traces – from the algorithms of social media platforms to open data acquired online – are implicated in *how news is transformed into knowledge*, and what that means for the social authority, trust and legitimacy that may (or may not) be invested in journalism going forward.

References

Abbott, A. D. (1988), *The System of Professions: An Essay on the Division of Expert Labor.* Chicago: University of Chicago Press.

Albarran, A. B. (2002), *Media Economics: Understanding Markets, Industries and Concepts.* Ames, IA: Blackwell.

Ananny, M. (2013), 'Toward an Ethics of Algorithms: Observation, Probability, and Time.' Paper presented at the Governing Algorithms Conference, New York, May 16–17.

Anderson, C. W. (2013a), 'Towards a sociology of computational and algorithmic journalism'. *New Media & Society*, 15(7), pp. 1005–21.

Anderson, C. W. (2013b), 'What aggregators do: Towards a networked concept of journalistic expertise in the digital age'. *Journalism*, 14(8), 1008–23.

Anderson, C. W. (2015), 'Between the unique and the pattern: Historical tensions in our understanding of quantitative journalism'. *Digital Journalism*, 3(3), 349–63.

Anderson, C. W., Bell, E. and Shirky, C. (2012), *Post-industrial Journalism: Adapting to the Present*. New York: Tow Center for Digital Journalism, Columbia University.

Bell, E. (2016), Facebook is eating the world. *Columbia Journalism Review*. http://www.cjr.org/analysis/facebook_and_media.php

boyd, d. and Crawford, K. (2012), 'Critical questions for big data: Provocations for a cultural, technological, and scholarly phenomenon'. *Information, Communication & Society*, 15(5), 662–9.

Borges Rey, E. (2017), Towards an epistemology of data journalism in the devolved nations of the UK: Changes and continuities in materiality, performativity and reflexivity, *Journalism*. Online-first version, February 2017. Retrieved from https://doi.org/10.1177/1464884917693864.

Braun, J. A. (2015), *This Program is brought to You By...: Distributing Television News Online*. New Haven, CT: Yale University Press.

Bruns, A. (2012), 'Reconciling community and commerce?' *Information, Communication & Society*, 15(6), 815–35.

Carlson, M. (2015), 'The robotic reporter: Automated journalism and the redefinition of labor, compositional forms, and journalistic authority'. *Digital Journalism*. 3(3), 416–31.

Carlson, M. (2017), *Journalistic Authority: Legitimating News in the Digital Era*. New York: Columbia University Press.

Collins, H. and Evans. R. (2007), *Rethinking Expertise*. Chicago, IL: University of Chicago Press.

Couldry, N. and Turow, J. (2014), 'Advertising, big data and the clearance of the public realm: Marketers' new approaches to the content subsidy'. *International Journal of Communication*, 8, 1710–26

Crawford, Kate. (2013), 'The Hidden Biases in Big Data'. *Harvard Business Review*, 1 April, http://blogs.hbr.org/2013/04/the-hidden-biases-in-big-data/

Crawford, K., Miltner, K. and Gray, M. L. (2014), 'Critiquing big data: Politics, ethics, epistemology'. *International Journal of Communication*, 8, 1663–72.

Deuze, M. (2007), *Media Work*. Boston: Polity Press.

Diakopoulos, N. (2015), 'Algorithmic accountability: Journalistic investigation of computational power structures'. *Digital Journalism*, 3(3), 398–415.

Domingo, D., Quandt, T., Heinonen, A., Paulussen, S., Singer, J. B. and Vujnovic, M. (2008), 'Participatory journalism practices in the media and beyond: An international comparative study of initiatives in online newspapers'. *Journalism Practice*, 2(3), 326–42.

Ekström, M. (2002), 'Epistemologies of TV-Journalism. A theoretical framework'. *Journalism: Theory, Practice and Criticism*, 3(3), 259–82.

Ekström, M. and Nohrstedt, S-A. (1996), *Journalistikens etiska problem* Rabén Prisma.

Ettema, J. and Glasser, T. (1987), 'On the epistemology of investigative journalism'. In: M. Gurevtich and M. Levy (eds), *Mass Communication Yearbook*, 6. London: Sage.

Ettema, J. S. and Glasser, T. L. (1998), *Custodians of Conscience: Investigative Journalism and Public Virtue*. New York: Columbia University Press.

Eyal, G. (2013), 'For a sociology of expertise: The social origins of the Autism epidemic.' *American Journal of Sociology*, 118(4),863–907.

Fairfield, J. and Shtein, H. (2014), 'Big data, big problems: Emerging issues in the ethics of data science and journalism'. *Journal of Mass Media Ethics*, 29(1), 38–51.

Gitelman, Lisa, (ed.) (2013), *'Raw Data' is an Oxymoron*. Cambridge, MA: MIT Press.

Godler, Y. and Reich, Z. (2013), 'How journalists think about facts: Theorizing the social conditions behind epistemological beliefs'. *Journalism Studies*, 14(1), 94–112

Howard, A. B. (2014), *The Art and Science of Data-Driven Journalism*. New York: Tow Center for Digital Journalism, Columbia University.

Howard, P. N. (2015), *Pax Technica: How the Internet of Things may Set Us Free or Lock Us Up*. New Haven: Yale University Press.

Karlsson, M. (2011), 'The immediacy of online news, the visibility of journalistic processes and a restructuring of journalistic authority'. *Journalism*, 12(3), 279–95.

Kitchin, R. and McArdle, G. (2016), 'What makes big data, big data? Exploring the ontological characteristics of 26 datasets'. *Big Data & Society*, 3(1), doi:10.1177/2053951716631130

Kraemer, F., K. van Overveld, and M. Peterson. (2011). 'Is There an Ethics of Algorithms?' *Ethics and Information Technology* 13(3), 251–260. doi:10.1007/s10676-010- 9233-7.

Lewis, S. C. (2012), 'The tension between professional control and open participation: Journalism and its boundaries'. *Information, Communication & Society*, 15(6), 836–66

Lewis, S. C. and Usher, N. (2013), 'Open source and journalism: Toward new frameworks for imagining news innovation'. *Media, Culture & Society*, 35(5), 602–19

Lewis, S. C and Usher, N. (2014), 'Code, collaboration, and the future of journalism: A case study of the Hacks/Hackers global network'. *Digital Journalism*, 2(3), 383–93.

Lewis, S. C. and Usher, N. (2016), 'Trading zones, boundary objects, and the pursuit of news innovation: A case study of journalists and programmers'. *Convergence: The International Journal of Research Into New Media Technologies*, 22(5), 543–60.

Lewis, S. C. and Westlund, O. (2015a), 'Big data and journalism: Epistemology, expertise, economics, and ethics'. *Digital Journalism*, 3(3), 447–66.

Lewis, S. C. and Westlund, O. (2015b), 'Actors, actants, audiences, and activities in cross-media news work: A matrix and a research agenda'. *Digital Journalism*, 3(1), 19–37

Lewis, S. C. and Carlson, M. (2016), 'The dissolution of news: Selective exposure, filter bubbles, and the boundaries of journalism'. In D. Lilleker, D. Jackson, E. Thorsen and A. Veneti (eds), *US Election Analysis 2016: Media, Voters and the Campaign*, 78. Poole, England: Center for the Study of Journalism, Culture and Community, Bournemouth University.

Lewis, S. C. and Westlund, O. (2016), 'Mapping the human–machine divide in journalism'. In T. Witschge, C. W. Anderson, D. Domingo and A. Hermida (eds), *The SAGE Handbook of Digital Journalism*, 341–53. London: SAGE.

Madrigal, A. C. (2014, June 10), Method journalism. *The Atlantic*. Retrieved from http://www.theatlantic.com/technology/archive/2014/06/method-journalism/372526/.

Manovich, L. (2012), 'Trending: The promises and the challenges of big social data'. In M. K. Gold (ed.), *Debates in the Digital Humanities*, 460–75. Minneapolis, MN: The University of Minnesota Press.

Matheson, D. (2004), 'Weblogs and the epistemology of the news: Some trends in online journalism'. *New Media & Society*, 6(4),443–468.

Mayer-Schönberger, V. and Cukier, K. (2013), *Big Data: A Revolution that will Transform how we Live, Work, and Think*. Boston: Houghton Mifflin Harcourt.

McChesney, R. (2012), 'Farewell to journalism?'. *Journalism Studies*, 13(5–6), 682–94.

Meikle, G. and Redden, G. (eds.) (2010), *News Online – Transformations and Continuities*. London: Palgrave Macmillan.

Meyer, P. (1973), *Precision Journalism: A Reporter's Introduction to Social Science Methods*. Bloomington: Indiana University Press.

Meyer, P. (1973), *Precision Journalism: A Reporter's Introduction to Social Science Methods*, 1st ed. Bloomington: Indiana University Press.

Nissenbaum, H. (2001), 'How computer systems embody values'. *Computer*, 34(3), 120–19.

O'Neil, C. (2016), *Weapons of Math Destruction: How Big Data Increases Inequality and Threatens Democracy*. New York: Random House.

Parasie, S. (2015), 'Data-driven revelation: Epistemological tensions in investigative journalism in the age of "Big data"'. *Digital Journalism*, 3(3),364–380. doi:10.1080/2167 0811.2014.976408.

Parasie, S. and Dagiral, E. (2013), 'Data-driven journalism and the public good: "Computer-assisted-reporters" and "programmer-journalists" in Chicago'. *New Media & Society*, 15(6), 853–71.

Pavlik, J. V. and Bridges, F. (2013), 'The emergence of augmented reality (AR) as a storytelling medium in journalism'. *Journalism & Communication Monographs*, 15(1), 4–59.

Picard, R. G. (2010), 'A business perspective on challenges facing journalism'. In David. A. L. Levy and Rasmus. K. Nielsen (eds), *The Changing Business of Journalism and its Implications for Democracy*. Reuters Institute for the Study of Journalism, University of Oxford.

Powers, M. (2012), '"In forms that are familiar and yet-to-be invented": American journalism and the discourse of technologically specific work'. *Journal of Communication Inquiry*, 36(1), 24–43.

Reich, Z. (2012), 'Journalism as bipolar interactional expertise'. *Communication Theory*, 22(4). 339–58

Schon, D. A. (1995), 'The new scholarship requires a new epistemology'. *Change* 27(6), 27–34.

Shah, D. V., Cappella, J. N. and Neuman, W. R. (2015), 'Big data, digital media, and computational social science: Possibilities and perils'. *The ANNALS of the American Academy of Political and Social Science*, 659(1), 6–13.

Smith, B. (2001), 'Objects and their environments: From aristotle to ecological ontology'. In *The Life and Motion of Socio-Economic Units*. In F. J. Raper and J.-P. Cheylan (eds), 79–97. London: Taylor and Francis.

Stavelin, E. (2014), *Computational Journalism: When Journalism Meets Programming*. Unpublished dissertation. Bergen: University of Bergen.

Storsul, T. and Krumsvik, A. H. (2013), *Media Innovations. A Multidisciplinary Study of Change*. Gothenburg: Nordicom.

Tandoc, E. C. (2014), 'Journalism is twerking? How web analytics is changing the process of gatekeeping'. *New Media & Society*, 16(4), 559–75.

Thurman, N. (2011), 'Making "the daily me": Technology, economics and habit in the mainstream assimilation of personalized news'. *Journalism*, 12(4), 395–415.

Tuchman, G. (1978), *Making News: A Study in the Construction of Reality*. New York: Free Press.

Tufekci, Z. (2014), 'Engineering the public: Big data, surveillance and computational politics'. *First Monday*, 19(7). doi:10.5210/fm.v19i7.4901

Usher, N. (2016), *Interactive Journalism: Hackers, Data, and Code*. Champaign, IL: University of Illinois Press.

Wahl-Jorgensen, K. (2015), 'Resisting epistemologies of user-generated content? Cooptation, segregation and the boundaries of journalism'. In M. Carlson and S. C. Lewis (eds), *Boundaries of Journalism: Professionalism, Practices, and Participation*. New York: Routledge.

Wahl-Jorgensen, K. (2016), 'Emotion and journalism'. In T. Witschge, C. W. Anderson, D. Domingo and A. Hermida (eds), *The SAGE Handbook of Digital Journalism*, 128–43. London: SAGE.

Ward, S. J. A. (2010), *Global Journalism Ethics*. Montreal: McGill-Queen's University Press.

Westlund, O. (2011), *Cross-Media News Work: Sensemaking of the Mobile Media (r)evolution*. Gothenburg: University of Gothenburg.

Westlund, O. (2012), 'Producer-centric versus participation-centric: On the shaping of mobile media'. *Northern Lights*, 10(1), 107–21.

Westlund, O. (2013), 'Mobile news: A review and model of journalism in an age of mobile media'. *Digital Journalism*, 1(1), 6–26.

Westlund, O. and Lewis, S. C. (2014), 'The agents of media innovation: Actors, actants, and audiences'. *The Journal of Media Innovations*, 1(2), 10–35.

Westlund, O and Lewis, S. C. (2017, forthcoming), 'Reconsidering news production: How understanding the interplay of actors, actants, and audiences can improve journalism education.' In R. Goodman and E. Steyn (ed.), *Global Journalism Education: Challenges and Innovations in the 21st Century*. Austin, TX: The Knight Center for Journalism in the Americas at The University of Texas at Austin.

Wiik, J. (2010), *Journalism in Transition. The Professional Identity of Swedish Journalists*, JMG Book Series no. 64. Gothenburg: University of Gothenburg.

Willig, Ida (2013), 'Newsroom ethnography in a field perspective'. *Journalism*, 14(3), 372–87.

Index